AF454463

Green, Closed Loop, Circular Bio-Economy

Green, Closed Loop, Circular Bio-Economy

Editors

Charisios Achillas
Dionysis Bochtis

MDPI • Basel • Beijing • Wuhan • Barcelona • Belgrade • Manchester • Tokyo • Cluj • Tianjin

Editors
Charisios Achillas
Department of Logistics,
Technical Educational Institute
of Central Macedonia
Greece

Dionysis Bochtis
Institute for Bio-economy and
Agri-Technology (IBO),
Centre for Research and
Technology—Hellas (CERTH)
Greece

Editorial Office
MDPI
St. Alban-Anlage 66
4052 Basel, Switzerland

This is a reprint of articles from the Special Issue published online in the open access journal *Sustainability* (ISSN 2071-1050) (available at: https://www.mdpi.com/journal/sustainability/special_issues/Green_Bio-economy).

For citation purposes, cite each article independently as indicated on the article page online and as indicated below:

LastName, A.A.; LastName, B.B.; LastName, C.C. Article Title. *Journal Name* **Year**, *Volume Number*, Page Range.

ISBN 978-3-0365-0210-6 (Hbk)
ISBN 978-3-0365-0211-3 (PDF)

Contents

About the Editors

Charisios Achillas is Assistant Professor at the International Hellenic University, Department of Supply Chain Management. He graduated in 1999 from the Department of Engineering (Aristotle University Thessaloniki) with a degree in Mechanical Engineering. He continued his studies with an MSc in Engineering Project Management in 2001 (UMIST, Manchester, UK). In 2009, he received his PhD Doctorate in the field of reverse logistics. Dr. Achillas is a senior researcher at the Institute of Bio-Economy and Agri-Technology, Centre for Research and Technology, Hellas (CERTH) and the Laboratory of Heat Transfer and Environmental Engineering, Department of Mechanical Engineering, Aristotle University of Thessaloniki. Dr. Achillas is the author of more than 180 scientific publications. His research interests span the fields of environmental management, sustainable development, and circular economy with a primary focus on reverse logistics and agri-chains.

Dionysis Bochtis works in the area of systems engineering focused on bioproduction and related provision systems with enhanced ICT and automation technologies up to fully robotized production systems. Throughout his career, he has held various positions, including Professor in Agri-Robotics at the Lincoln Institute for Agri-Food technologies, University of Lincoln, UK, and Senior Scientist in Operations Management at the Department of Engineering of Aarhus University, Denmark. Currently, he is the Director of the Institute for Bio-economy and Agri-technology (IBO), Centre for Research and Technology, Hellas (CERTH). He runs numerous research projects on ICT and robotics in agricultural production. He is author of more than 300 articles (more than 100 in peer-reviewed journals) and has been invited for more than 30 keynote speeches around the globe. He is involved in the board of a number of international scientific organizations, holding positions including the following:

- President of the Commission Internationale de l'Organisation Scientifique du Travail en Agriculture, (CIOSTA, founded in Paris in 1950), for the periods 2011–2013 and 2017–2019;

- Chair of Section V (Systems Management) of the International Commission of Agricultural and Biosystems Engineering (CIGR), 2020–2022;

- Chair of the European Federation for Information Technologies in Agriculture (EFITA), 2019–2021.

Preface to "Green, Closed Loop, Circular Bio-Economy"

In recent years, bioeconomy strategies have been implemented and adapted internationally. In the bioeconomy, materials are to a certain extent circular by nature. However, biomaterials may also be used in a rather linear way. Lately, a transition towards a circular economy, a more restorative and regenerative economic model, is being promoted worldwide. A circular economy offers an alternative model aiming at "doing more and better with less". It is based on the idea that circulating matter and energy will diminish the need for new input. Its concept lies in maintaining the value of products, materials, and resources for as long as possible and at the same time minimizing or even eliminating the amount of waste produced. Focused on "closing the loops", a circular economy is a practical solution for promoting entrepreneurial sustainability, economic growth, environmental resilience, and a better quality of life for all. The most efficient way to close resource loops is to find value in the waste. Different modes of resource circulation may be applied, e.g., raw materials, by-products, human resources, logistics, services, waste, energy, or water.

To that end, this Special Issue seeks to contribute to the circular bioeconomy agenda through enhanced scientific and multidisciplinary knowledge to boost the performance efficiency of circular business models and support decision-making within the specific field. The Special Issue includes innovative technical developments, reviews, and case studies, all of which are relevant to green, closed-loop, circular bioeconomy.

Charisios Achillas, Dionysis Bochtis
Editors

Editorial

Toward a Green, Closed-Loop, Circular Bioeconomy: Boosting the Performance Efficiency of Circular Business Models

Charisios Achillas [1],* and Dionysis Bochtis [2]

[1] Department of Supply Chain Management, School of Economics and Business Administration, International Hellenic University, GR 60100 Katerini, Greece

[2] Institute for Bio-Economy and Agri-Technology (iBO), Center for Research and Technology–Hellas (CERTH), 6th km Charilaou-Thermi Rd, GR 57001 Thermi, Greece; d.bochtis@certh.gr

* Correspondence: c.achillas@ihu.edu.gr; Tel.: +30-23510-47860

Received: 2 December 2020; Accepted: 3 December 2020; Published: 4 December 2020

In recent years, bioeconomy strategies have been successfully implemented and widely adopted internationally. Such strategies have been promoted, mostly in an effort to combat and confront climate change, which is considered not only an ecological question, but "the most systemic threat to humankind", as characterized by the UN Secretary-General in 2018, and also a critical point towards global security and freedom [1,2]. Moreover, resource scarcity is also considered an additional critical issue at present. The pressures imposed by economic development and the expansion of the global population places the problem of the intensification of raw material use and resource scarcity at the epicenter of the discussion on global sustainability for generations to come [3].

Within the bioeconomy, materials are, to a certain extent, circular by nature. However, biomaterials may be also used in a rather linear way. Lately, a transition towards a circular economy has been promoted worldwide. In brief, circular business models are targeting the establishment of a more restorative and—at the same time—regenerative system, where material inputs, energy use, waste production, and emissions to the natural environment are minimized, if not eradicated, by narrowing and closing material and energy loops. The aim is to efficiently utilize materials by extracting the maximum value from them whilst in use, providing yield for as long as possible, and recovering their residual values at the end of their useful lifetimes [4,5]. In this context, a circular economy offers an alternative model aimed at "doing more and better with less" on a life cycle basis (life cycle thinking), in contrast to the traditional linear model that is based on the "take—make—dispose" production and consumption strategies.

More specifically, circularity is based on the idea of redistributing matter and energy in an effort to diminish the need for new inputs in supply chains. The concept lies in maintaining the value of products, materials, and resources for as long as possible and, at the same time, minimizing, or even eliminating, the amount of waste produced. This can be achieved only through the application of modifications to the design of products or services, so as to increase their useful lifetimes, upgrade their maintainability (e.g., through the adoption of the Design for Disassembly approach), support their reuse potential, either as a full product or partially, with the refurbishment and reuse of their components, while also improve recyclability and upcycling.

Focused on the "closing the loop" philosophy, the circular economy is a practical solution for promoting entrepreneurial sustainability, economic growth, environmental resilience, and a better quality of life for all. The most efficient way to close resource loops is to find value in the waste [6]. In this light, nothing is regarded as waste; on the contrary, everything is viewed as a resource with a residual value that can be reintroduced and exploited within the same or a different supply chain. In practice, a number of different modes of resource circulation may be applied. Closing the loop can be applied to raw materials, by-products, human resources, logistics, services, waste, energy, and water.

The benefits that derive from the adoption of circular economy models are critical towards economic development and public prosperity. In particular, the application of circular economic models reduces the requirements for raw materials, water use, and energy dependency, creating significant savings by better controlling and reducing operational costs, thus improving business competitiveness. Moreover, such models typically result in the upgrade of supply chain security since organizations are far less dependent on external inputs. Apart from the economic benefits, the adoption of circular economy models also results in significant environmental and social benefits, enhancing sustainability. The extraction of raw materials, waste quantities, and emissions are minimized, decreasing the pressure imposed on the natural environment; greenhouse gas emissions are abated, contributing to internationally agreed-upon policies against climate change and global warming; employment opportunities arise in newly established sectors, while new business opportunities are formed through the development of novel products and services that link the output and waste of one entity to the required input for another [7–11].

At an international level, circularity has been expanded relatively recently. Although initially coined as a term in the 1970s, the circular economy attracted the intensified interest of major, multinational producers and policymakers only during recent years. Across the world, key stakeholders are eager to shift from linearity to circularity, primarily in order to enhance the competitiveness of the business environment under strict restrictive environmental conditions and a fast technological transformation [12,13]. The introduction of circular economy models lies mostly on four key action areas, namely primary and industrial production, consumption of goods, management of waste, and exploitation of secondary raw materials. Currently, circularity is widely applied primarily to business sectors, such as the plastics industry, the raw materials sector, the construction (and demolition) industry, as well as to the food industry, and fields related to the bioeconomy, such as the production of biomass and bio-based products [14]. The increasingly stricter environmental legislation and policies at an international level, together with prevailing "green" initiatives worldwide, have triggered the necessity for the development of a circular ecosystem on a global scale. To that end, policy and legal frameworks have been established in more and more countries in the developed and the developing world, in the effort to support the exploitation of waste, its reuse, refurbishment, or recycling and reintroduction to similar or other supply chains and production processes.

Of course, the adoption of circular economy models requires collective effort and teamwork. No single organization or enterprise can efficiently apply circular economy practices on their own. The circular economy is highly dependant on the exchange of knowledge and experiences between stakeholders, while thriving on business networks and the application of industrial symbiotic practices. In this context, the only path for each entity to succeed in minimizing waste and exploiting secondary resources includes cooperation with other players who have a similar mentality and seek work in the same direction, or better yet, are already a few steps ahead.

Undoubtedly, the predominance of the circular economy in the field of bioeconomy requires the active involvement of all types of stakeholders within the quadruple helix, namely government, industry, academia, and civil society [15–17]. Moreover, the prevalence of circular business models over linear ones necessitates the adoption of multidisciplinary scientific and technical approaches [18]. This Special Issue aims to bring to light novel production systems, innovative design processes, and pioneering engineering advancements. The thirteen (13) articles that are published in the Special Issue efficiently contribute to the circular bioeconomy agenda through enhanced scientific and multidisciplinary knowledge. It is evident that contributions related to innovative technical developments, reviews, and case studies effectively boost the performance efficiency of circular business models, and support decision-making within the field of green, closed-loop, circular bio-economies. In brief, the contributions span around topics related to sustainability, agri-business, life cycle thinking, green production, reverse logistics, waste management, bioresources, bioenergy, and a bioeconomy. The thematic areas that are discussed within the Special Issue cover views and attitudes on a bioeconomy, policymaking and decision-making in the circular economy, social issues related to the topic, and also reviews and secondary sources that synthesize and analyze research already conducted in the field.

More specifically, three (3) articles included in the Special Issue deal with "views and attitudes on the bioeconomy". In particular, Purwestri et al. conducted a preliminary assessment of a survey utilized in the Czech Republic on the bioeconomy, as part of a nationwide relevant survey [19]. The article focuses on wood and non-wood forest products. The work aims to provide an initial evaluation regarding the use of forest products and related factors, and to make recommendations on developing wood consumption and promoting other forest ecosystem services for the adoption of a forest bioeconomy strategy in the Czech Republic. The survey revealed the challenge to switch the practice from using fossil-based heating to a wood boiler energy source. Moreover, turning wood into products with a high added-value is highly recommended. The promotion of wood and non-wood forest products is encouraged, starting with increasing awareness and knowledge of the strength of the forest-based sector as a renewable energy resource and the importance of forest ecosystem services, using different channels as sources of information. On a similar topic, Priefer and Meyer examine the way that scientists think about the concept of a bioeconomy in Germany [20]. In this work, an online survey was carried out among scientists involved in a regional bioeconomy research program in southern Germany in order to gain insight into their understanding of a bioeconomy. Moreover, the survey provided information about cooperation and major challenges in the future development of three biomass utilization pathways: biogas, lignocellulose, and microalgae. The analysis showed that a resource-oriented understanding of a bioeconomy is favored. The political objectives for a European bioeconomy are widely accepted, and it is expected that ongoing research can significantly contribute to achieving these goals. The two different pathways for shaping the bioeconomy that are discussed in the debate—the technology-based approach and the socio-ecological approach—are considered compatible rather than contrary. Within the same topic, Fiore et al. present the key findings of the InnovaEcoFood project, which focuses on the use of by-products of the Piedmontese rice and wine production chains to valorize their untapped potential in the food sector by identifying how value could be created from the waste [21].

An additional six (6) contributions are related to the thematic area of "policymaking and decision-making" in the field of the circular economy. In particular, Singh et al. evaluate the effectiveness of climate change adaptations in the Sundarbans, the world's largest coastal river delta and the largest uninterrupted mangrove ecosystem [22]. With the use of a fuzzy cognitive maps-based approach, the authors aim to elicit and integrate stakeholders' perceptions regarding current climate forcing, consequent impacts, and efficacy of the existing adaptation measures. The simulations revealed that while existing adaptation practices provide resilience to an extent, they are grossly inadequate in the context of providing future resilience, and adaptations may not be entirely transformative in such a fragile ecosystem. According to the authors, measures that are likely to enhance adaptive capabilities of the local communities include those involving gender-responsive and adaptive governance, human resource capacity building, commitments of global communities for adaptation financing, education and awareness programs, and embedding indigenous and local knowledge into decision making.

Ovezikoglou et al. focus on the funding mechanisms of circular interventions and specifically on the development of a novel methodological scheme for the sustainability assessment of alternative investments with the use of indicators, based on a Multiple Criteria Methodological Framework [23]. A number of alternative indicators are proposed, assessed over four criteria (environment, society, economy, and technology), and hierarchized in order to select the optimal bundle to be used for the assessment of future industrial investments. The analysis results in the fact that "Resource Savings" is the criterion that optimally characterizes the sustainability of an investment in a bioeconomy, coupled with "Recycling" and "Research, Innovation, Development" indicators.

Anagnostis et al. turned their attention to the application of artificial neural networks for natural gas consumption forecasting [24]. The study explores different types of neural network approaches for forecasting natural gas consumption in cities across Greece. A comparative analysis is conducted between the selected networks, in an effort to better understand individual locations' characteristics and public habitual patterns.

Kyriakarakos et al. propose a paradigm shift in rural electrification investments in Sub-Saharan Africa through agriculture [25] Since a large number of people around the world still do not have access to electricity, private sector investments and new innovative business models are required. In this direction, microgrid electrification in rural areas with the active involvement of agriculture-related businesses is proposed by the authors, and financing of such activities is investigated in the specific work. The analysis results in the fact that the high costs of rural electrification can be met through the increased value of locally produced products. At the same time, cross-subsidization can take place to decrease the cost of household electrification.

The work of Katchova and Sant'Anna puts emphasis on the location of ethanol plants and the impact of that decision on corn revenues for U.S. farmers [26]. Specifically, the opening of an ethanol plant may increase local demand for corn, pressuring increases in the local corn basis. The question that is answered in this manuscript is what the effect of location on corn contract prices and revenues is. The authors conclude that policymakers should focus their resources on promoting greater efficiency in corn production to boost farmers' revenues.

Chastain concentrates on ammonia volatilization losses during the irrigation of liquid animal manure [27]. Ammonia loss resulting from land application of liquid animal manure varies depending on the composition of the manure and the method used to apply manure to cropland. High levels of ammonia volatilization result in an economic loss to the farmer, based on the value of the nitrogen, and have also been shown to be a source of air pollution. Using irrigation as a method of applying liquid manure to cropland has generally been accepted as a method that increases the volatilization of ammonia. Through measurements, and in line with previous studies on irrigation performance, Chastain concludes with the fact that volatilization losses during irrigation were not found to be statistically significant, and evaporation losses were small.

Within the topic of "social issues" related to bioeconomy, two (2) articles have been published. Papageorgiou et al. propose a fuzzy cognitive map-based sustainable socio-economic development plan for rural communities [28]. Specifically, fuzzy cognitive maps are recommended to be employed by policymakers—as a powerful and easy-to-use tool for representing complex systems—in order to keep human activities within a safe limit of the planetary boundaries. The work studies a complex phenomenon of poverty eradication and socio-economic development strategies in rural areas in India, and various scenarios examining the economic sustainability and livelihood diversification of poor women in rural areas were performed. With a similar concept, Bochtis et al. extensively discuss the agricultural workforce crisis in light of the recent COVID-19 pandemic, which seriously affected the agricultural workforce and jeopardized food security [29]. The work aims at assessing the COVID-19 pandemic impacts on agricultural labor and suggests strategies to mitigate them. Following an analytic research methodology related to the benchmarking of risks to agricultural workers by individual tasks within agricultural production, depending on a number of key criteria, the authors conclude that a series of control measures need to be adopted so as to enhance the resilience and sustainability of the sector, as well as to protect farmers with physical distancing, hygiene practices, and personal protection equipment.

There are also two (2) review articles that contribute to the topic. Lampridi et al. propose a methodological framework for the systematic literature review of agricultural sustainability studies [30]. The framework synthesizes all the available literature review criteria and introduces a two-level analysis facilitating systematization, data mining, and methodology analysis. The proposed framework is validated for the systematic review of 38 crop agricultural sustainability assessment studies at the farm level within the last decade. Moreover, Angelopoulou et al. review the progress made in the last decade in respect to laboratory and proximal sensing spectroscopy in the visible and near infrared and shortwave infrared (VNIR–SWIR) wavelength region for soil organic carbon and soil organic matter estimation as an alternative to analytical chemistry measurements [31]. The results of over fifty selective studies are critically discussed and the factors that affect the accuracy of spectroscopic measurements for both laboratory and in situ applications are analyzed.

It is the editors' strong belief that the material included in the present Special Issue bridges part of the gap between the bioeconomy on one hand and green supply chains, closed-loop systems, and the circular economy on the other. As thoroughly discussed in this editorial, as well as in the contributing manuscripts that are included in the Special Issue, the only efficient way to close resource loops is to find value in the waste. It is evident that shifting from conventional, linear business modules to circular ones is a prerequisite for sustainable development and the achievement of the internationally acknowledged Sustainable Development Goals.

Author Contributions: Conceptualization: C.A. and D.B.; Writing—original draft preparation C.A. and D.B.; Writing—review and editing: C.A. and D.B. All authors have read and agreed to the published version of the manuscript.

Funding: This research received no external funding.

Conflicts of Interest: The authors declare no conflict of interest.

References

1. Aguiar, A.P.D.; Collste, D.; Harmáčková, Z.V.; Pereira, L.; Selomane, O.; Galafassi, D.; Van Vuuren, D.; Van Der Leeuw, S. Co-designing global target-seeking scenarios: A cross-scale participatory process for capturing multiple perspectives on pathways to sustainability. *Glob. Environ. Chang.* **2020**, *65*, 102198. [CrossRef]
2. Shrivastava, P.; Stafford Smith, M.; O'Brien, K.; Zsolnai, L. Transforming Sustainability Science to Generate Positive Social and Environmental Change Globally. *One Earth* **2020**, *2*, 329–340. [CrossRef]
3. Nyam, Y.S.; Kotir, J.H.; Jordaan, A.J.; Ogundeji, A.A.; Adetoro, A.A.; Orimoloye, I.R. Towards Understanding and Sustaining Natural Resource Systems through the Systems Perspective: A Systematic Evaluation. *Sustainability* **2020**, *12*, 9871. [CrossRef]
4. Banias, G.; Achillas, C.; Vlachokostas, C.; Moussiopoulos, N.; Stefanou, M. Environmental impacts in the life cycle of olive oil: A literature review. *J. Sci. Food Agric.* **2017**, *97*, 1686–1697. [CrossRef] [PubMed]
5. Tsarouhas, P.; Achillas, C.; Aidonis, D.; Folinas, D.; Maslis, V. Life Cycle Assessment of olive oil production in Greece. *J. Clean. Prod.* **2015**, *93*, 75–83. [CrossRef]
6. Achillas, C.; Moussiopoulos, N.; Karagiannidis, A.; Banias, G.; Perkoulidis, G. The use of multi-criteria decision analysis to tackle waste management problems: A literature review. *Waste Manag. Res.* **2013**, *31*, 115–129. [CrossRef]
7. Jarosch, L.; Zeug, W.; Bezama, A.; Finkbeiner, M.; Thrän, D. A Regional Socio-Economic Life Cycle Assessment of a Bioeconomy Value Chain. *Sustainability* **2020**, *12*, 1259. [CrossRef]
8. Bochtis, D.D.; Sørensen, C.G.; Busato, P.; Berruto, R. Benefits from optimal route planning based on B-patterns. *Biosyst. Eng.* **2013**, *115*, 389–395. [CrossRef]
9. Rodias, E.; Berruto, R.; Busato, P.; Bochtis, D.; Sørensen, C.G.; Zhou, K. Energy savings from optimised in-field route planning for agricultural machinery. *Sustainability* **2017**, *9*, 1956. [CrossRef]
10. de Schutter, L.; Giljum, S.; Häyhä, T.; Bruckner, M.; Naqvi, A.; Omann, I.; Stagl, S. Bioeconomy Transitions through the Lens of Coupled Social-Ecological Systems: A Framework for Place-Based Responsibility in the Global Resource System. *Sustainability* **2019**, *11*, 5705. [CrossRef]
11. Rodias, E.C.; Lampridi, M.; Sopegno, A.; Berruto, R.; Banias, G.; Bochtis, D.D.; Busato, P. Optimal energy performance on allocating energy crops. *Biosyst. Eng.* **2019**, *181*, 11–27. [CrossRef]
12. Duquet, B.; Brunelle, C. Subcentres as Destinations: Job Decentralization, Polycentricity, and the Sustainability of Commuting Patterns in Canadian Metropolitan Areas, 1996. *Sustainability* **2020**, *12*, 9966. [CrossRef]
13. Marinoudi, V.; Sørensen, C.G.; Pearson, S.; Bochtis, D. Robotics and labour in agriculture. A context consideration. *Biosyst. Eng.* **2019**, *184*, 111–121. [CrossRef]
14. Tomaszewska, J. Polish Transition towards Circular Economy: Materials Management and Implications for the Construction Sector. *Materials (Basel)* **2020**, *13*, 5228. [CrossRef]

15. Feleki, E.; Vlachokostas, C.; Achillas, C.; Moussiopoulos, N.; Michailidou, A.V. Involving decision-makers in the transformation of results into urban sustainability policies. *Eur. J. Environ. Sci.* **2016**, *6*, 7–10. [CrossRef]

16. Marting Vidaurre, N.A.; Vargas-Carpintero, R.; Wagner, M.; Lask, J.; Lewandowski, I. Social Aspects in the Assessment of Biobased Value Chains. *Sustainability* **2020**, *12*, 9843. [CrossRef]

17. Dieken, S.; Venghaus, S. Potential Pathways to the German Bioeconomy: A Media Discourse Analysis of Public Perceptions. *Sustainability* **2020**, *12*, 7987. [CrossRef]

18. Barragán-Ocaña, A.; Silva-Borjas, P.; Olmos-Peña, S.; Polanco-Olguín, M. Biotechnology and Bioprocesses: Their Contribution to Sustainability. *Processes* **2020**, *8*, 436. [CrossRef]

19. Purwestri, R.C.; Hájek, M.; Šodková, M.; Jarský, V. How arewood and non-wood forest products utilized in the Czech Republic? A preliminary assessment of a nationwide survey on the bioeconomy. *Sustainability* **2020**, *12*, 566. [CrossRef]

20. Priefer, C.; Meyer, R. One concept, many opinions: How scientists in Germany think about the concept of bioeconomy. *Sustainability* **2019**, *11*, 4253. [CrossRef]

21. Fiore, E.; Stabellini, B.; Tamborrini, P. A systemic design approach applied to rice and wine value chains. The case of the innovaecofood project in piedmont (italy). *Sustainability* **2020**, *12*, 9272. [CrossRef]

22. Singh, P.K.; Papageorgiou, K.; Chudasama, H.; Papageorgiou, E.I. Evaluating the effectiveness of climate change adaptations in the world's largest Mangrove Ecosystem. *Sustainability* **2019**, *11*, 6655. [CrossRef]

23. Ovezikoglou, P.; Aidonis, D.; Achillas, C.; Vlachokostas, C.; Bochtis, D. Sustainability assessment of investments based on a multiple criteria methodological framework. *Sustainability* **2020**, *12*, 6805. [CrossRef]

24. Anagnostis, A.; Papageorgiou, E.; Bochtis, D. Application of artificial neural networks for natural gas consumption forecasting. *Sustainability* **2020**, *12*, 6409. [CrossRef]

25. Kyriakarakos, G.; Balafoutis, A.T.; Bochtis, D. Proposing a paradigm shift in rural electrification investments in Sub-Saharan Africa through Agriculture. *Sustainability* **2020**, *12*, 3096. [CrossRef]

26. Katchova, A.L.; Sant'Anna, A.C. Impact of Ethanol plant location on corn revenues for U.S. Farmers. *Sustainability* **2019**, *11*, 6512. [CrossRef]

27. Chastain, J.P. Ammonia volatilization losses during irrigation of liquid animal manure. *Sustainability* **2019**, *11*, 6168. [CrossRef]

28. Papageorgiou, K.; Singh, P.K.; Papageorgiou, E.; Chudasama, H.; Bochtis, D.; Stamoulis, G. Fuzzy Cognitive Map-Based Sustainable Socio-Economic Development Planning for Rural Communities. *Sustainability* **2019**, *12*, 305. [CrossRef]

29. Bochtis, D.; Benos, L.; Lampridi, M.; Marinoudi, V.; Pearson, S.; Sørensen, C.G. Agricultural workforce crisis in light of the COVID-19 pandemic. *Sustainability* **2020**, *12*, 8212. [CrossRef]

30. Lampridi, M.; Sørensen, C.; Bochtis, D. Agricultural Sustainability: A Review of Concepts and Methods. *Sustainability* **2019**, *11*, 5120. [CrossRef]

31. Angelopoulou, T.; Balafoutis, A.; Zalidis, G.; Bochtis, D. From Laboratory to Proximal Sensing Spectroscopy for Soil Organic Carbon Estimation—A Review. *Sustainability* **2020**, *12*, 443. [CrossRef]

Publisher's Note: MDPI stays neutral with regard to jurisdictional claims in published maps and institutional affiliations.

 *sustainability*

Article

How Are Wood and Non-Wood Forest Products Utilized in the Czech Republic? A Preliminary Assessment of a Nationwide Survey on the Bioeconomy

Ratna C. Purwestri, Miroslav Hájek *, Miroslava Šodková and Vilém Jarský

Faculty Forestry and Wood Sciences, Czech University of Life Sciences Prague, Kamýcká 129, 16500 Praha 6–Suchdol, Czech Republic; purwestri@fld.czu.cz (R.C.P.); sodkova@fld.czu.cz (M.Š.); jarsky@fld.czu.cz (V.J.)
* Correspondence: hajek@fld.czu.cz; Tel.: +420-224-383-707

Received: 13 November 2019; Accepted: 8 January 2020; Published: 11 January 2020

Abstract: The Czech forests occupy 33.7% of the total country area; thus, wood and non-wood forest products (NWFPs) are important resources for the country. To date, the country has not adopted a forest bioeconomy strategy. A forest bioeconomy is defined as all activities that relate to the forest ecosystem services (FES). This study aimed to provide an initial evaluation regarding the use of forest products and related factors, and to make recommendations on developing wood consumption and promoting other FES for the adoption of a forest bioeconomy strategy in the country. The research study was part of a nationwide survey in June 2019. An online panel of 1050 respondents aged 18–65 years old was recruited based on a quota sampling procedure. Wood products were the most preferred material for furniture (96.3%) and building materials (46.3%). In total, 38.6% of Czech residents used wood as a source of energy, mostly in the form of firewood. It is challenging to switch the practice from using fossil-based heating to wood boiler energy source. The further development of wood into products with a high added value is recommended. Picking mushrooms and berries were among the popular activities in relation to NWFPs. The promotion of wood and NWFPs is encouraged, starting with increasing awareness and knowledge of the strength of the forest-based sector as a renewable energy resource and the importance of FES, using different channels as sources of information.

Keywords: forestry; wood; non-wood forest products; bioeconomy

---

1. Introduction

As a result of growing concerns about dependency on fossil-based energy-sources and their impact on climate change, as well as increasing awareness of and preference for sustainable production and consumption patterns, bioeconomy has become a significant solution. The European Commission (EC) defined the bioeconomy as an economy that *"encompasses the production of renewable biological resources and their conversion into food, feed, bio-based products, and bioenergy"*. Agriculture, forestry, fisheries, food, pulp, and paper production, as well as some of the chemical, biotechnological and energy industries, are expected to contribute to bioeconomy activities [1,2]. Many of the strategies were further developed to improve the national economy and create job opportunities, and at the same time manage the forest sustainably. In addition to the above definitions, Winkel [3] described the forest-based bioeconomy as *"all economic activities that relate to forests and forest ecosystem services, including biomass-based value chains and the economic utilization of other types of forest ecosystem services (FES)"*.

Forest-based wood production is leading the way to renewable energy sources, which are part of a long tradition in European countries [4,5]. In addition to wood forest products, forests offer valuable forest ecosystem services and other benefits for the well-being of the people [6]. The services provided include provisioning and regulating, as well as basic, supportive and cultural services. Provisioning services cover the products obtained from ecosystems, e.g., food, water, construction and firewood, and fiber, while regulating services cover the benefits obtained from the regulation of ecosystem services, such as erosion control, climate regulation, and precipitation. Cultural services are defined as all nonmaterial benefits obtained from the forest, including spiritual, aesthetic, religious, and recreational values, all of which contribute to our well-being, social and cultural functions. Forests also provide supporting services that are necessary for the production of all other ecosystem services, e.g., nutrient and water cycling, soil formation and retention, and photosynthesis. [6–14].

In the Czech Republic, forests cover about 2.7 million ha (33.7%) of the total country area, lower than all European Union (EU) countries together (40.3%) and Austria (45%), but comparable to the German forested landscape (±31%) [15–18]. Forests are an important part of Czech history and culture. Based on the 2017 Czech forest report, forestry (forestry and logging) and the wood processing industry's shares accounted for 1.180% of the gross value added (GVA) at basic prices, not including the paper and furniture industries, which would add a contribution up to 2.018% of the GVA. The share of the forestry and wood processing industry alone was slightly lower than agriculture's share (1.713%), indicating the importance of the forest-based sector in the country [19].

Due to a growing global concern to replace fossil-based fuels with renewable energy sources, the forest-based sector has become a backbone for bioeconomy strategy. A shift in wood production from weakly regulated forests toward sustainable forest management is accompanied by third-party certification, as promoted in forest strategies in EU countries like Finland, Sweden, Germany, and Austria, that has changed the demand for wood in these regions. In 2014, Sweden became the top producer of primary wood products among EU countries by approximately 70 million m^3, followed by Finland (± 57 million m^3) and Germany (about 54 million m^3), while the Czech Republic and Austria contributed about 15 and 17 million m^3, respectively [20]. The Czech Republic was also named as one of the main roundwood exporter countries in 2016 [21]. In 2017, timber production in the Czech Republic resulted in 19,387 million m^3, of which roundwood production amounted to 11,488 million m^3. Most of timber production, including softwood–roundwood, and pulpwood, is exported, mainly to Austria and Germany, for further processing. However, the supply for sawmills and pulp mills in some regions is still insufficient, and this has caused the country to import from Slovakia, Germany, and Poland [19,22].

Although the bioeconomy strategy has not been mentioned in the Czech National Forest Programme (NFP) [16], bioeconomy has been mentioned in the 2018 draft strategy of the Ministry of Agriculture (MoA). In addition to timber production as one of the fundamental priorities in the Czech forest-based sector, non-wood products, like forest fruits and mushrooms, are also considered important FES [22]. Thus, it is important to provide a view of the current situation of forest products' utilization and preferences by the Czech public. The results can be used to inform policymakers and other stakeholders to offer better understanding, and as a baseline to make recommendations on further actions for the adoption of the forest bioeconomy strategy and the promotion of FES.

2. Materials and Methods

2.1. Study Area

The Czech forests cover about 33.7% of the total country area. In 2017, 71.9% of the total Czech forests consisted of coniferous trees, 50.3% of them being Norway spruce (*Picea Abies*). Deciduous trees, such as beech (8.4%) and oak (7.2%), covered 27% of the total forested region, and the rest (1.1%) was forested land without trees [19].

2.2. Design of the Study

The research study was part of a nationwide survey. The survey itself was part of the "Advanced research supporting the forestry and wood-processing sector's adaptation to global change and the 4th industrial revolution" and the "Diversification of the Impact of the Bioeconomy on Strategic Documents of the Forestry-Wood Sector as a Basis for State Administration and the Design of Strategic Goals" research project. The study was carried out in June 2019 in co-operation with an external market research company, REMMARK, a.s (Prague, Czech Republic). The company used the computer-assisted web interviewing (CAWI) technique to recruit the online respondents. No private information was required, and the respondents were anonymous. The online participants aged 18–65 years were recruited proportionally based on age, sex, education level, region, and village size. This technique generates emails and sent the questionnaires to the potential respondents based on the company's list through different online platforms, (e.g., Yahoo email). We have no information on the number of sent-out questionnaires. The survey was terminated after reaching the minimum required sample size. All returned questionnaires were included in the analysis (100%). The respondents were asked to answer a closed-ended questionnaire consisting of socio-demography characteristics and information on FES utilization. Additional information could be written/typed, in order to explain the answer option "others". The answers were later grouped and coded for further analysis.

2.3. Data Analysis

Descriptive data for the general characteristics of the respondents were used for single traits. Frequencies were presented by absolute numbers and their proportions. A group comparison of traits that determine the FES was made via a chi's square test or the Fischer exact test for categorical data. The age of the respondents was checked against an expected normal distribution using quantile-quantile (Q-Q) plot.

The target respondents of the survey were within a productive age. The age of the respondents was categorized as 18–24 (youth employment), while 25–54 and 55–65 years were defined as prime and mature working age, respectively [23,24]. The education levels of the respondents were categorized as elementary school, secondary school without official graduation or with vocational training (secondary school/vocational training), graduated from high school, or university level. In this article, the place of residence was grouped based on region. To fulfill further data analysis, we combined the respondents that never visited the forest with those who went one or two times per year and created a dummy variable of 0 = frequent visitors and 1= never/rarely. The frequency of forest visits is considered as one of the indicators of utilization of FES.

Two scoring systems were used to define the preferences and opinions of the respondents. The first method used five categorizations of opinions as follows:

(1)   Certainly not;
(2)   Rather not;
(3)   Neither;
(4)   Rather yes;
(5)   Certainly yes.

The second method applied five degree of preferences, in which 1 (one) is most preferred and 5 (five) is least preferred.

Binary logistic regression with a forward stepwise approach was applied to identify potential predictors of the frequency of forest visits and utilization of forest products and ecosystem services. The following covariates associated with the dependent variables were included in the initial model: age, education level, and characteristics of the place of residence.

To designate the statistical significance in all analyses, a p-value of less than 0.05 was used. Statistical analysis was performed using IBM SPSS statistics version 25 (IBM Corp., Armonk, NY, USA).

3. Results

In total, 1050 respondents age 18–65 years were recruited in the study. As the samples were proportionally drawn by population size per region and village, the proportions of the residential locations of our respondents were similar to the national statistics office [25].

3.1. Forest Visitor General Characteristics

Respondents to the survey were from different education levels and age groups, of people aged between 18 and 65 years (Table 1). Respondents at the elementary school level were mostly from the younger age group. The respondents' data on education levels were similar ($p = 0.373$) to those from the Czech statistics office in 2011. The proportion of Czech citizens who graduated from elementary, secondary school without official graduation, high school, and university level that were reported by the national statistics office were 17.2%, 33%, 27.1%, and 12.5%. [26].

Table 1. General characteristics of the respondents (N = 1050).

Characteristics	Percentage (n) or Mean ± sd (Min–Max)
Gender (female)	49.2 (517)
Age (years),	42.4 ± 13.5 (18–65)
Age group	
- 18–24 years	10.0 (105)
- 25–54 years	65.4 (687)
- 55–65 years	27.0 (283)
Education level	
- Elementary school	10.4 (109)
- Secondary school/vocational training	37.7 (396)
- High school graduates	36.3 (381)
- University	15.6 (164)
Place of residential (region)	
- Prague (capital city)	13.9 (146)
- Bohemia	53.5 (562)
- Moravia	32.6 (342)
Size of the city	
- up to 1000 inhabitants	16.2 (170)
- 1001–5000 inhabitants	20.8 (218)
- 5001–20,000 inhabitants	18.2 (191)
- 20,001–100,000 inhabitants	21.5 (226)
- > 100,000 inhabitants	23.3 (245)
Frequency of forest visits	
- Several times/week	17.0 (178)
- Once a week	27.0 (283)
- Once a month	33.2 (349)
- One or two times/year	21.8 (229)
- Never	1.0 (11)

The majority of respondents reported that they visited the forests regularly (about 77.1%), 21.8% of them admitted they rarely went to a forested region (one or two times per year), and only 1.0% of them never visited the forest. Those who had never visited the forest came from lower education levels (three from elementary and eight persons from secondary school/vocational training), aged between 25–54 years.

When we used binomial groups of forest visitors (0 = frequently, 1 = never and one or two times per year) and correlated the groups with other characteristics of the respondents, we found that significantly more people visited forests regularly across the regions than those who never went or went once or twice per year ($p = 0.048$). However, 46.7% of respondents from secondary school/vocational training graduates were found to be in the group of never to rarely visiting the forest, which was significantly higher than other groups of education levels ($p = 0.004$). Binary logistic regression revealed the predictor of secondary school/vocational training were about 1.6-fold more likely (95% confidence limit of 1.21–2.17) to rarely to never visit the forest ($p = 0.001$).

3.2. Wood-Based Forest Products

In total, 46.3% of respondents considered that wood is a better material for building construction than other non-wood materials. The reasons for selecting wood as a material used in building construction are presented in Figure 1. The major positive reasons were its short preparation and environmentally-friendly material, followed by it being an energy saver and having a reasonable price. In contrast, its short lifespan and non-fireproof material prevented respondents from selecting wood as a building material.

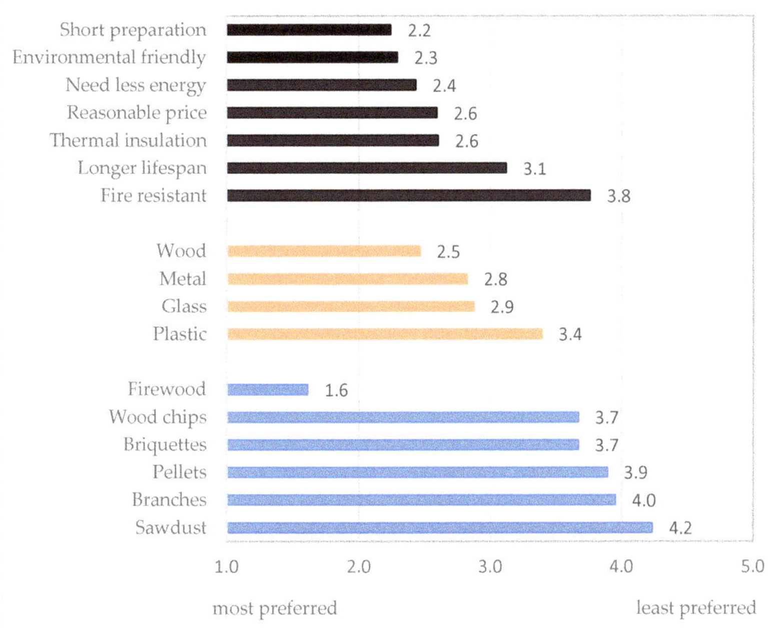

Figure 1. Reasons for selecting wood as a building materials (n = 1050), preferences of furniture material (n = 1050), and type of wood preferences for source of energy (n = 450).

Woods are the most preferred material for furniture over metal, glass, and plastic. As many as 96.3% of the respondents answered with a score of one (1) or two (2) for a preference of wood materials.

In total, 38.6% (n = 405) of the respondents were users of fireplaces, wood stoves/burners, or wood boilers. Firewood was the most favoured, while sawdust was the least preferred compared to other types of fuelwood (Figure 1).

3.3. Selected Non-Wood Forest Products Utilization

In this study we used the terminology of non-wood forest products based on the Food and Agriculture Organization (FAO) definition that excludes all woody raw materials [27]. Table 2 presents information about the NWFPs' utilization by the respondents and/or their family members. Mushrooms were the most favoured NWFPs (58.5% and 27.4% of the respondents certainly and rather used them, respectively), followed by forest berries. Forest herbs and flowers were among the least utilized NWFPs. The preferences were similar across age groups and education levels.

Table 2. Utilization of non-wood forest products.

Non-Wood Forest Products (N = 1050)	Utilization % (n)				
	1 (Certainly Not)	2 (Rather Not)	3 (Neither)	4 (Rather Yes)	5 (Certainly Yes)
Mushrooms	2.5 (26)	3.5 (37)	8.1 (85)	27.4 (288)	58.5 (614)
Berries	2.8 (29)	6.6 (69)	12.1 (127)	30.7 (322)	47.9 (503)
Forest honey	15.6 (164)	15.6 (164)	19.1 (201)	27.0 (284)	22.6 (237)
Forest herbs	9.1 (96)	23.7 (249)	24.6 (258)	23.5 (247)	19.0 (200)
Forest flowers	26.6 (279)	27.1 (285)	18.8 (197)	16.5 (173)	11.0 (116)

In total, 46.7% of the respondents reported that they did not consume meat from game animals, followed by 18.6% and 21.4% of them, who consumed it less than two times and two to four times per year, respectively. A few of the respondents reported the frequent consumption of game animals of five to eleven times per year (7.1%), once a month (3.7%), and more than once a month (2.5%).

Among respondents who utilized NWFPs, gender played a role in their preferences. Female respondents utilized herbs (54.6%) and flowers (54.7%) significantly more than male respondents (45.4% and 45.3%, respectively). Additionally, more male respondents were engaged in the consumption of game animals. The youngest age group and elementary school graduates were major forest flower collectors for decoration.

3.4. Sources of Information

The main sources of information about forests were collected to identify the means of communication utilized by the respondents. We discovered that television (55%), followed by friends and families (39.6%), were the main sources of information. Online news (32.2%) was used as a source of information in almost the same proportion as social media (29.6%). Radio and printed media as two conventional sources of news were still used by 14.6% and 13.3% of people in the country, respectively. Out of 4.4% (n = 46) of respondents who reported gathering information about forests from other sources, almost half of them (n = 20) said they had gained it from their own experiences after visiting the forest, followed by seven respondents who had obtained it from an official forestry website.

When correlating age and education level with source of information, the proportion of TV viewers (n = 578) was dominant across all age groups and education levels (Figure 2). The young age group of 18–24 years old, and respondents with a higher education level, were the major users of all of the provided sources of information.

Especially within older age groups of respondents, the preference for receiving information from TV was more than 60%. A high proportion of respondents of secondary school/vocational training were TV viewers (61.1%). The majority of respondents within the age group of 18–54 years utilized online news and social media as their primary sources of news. Among users, male respondents (56.5%) were online media readers ($p = 0.012$), while the women were social media users (55%, $p = 0.018$). Peer groups and families were considered an important source of information, especially within the age group of 18–24 years old. Although radio and printed media were at the bottom two of the sources, retired respondents still considered radio as an important information provider.

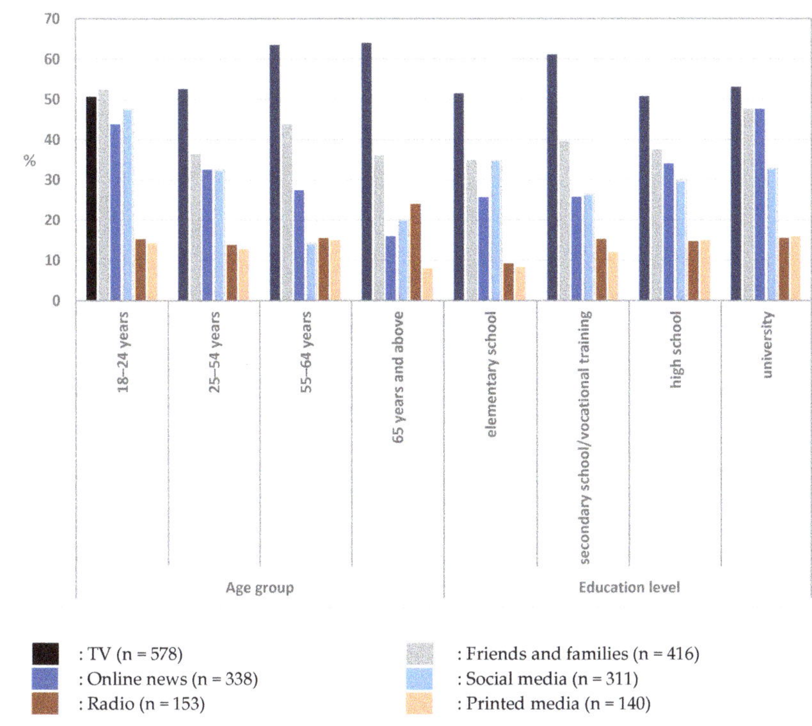

Figure 2. Sources of information based on age group and education level.

4. Discussion

As the main forest product, woods are utilized for different purposes, for instance roundwood for industrial purposes, pulp and paper, housing, and furniture material. [28]. Wood for building materials, furniture, and energy sources were considered to be commonly used in the Czech Republic [19]. Between the years 2000 and 2018, the Czech statistical office reported an increased trend (up to 15%) in wooden house/building constructions [29]. In this survey, 46.3% of the respondents reported that they considered wood a better material for housing than other materials, due to the short duration of material preparation, its environmentally friendly factor, the reasonable price, and its thermal insulating character. In contrast, wood's short lifespan and non-fireproof characteristics were in the bottom two of the reasons. Our results revealed that the strength and weakness of wood could either discourage or encourage customers to buy wood materials. By aiming to find solutions that could produce wood materials with a long lifespan, that are fireproof and sustainable at a reasonable price, wood consumption in the country can be improved. Additionally, the Czech statistical office also reported that utilization of wood as a furniture material (excluding kitchen furniture) in the country was similar over the years, especially 2017–2018 [30]. Our survey presented a preference for wood for furniture materials, which gives an opportunity for the development of value-added products in this sector, too. This can be seen also as a potential market for producers of the local wood furniture, with the caution of a potential change in consumption patterns from roundwood to, e.g., plywood furniture.

In 2015, the Czech statistical office reported that about 25% of utilized energy was from renewable resources. From 1995 to 2015, the trend of using energy from renewable resources nearly doubled. Use of renewable energy source was mostly at household levels (66.5%), followed by industry (25.2%) and other sectors (8.3%) [31]. Wood as a renewable energy resource was used by 38.6% of our respondents. Firewood was the most preferred type of wood as a source of energy. The results imply that the simple use of wood as an energy source was still favourable. In this survey, we did not ask further questions regarding the respondents' reasons for selecting an energy source at home. About two million m^3 per capita of wood was used as a source of energy in the country, which was lower compared to the roughly twelve and six million m^3 users in Germany and Austria, respectively, indicating that it is challenging to change the practice of using a fossil-based heater to a wood boiler energy source.

Studies from Finland, Germany, and Austria reported growing economic activity in rural areas, which includes biomass-based value chains and the economic utilization of other types of FES [32–34]. In comparison to Austria and Germany (about 14,800 and 19,900 full-time equivalent (FTE), respectively), the country had 42,500 FTE employmen in forest-based sectors. However, the annual change rate of FTE decreased in the Czech Republic (− 6.85%) [35]. In the Czech Republic, the number of employees in the forestry sector has systematically decreased in the last two decades, mainly due to low wages in this sector. This mainly concerns workers, and started to manifest itself negatively in the last two years in connection with the bark beetle outbreak. There is a lack of staff both in logging activities (inability to remove infested trees within the legal deadline), and especially in planting activities (there are not enough workers to plant trees). Therefore, the promotion of production in the bioeconomy, and changing the consumption pattern of the Czech people in the future, could potentially improve the employment situation in the forest-based sector and its value chains. The new challenge is to launch a better use of biological renewable resources in the Czech Republic in a suitable form for the future of society in sustainable manner.

To date, only wood forest products are considered to provide significant economic value for the Czech forest owners. Through the adoption of the forest bioeconomy strategy, it is expected that other FESs will also be promoted. NWFPs, especially mushrooms and different varieties of berry, were important for the recreational activity and socio-economic value of the Czech public [36,37]. The Czech MoA [22] reported that mushrooms were the most picked NWFPs (21,900 kg per year), and, altogether, collected berries amounted to 17,000 kg per year. Our current study also found a similar trend in preferred NWFPs (mushrooms and berries), which suggests the importance of further promoting this FES. In this survey, we did not ask how the respondents obtained the NWFPs (e.g., self-picking or buying the berries, honey, etc.). However, we discovered the importance of the respondent's gender and their preferred NWFPs, with women preferring more diverse NWFPs than men, such as collecting forest flowers and honey. In Switzerland, women were also reported as potential consumers of NWFPs [38].

In contrast, hoof game meat seemed to be of more interest for males compared to females. The Czech statistical office reported that consumption of game meat was still considered to be low (1.2% from the total consumed meat per year per capita). Additionally, the statistical office also reported an increasing trend of game animal consumption, from 0.5 kg per capita per year in 2006 to 0.9 kg in 2013 [39]. Game animals, such as wild boars and deer, are raised more naturally than livestock on a farm. The potential increase in game meat consumption due to hunting activities could cause problems in the forest, such as damage to the growth of young shoots. Additionally, the argument surrounding the function of hunting game as a recreational activity (e.g., getting a trophy, as a type of sport) and wildlife management [40] is an ongoing discussion among associated stakeholders in the Czech Republic, in addition to the problem of too high a stock of hoofed game [22].

By visiting the forests, the people utilize the recreation services of the forests. Access to the Czech forest is a public right. At the moment, there are no official data regarding the frequency of the forest visits of the Czech general population. However, concerning mushroom and berry picking, 90% of Czech visitors visited the forests at least once a year, of which 20% visited on a weekly basis [36]. Our study reported that a very high proportion (99%) of respondents aged 18–65 years visited the forest at least once per year (Table 1), which was higher than the 90% of Czech forest visitors in 2005 [36]. The reason for this was probably due to the age range of our respondents, which was at a productive age, and not including those below 15 and above 65 years, who are less active and more dependent. Meanwhile, 66.67% of the German general population went to forests at least once a year [17], while 40% of Austrians visited forests every week [41]. Out of the 99% of forest visitors in our survey, 77.1% of them visited the Czech forest frequently. However, detailed reasons for visiting the forest were not asked for in this survey. In another national survey (The Market & Media & Lifestyle), walking and doing sport were the largest drivers of the Czech forest visits [42]. By understanding the drivers of forest visits, programs concerning the promotion of FES could be well targeted and developed by the respective forest owners and enterprises.

Our results also revealed that the proportion of respondents that have graduated from secondary school/vocational training (46.7%) was significantly higher in the group of never or rarely visited the forests than other education levels ($p = 0.004$). Based on the results of binary logistic regression analysis, respondents that have graduated from secondary school/vocational training had a 1.6 times odds ratio of not visiting or rarely going to the forest ($p = 0.001$). The education plans of the secondary vocational schools involve more practical and physical work than regular high schools, which lead to more physical jobs and rotating work schedules (shift). In this study, we did not collect information on the type of occupation of the respondents. It was likely that the respondents who had graduated from the secondary school/vocational training had time constraints to do recreational activities due to the type of occupation (excluding those who graduated from a forestry or agriculture secondary vocational school).

In order to increase public awareness with regards to the adoption of a bioeconomy strategy, it is important to select an appropriate information provider, since changes in consumer behavior are expected. Respondents in this national survey still favoured TV and the peer group/family as major information channels. Online news and social media were the next most frequently selected information providers, especially for the respondents below 55 years old. We also found that gender influenced preferences in utilizing online news (male) or social media (female). The internet has been predicted to win over more traditional information providers [43]. But, since a high proportion of respondents in this survey selected TV as the source for information, therefore, TV should still be considered as an important channel, together with other selected mediums.

5. Conclusions and Recommendations

The results of the survey presented wood products as the most preferred material for furniture (96.3%) and building materials (46.3%). We found that 38.6% of Czech residents used wood as a source of energy, mostly in the form of firewood. It is challenging to change the practice of using fossil-based heaters to wood boiler energy sources. However, by addressing the positive attributes of wood and their impact on the future environment, it is likely that an increase in awareness and changes in consumer patterns can be expected. As the country has not yet adopted a bioeconomy strategy, a review study on the forest bioeconomy in other European countries is recommended, in order to give a better understanding of the impacts of bioeconomy strategies on a country's economic growth, particularly in the forest-based sector. The promotion of wood and non-wood forest products is encouraged, starting with increasing awareness and knowledge of the strength of the forest-based sector as a renewable energy resource and the importance of forest ecosystem services for recreation, health, and the well-being of the people. The increasing trends in NWFPs' utilization can be further promoted by creating different events in relation to respective FESs, while being careful to regulate the sustainable and shared responsibility aspects of protecting the forest. We also propose to investigate the reasons for visiting the forest and design a targeted program for a specific population (based on age, education level, type of occupation, or gender) in the country. It is important to utilize all channels as an information source for the importance of FES and the bioeconomy, depending on the target groups. We also propose to include forestry extension programs at school, especially in secondary vocational schools, aiming to inform people about and link people with forests, and to improve their participation in safeguarding the environment.

Author Contributions: Conceptualization, methodology, validation, M.H.; writing—original draft preparation, R.C.P. and M.H.; writing—review and editing, R.C.P., M.Š., M.H. and V.J., visualization, R.C.P., V.J. and M.H.; supervision, M.H.; project administration, M.H.; funding acquisition, M.H. All authors have read and agreed to the published version of the manuscript.

Funding: This research was financed by the Operational Program Research, Development and Education (OP RDE), the Ministry of Education of the Czech Republic, grant no. CZ.02.1.01/0.0/0.0/16_019/0000803 and by the Ministry of Agriculture of the Czech Republic, grant no. QK1920391.

Acknowledgments: The authors thank the support from the project "Advanced research supporting the forestry and wood-processing sector´s adaptation to global change and the 4th industrial revolution" and the project "Diversification of the Impact of the Bioeconomy on Strategic Documents of the Forestry-Wood Sector as a Basis for State Administration and the Design of Strategic Goals".

Conflicts of Interest: The authors declare no conflict of interest.

References

1. European Commission. *Innovating for Sustainable Growth: A Bioeconomy for Europe*; European Commission: Brussels, Belgium, 2012; p. 16. ISBN 978-92-79-25376-8.
2. European Commission. *A Sustainable Bioeconomy for Europe: Strengthening the Connection between Economy, Society and the Environment. Updated Bioeconomy Strategy*; European Commission: Brussels, Belgium, 2018; p. 4. ISBN 978-92-79-94144-3.
3. Winkel, G. Introduction. In *Towards a Sustainable European Forest-Based Bioeconomy. Assessment and the Way forward. Winkel, G. Ed.*; European Forest Insitute: Joensuu, Finland, 2017; pp. 15–18. ISBN 978-952-5980-42-4.
4. Glück, P. Social Values in Forestry-Synopsis. *Ambio* **1987**, *16*, 158–160. Available online: https://www.jstor.org/stable/4313346?seq=1 (accessed on 3 May 2019).
5. Giurca, A.; Späth, P. A Forest-Based Bioeconomy for Germany? Strengths, Weaknesses and Policy Options for Lignocellulosic Biorefineries. *J. Clean. Prod.* **2017**, *153*, 51–62. [CrossRef]
6. Millennium Ecosystem Assessment. *Ecosystems and Human Well-Being: Synthesis*; Island Press: Washington, DC, USA, 2005; p. VI. ISBN 1-59726-040-1.
7. Nowak, D.; Noble, M.H.; Sisinni, S.M.; Dwyer, J.F. People and Trees: Assessing the US Urban Forest Resource. *J. For. Res.* **2001**, *99*, 37–42.
8. Krieger, D. *Economic Value of Forest Ecosystem Services: A Review*; The Wilderness Society: Washington, DC, USA, 2001; pp. 1–31.
9. Millennium Ecosystem Assessment. *Ecosystems and Human Wellbeing: Opportunities and Challenges for Business and Industry*; World Resources Institute: Washington, DC, 2005; pp. 3–5.
10. Pettenella, D.; Secco, L.; Maso, D. NWFP&S Marketing: Lessons Learned and New Development Paths from Case Studies in Some European Countries. *Small-scale For.* **2007**, *6*, 373–390.
11. Nijnik, M.; Nijnik, A.; Brown, I. Exploring the Linkages between Multi-Functional Forestry Goals and the Legacy of Spruce Plantations in Scotland. *Can. J. For. Res.* **2016**, *46*, 1247–1254. [CrossRef]
12. Grêt-Regamey, A.; Altwegg, J.; Sirén, E.; van Strien, M.; Weibel, B. Integrating Ecosystem Services into Spatial Planning-A Spatial Decision Support Tool. *Landsc. Urban Plan.* **2016**, *165*, 206–219. [CrossRef]
13. de Arano, I.M.; Muys, B.; Topi, C.; Petenella, D.; Feliciano, D.M.S.; Rigolot, E.; Lefevre, F.; Prokofieva, I.; Labidi, J.; Carnus, J.M.; et al. *A Forest-Based Circular Bioeconomy for Southern Europe: Visions, Opportunities and Challenges. Reflections on the Bioeconomy*; European Forest Institute: Joensuu, Finland, 2018; pp. 1–119.
14. Marusakova, L.; Sallmannshofer, M.; Kaspar, J.; Schwarz, M.; Tyrvainen, L.; Bauer, N. Human Health and Sustainable Forest Management. In *Human Health and Sustainable Forest Management*; Marusakova, L., Sallmannshofer, M., Eds.; Forest Europe—Liaison Unit Bratislava: Zvolen, Slovakia, 2019; pp. 58–97. ISBN 978-80-8093-266-4.
15. Eurostat-European Commission. Agriculture, fishery and forestry statistics, Main results—2010-11. 2012. Available online: https://ec.europa.eu/eurostat/documents/3930297/5967972/KS-FK-12-001-EN.PDF/0de35d0b-aad0-4cfa-9319-c30f05d46ace (accessed on 10 January 2020).
16. Ministry of Agriculture of the Czech Republic. *National Forest Programme for the Period until 2013*; Ministry of Agriculture of the Czech Republic: Praha, Czech Republic, 2008; p. 7. ISBN 978-80-7084-758-9.

17. Federal Ministry of Food, Agriculture and Consumer Protection (Bundesministerium für Ernährung, Landwirtschaft und Verbraucherschutz/BMELV). *Forest Strategy 2020 Sustainable Forest Management—And Opportunity and Challenge for Society*. Available online: https://www.bmel.de/SharedDocs/Downloads/EN/Publications/ForestStrategy2020.pdf?__blob=publicationFile (accessed on 3 May 2019).

18. Federal Ministry for Sustainability and Tourism. *Austrian Forest Strategy 2020+*; Federal Ministry for Sustainability and Tourism: Wien, Austria, 2018; p. 46.

19. Ministry of Agriculture of the Czech Republic (MoA). Information on Forests and Forestry in The Czech Republic by 2017. Available online: http://eagri.cz/public/web/file/615927/Zprava_o_stavu_lesa_2017_ENG.pdf (accessed on 10 October 2019).

20. Eurostat-European Commission. Forestry statistics in detail. Statistics Explained. Available online: https://ec.europa.eu/eurostat/statistics-explained/pdfscache/29576.pdf (accessed on 10 May 2019).

21. FAO. Global Forest Products Facts and Figures 2016. Available online: http://www.fao.org/3/i7034en/i7034en.pdf (accessed on 3 May 2019).

22. Ministry of Agriculture of the Czech Republic (MoA). Information on Forests and Forestry in The Czech Republic by 2012. Available online: http://eagri.cz/public/web/file/272639/ZZ_2012_ENG.pdf (accessed on 10 January 2020).

23. Gruber, J.; Milligan, K.; Wise, D. *Social Security Programs and Retirement Around the World: The Relationship to Youth Employment, Introduction and Summary*; NBER Working Papers; National Bureau of Economic Research, Inc., 2009; no. wp. 14647; Available online: https://doi.org/10.3386/w14647 (accessed on 10 October 2019).

24. Chomik, R.; Piggott, J. Mature-age labour force participation: Trends, barriers, incentives, and future potential. Available online: http://cepar.edu.au/sites/default/files/Mature-age_labour_force_participation.pdf (accessed on 10 October 2019).

25. Czech Statistical Office. Population Statistics Department (Kde a jak bydlí české domácnosti?). Available online: https://www.czso.cz/csu/czso/kde-a-jak-bydli-ceske-domacnosti-p2eqbgktkl (accessed on 11 November 2019).

26. Czech Statistical Office. Population Statistics Department. Educational Level of Population According to Census Results. Available online: https://www.czso.cz/documents/10180/20536250/17023214.pdf/7545a15a-8565-458b-b4e3-e8bf43255b12?version=1.1 (accessed on 9 November 2019).

27. Dembner, S.A.; Perlis, A. Towards a Harmonized Definition of Non-Wood Forest Products. *Unasylva* **1999**. Issue No. 198. Available online: http://www.fao.org/3/x2450e/x2450e0d.htm#fao%20forestry (accessed on 9 January 2020).

28. FAO; UNECE. Forest Products Annual Market Review 2017–2018. Available online: https://www.unece.org/fileadmin/DAM/timber/publications/FPAMR2018.pdf (accessed on 11 November 2019).

29. Czech Statistical Office. Press Release - Building Construction Has Been Successful in Recent Years (Stavebnictví Se v Posledních Letech Daří). Available online: https://www.czso.cz/csu/czso/stavebnictvi-se-v-poslednich-letech-dari (accessed on 12 December 2019).

30. Czech Statistical Office. Manufacture of Selected Products from Industry - 2018 (Výroba Vybraných Výrobků v Průmyslu - 2018). Available online: https://www.czso.cz/csu/czso/vyroba-vybranych-vyrobku-v-prumyslu-2018 (accessed on 12 December 2019).

31. Czech Statistical Office. Fuel and Energy Consumption in Households. Department of Industry, Construction and Energy Statistics (Spotřeba Paliv a Energií v Domácnostech). Available online: https://www.czso.cz/documents/10180/50619982/ENERGO_2015.pdf/86331734-a917-438a-b3c2-43a5414083fc?version=1.4 (accessed on 12 December 2019).

32. Finnish Environment Institute. Finnish Environment Institute Reports 13 | 2017. Renewal of forest based manufacturing towards a sustainable circular bioeconomy. 2017. Available online: https://pdfs.semanticscholar.org/c5e9/20375fde67380d02f152a505f01352768931.pdf (accessed on 13 December 2019).

33. Biobased Industries Consortium. Bioeconomy Regions in Europe. 2017. Available online: https://biconsortium.eu/sites/biconsortium.eu/files/publications/BIC_GA_Brochure_Bioeconomy_regions_in_Europe_Nov_2017.pdf (accessed on 9 January 2020).

34. BIOPRO. Country Report. Cross-Clustering Partnership for Boosting Eco-Innovation by Developing a Joint Bio-Based Value-Added Network for the Danube Region. Framework Conditions for Cluster Development in Bio-Based Industry in the Region of Baden-Württemberg, Germany. 2018. Available online: http://www.ipe.ro/Country%20Report%20Baden%20W.pdf (accessed on 5 December 2019).

35. FAO. Global Forest Resources Assessment 2015: Desk Reference. Rome. 2017. Available online: http://www.fao.org/forest-resources-assessment/past-assessments/fra-2015/en/ (accessed on 5 December 2019).

36. Šišák, L. Importance of Non-Wood Forest Product Collection and Use for Inhabitants in the Czech Republic. *J. For. Sci.* **2006**, *52*, 417–426. [CrossRef]

37. Šišák, L.; Riedl, M.; Dudik, R. Non-Market Non-Timber Forest Products in the Czech Republic-Their Socio-Economic Effects and Trends in Forest Land Use. *Land Use Policy* **2016**, *50*, 390–398. [CrossRef]

38. Seeland, K.; Kilchling, P.; Hansmann, R. Urban Consumers' Attitudes Towards Non-Wood Forest Products and Services in Switzerland and an Assessment of Their Market Potential. *Small-scale For.* **2007**, *6*, 443–452. [CrossRef]

39. Czech Statistical office (CZSO). Consumption of Food and Non-Alcoholic Beverages (Annual Per Capita Averages). 2015. Available online: https://www.czso.cz/documents/10180/20562003/2701391501.pdf/1547f1b0-eeac-482f-8ea0-2289d3b4ed3e?version=1.1 (accessed on 25 November 2019).

40. Fischer, A.; Sandström, C.; Delibes-Mateos, M.; Arroyo, B.; Tadie, D.; Randall, D.; Hailu, F.; Lowassa, A.; Msuha, M.; Kereži, V.; et al. On the Multifunctionality of Hunting—An Institutional Analysis of Eight Cases from Europe and Africa. *J. Environ. Plan. Manag.* **2013**, *56*, 531–552. [CrossRef]

41. Federal Ministry for Agriculture, Forestry Environment and Water Management. *Forest in Austria*; Federal Research Centre for Forests, Natural Hazards and Landscape: Vienna, Austria, 2017; p. 24. ISBN 978-3-902762-73-3.

42. Šodková, M.; Purwestri, R.C.; Riedl, M.; Jarský, V.; Hájek, M. What Drives the Forest Visit? Results of a National Survey in Czech Republic. unpublished work, manuscript in preparation. 2020.

43. Havick, J. The Impact of the Internet on a Television-Based Society. *Technol. Soc.* **2000**, *22*, 273–287. [CrossRef]

 *sustainability* 

Article

One Concept, Many Opinions: How Scientists in Germany Think About the Concept of Bioeconomy

Carmen Priefer [1,2] **and Rolf Meyer** [1,*]

[1] Institute for Technology Assessment and Systems Analysis (ITAS), Karlsruhe Institute of Technology (KIT), 76133 Karlsruhe, Germany

[2] Department of Nutritional Behaviour, Federal Research Institute of Nutrition and Food (Max Rubner-Institut), 76131 Karlsruhe, Germany

* Correspondence: rolf.meyer@kit.edu

Received: 19 June 2019; Accepted: 1 August 2019; Published: 6 August 2019

Abstract: The official bioeconomy strategies in Europe and Germany pursue a technology-based implementation pathway and stipulate a wide range of objectives to be achieved with a bio-based economy. Reviews of the scientific and societal debate have shown that the technology fix meets criticism and that there is a controversial discussion about possible ways to shape the transition process. Against this background, an online survey was carried out among scientists involved in a regional bioeconomy research program in southern Germany in order to gain insight into their understanding of a bioeconomy. Moreover, the survey provides information about cooperation and major challenges in the future development of three biomass utilization pathways: biogas, lignocellulose, and microalgae. The analysis showed that a resource-oriented understanding of a bioeconomy is favored. The political objectives for a European bioeconomy are widely accepted, and it is expected that ongoing research can significantly contribute to achieving these goals. The two different pathways for shaping the bioeconomy that are discussed in the debate—the technology-based approach and the socio-ecological approach—are considered compatible rather than contrary. Up to now, scientific cooperation has prevailed, while cooperation with societal stakeholders and end-users has played a minor role.

Keywords: bioeconomy; survey; strategies; research program; biogas; lignocellulose; microalgae

---

1. Introduction

Over the last 10 years, bioeconomy has become an important issue in research and innovation policy, especially in industrialized countries. The core idea of the concept is to replace nonrenewable fossil fuel resources with renewable biogenic feedstock in industrial and energy production. This is meant to pave the way for a more sustainable and eco-efficient economy [1]. Worldwide, a number of countries and international organizations have already developed dedicated bioeconomy strategies, such as the European Union (EU), the Organization for Economic Co-Operation and Development (OECD), the West Nordic countries (Iceland, Greenland, and the Faroe Islands), Australia, Finland, France, Germany, Japan, Malaysia, South Africa, Spain, Sweden, and the United States [2–4]. In other countries, strategies are currently under development. Besides international and national activities, a number of regional bioeconomy initiatives and innovation networks have been established [5]. The Bio-Based Industries Joint Undertaking (BBI JU), a public–private partnership between the European Commission and the Bio-Based Industries Consortium started in 2014 under Horizon 2020, is a key activity at the EU level for implementing the EU bioeconomy strategy [6].

These official strategies use different definitions of bioeconomy, ranging from equalization with biotechnology to a broad societal transition involving a variety of technologies and economic sectors [7,8]. In the literature, it has been pointed out that the terms "bioeconomy" and "bio-based economy" have no clear and binding definition [9]. Definitions strongly depend on the stakeholders involved and their interests, but also show some similarities [10]. Overall, heterogeneous bioeconomy definitions are used for similar contexts.

Various analyses of bioeconomy debates have worked out that different and even contrasting visions and transition pathways are envisioned [7,8,11–14]. Official bioeconomy strategies pursue either a biotechnology vision or a transformation respectively bioresource vision [7,11]. Despite their different foci, all official strategies envision a technology-based transition to a bioeconomy [7]. In the biotechnology vision, economic growth is based on biotechnologies, and the valorization of bioresources is expected to drive economic growth in the bioresource vision [12]. Life sciences and biotechnology and new biomass conversion technologies are seen as enabling technologies in various sectors. A strong partnership between policy, science, and industry; the promotion of international cooperation; the establishment of global value chains; and the granting of patents should improve international competitiveness and contribute to economic growth and employment [8].

In criticism of this understanding, an alternative vision of bioeconomy has been developed in the scientific and societal debate, focusing on socio-ecological aspects and alternatives to the currently dominant model of industrialized agriculture [8]. Rather than concentrating on biotechnology and/or conversion technologies for new value chains, the focus is on agroecological techniques and methods such as increasing plant genetic diversity, improving nutrient recycling, enhancing biodiversity, and improving the health of soil, crops, and livestock. This vision gives high priority to sustainability concerns and calls for a public goods-oriented bioeconomy [14,15]. The contrasting visions have different underlying objectives to be achieved in the context of bioeconomy (See Section 3.4).

The common understanding is that the bioeconomy should contribute to a more sustainable future. However, sustainability is understood differently in the two visions, and the sustainability requirements are often not clearly specified [8]. Common weaknesses and gaps in current bioeconomy strategies are seen in terms of the sound use of land and water resources, waste management along the whole value chain, possible competition between the different biomass end-use sectors, and mechanisms to benefit farmers [1]. Different understandings of the relationship between bioeconomy and sustainability have been identified, ranging from sustainability being an inherent characteristic of the bioeconomy to the bioeconomy having no positive impact and the bioeconomy being disadvantageous for sustainability [16]. This reflects different expectations about the future development of bio-based value chains and especially their environmental impacts.

Regional bioeconomy concepts and activities are expected to play a significant role. Regions are considered to be important in facilitating collaborations between industries and research institutions [17]. In contrast, developing international cooperation is an important objective of national bioeconomy strategies with their technology focus. The desired international cooperation involves different aspects, such as the exchange of information and knowledge transfer, international coordination between research and innovation activities, and the setting up of joint multinational research and technology activities [8]. In addition, a successful transition to bio-based feedstock would require huge volumes of biomass, which could not be mobilized sufficiently in Europe. The consequences could be an expanded international biomass trade and/or the relocation of processing capacities to regions with high biomass potential [18]. Instead, the alternative vision favors home-grown biomass that should be regionally traded and processed [19].

A close interaction between research and industry is a central element in most bioeconomy strategies [7]. The proposed thinking in value chains demands cooperation between different disciplines and actors. However, the establishment of new networks cannot be easily achieved [20] (p. 294).

Against this backdrop of bioeconomy strategies and controversies, the overall aim of our research was to analyze researchers' views on the bioeconomy by conducting an online survey among scientists involved in a regional bioeconomy research program using an explorative approach. The investigation was carried out in the context of the Bioeconomy Research Program Baden-Württemberg, which is funded by the Baden-Württemberg Ministry of Science, Research, and the Arts (MWK). The program aims to improve the strategic position of Baden-Württemberg's research institutions in the field of bioeconomy, provide support in the acquisition of third-party funding, and draw industry's attention to the possibilities of new bio-based products and processes in order to accelerate the transfer of research results into practice [21] (p. 9). The federal state's research landscape is characterized by agricultural and forestry science, biology, biotechnology, chemistry, and process engineering [21] (p. 50).

In the first funding round of the Bioeconomy Research Program Baden-Württemberg, around 50 projects were funded. The research program is thematically divided into the following three fields (hereafter also referred to as clusters or networks):

- Sustainable and flexible value chains for biogas in Baden-Württemberg;
- Lignocellulose as an alternative resource platform for new materials and products; and
- The Integrated use of microalgae for nutrition.

The production of electricity, heat, and methane in biogas plants is a long-established technology in Baden-Württemberg. Biogas is intended to play an important role in Baden-Württemberg's future energy mix, especially for decentralized electricity and a heat supply in rural areas and as control energy to compensate for the fluctuating supply of solar and wind energy. In line with this goal, the projects in this research field explore options for increasing performance and optimizing the process and system integration of existing plants; estimate the potential of alternative substrates such as sewage sludge, residual, and waste materials; and deal with the further technical development of methane production. In contrast to biogas production, the use of lignocellulosic biomass in biorefineries and the industrial use of microalgae are still in the research and development phase. Accordingly, the projects in these two fields are more basic research-oriented. In the lignocellulose cluster, the focus is on improving pulping processes; optimizing synthesis routes; and assessing the availability, combinability, and improvement of feedstocks and the ecological and economic effects of the cultivation and use of lignocellulose. The microalgae research field is mainly concerned with improving extraction methods, optimizing cultivation processes and product quality, and establishing production systems.

The following sections describe the methodology used and present the results of the survey, also addressing similarities and major differences between the clusters. Finally, the main findings are summarized and the results are compared to existing investigations.

2. Materials and Methods

The aim of the survey was to analyze the scientists' understanding of the concept of bioeconomy and their attitudes toward the objectives of selected bioeconomy strategies, to enquire about the current status of the projects (e.g., existing collaborations), to gain insight into certain problem areas (e.g., competing uses and environmental impacts), and to identify similarities and differences between the three investigated biomass value chains. In order to gather this information, an online survey was conducted. All researchers involved in the biogas, lignocellulose, and microalgae networks within the Bioeconomy Research Program Baden-Württemberg (cluster spokespersons, cluster coordinators, project managers, project staff members, doctoral students) were interviewed. Prior to the survey being carried out, the inquiry was announced at the respective cluster meetings in order to attract the interest of the scientists. The survey started in mid-March 2016, and the online questionnaire was open for a period of six weeks. Due to the high number of international scientists in the lignocellulose and microalgae clusters, these groups were also offered an English version of the questionnaire. The survey was developed based on the results of literature reviews on existing strategies [7] and scientific and societal discourse [8]. Before starting the survey, the questionnaire was tested on a group of experts from different scientific disciplines at the Institute for Technology Assessment and Systems Analysis (ITAS) and with spokespersons from the three thematic clusters. The questionnaire (Supplement S1) consisted of five sections:

- Topic 1. Perspectives on bioeconomy;

- Topic 2. Value chains;
- Topic 3. Potentials and competing uses;
- Topic 4. Effects on the environment;
- Topic 5. Collaborations.

Topics 1 and 5 were identical in content for all participants. The questions in Topic 1 addressed the level of agreement with definitions, objectives, and implementation pathways outlined in the bioeconomy strategies and debates. The purpose was to gain insight into the extent to which the bioeconomy strategies serve as a guiding function for researchers in the field. Topic 5 included questions on partners and issues of collaboration in order to identify the prevailing form of cooperation and to assess the extent to which the desired cooperation and networking is realized in the projects of the Bioeconomy Research Program Baden-Württemberg.

The questions in Topics 2 to 4 were adapted to the specific characteristics of the individual clusters. The purpose of the questions was not to obtain exact scientific data, but to elicit the scientists' assessments and personal opinions on the opportunities and limits of the individual biomass value chains based on their experience. The questionnaire included closed and open questions. Depending on the content of the question, multiple answers were possible. This is indicated in the figures, where applicable.

In total, 81 persons participated in the survey, of which only 49 completed the entire questionnaire. In relation to the total number of persons invited to the survey (144), this corresponded to a net response rate of 34%. The response rate was highest in the microalgae research cluster (44%), medium in the lignocellulose cluster (32%), and lowest in the biogas cluster (29%). Nevertheless, the response rates can be considered rather high for a social scientific survey. This success can be explained by the previous promotion of the survey in the cluster meetings. The response rate to questions in Topics 3 and 4 was lower than to those in the first parts of the questionnaire (about 24% on average compared to 34%). A possible reason could be that these questions were primarily related to accompanying research, and not all scientists have expertise in this field. In addition, these sections dealt with future-oriented issues that are still under research. Thus, scientists are not yet in a position to make an assessment. These parts are therefore presented in a qualitative and concise way without the claim of being representative of the opinion of all respondents.

Representativeness for the entire research community cannot be answered conclusively due to a lack of information on personal characteristics (such as age, qualification, position) of the participants in the Bioeconomy Research Program Baden-Württemberg. However, the responses related to scientific background (Section 3.1) indicated that the respondents represented a cross-section of scientific career stages.

The answers were evaluated quantitatively and qualitatively. The open questions were evaluated using the quantitative content analysis according to Mayring [22] in order to build content categories and group the answers into these content categories. The results presented in this paper focus on the scientists' understandings of bioeconomy and on existing collaborations. In addition, the "Results" section sketches challenges for the further development of the three research fields as indicated by the respondents under the topics "competing uses" and "environmental impacts".

3. Results

The following section summarizes the results for all three clusters and identifies the main differences between them. The results are not necessarily presented according to the questionnaire's sequence. Instead, particularly interesting statements, points of clear consensus, or opposing positions are outlined.

3.1. Scientific Background of Respondents

The scientists were asked about which field of activity their project could be assigned to. Multiple answers were possible. According to the responses, the majority (53%) related their projects to biomass production, conversion, and product development. The remaining 47% were assigned to evaluation, systems analysis, and societal framework conditions (Figure 1). Some respondents gave multiple answers, which indicated that their projects deal with different technical or different assessment aspects at the same time or even combine technical and nontechnical research areas such as sustainability assessment. In the biogas cluster, the share of assessment-related research activities was highest, with 58% of the responses. In the microalgae cluster, it was lowest, with 30%. This can be explained by the different development stages of the fields. While the biogas sector is well elaborated from a technical point of view and challenges arise from the lack of competitiveness and acceptance by local residents, microalgae production and conversion is a rather new approach that requires further technical research, with the consequence that an assessment of environmental and societal issues fades into the background.

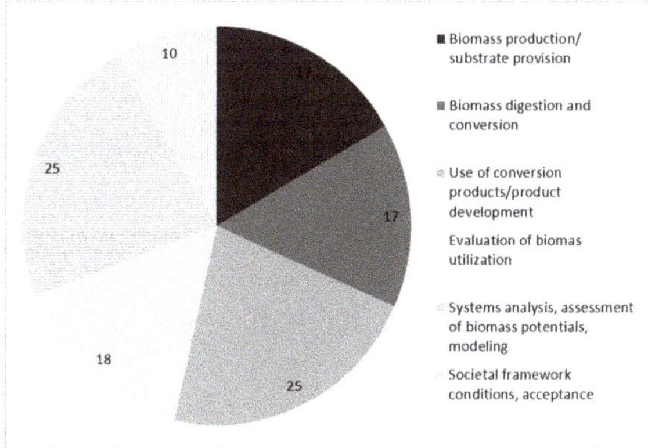

Figure 1. Respondents' assignment of projects to fields of activity (*n* = 79, n = number of answers, multiple answers possible).

Thirty-one percent of the respondents assigned their project to basic research and 69% to applied research (Figure 2). This relation was quite different for each cluster. In biogas research, 96% of the responses referred to applied research, while this share was only 55% in the lignocellulose cluster and 62% in the microalgae cluster. In all three clusters, a limited number of multiple answers were given, i.e., the respective research projects were assigned to both basic and applied research. The number of multiple entries was highest in the lignocellulose cluster, which could indicate that some research projects are in a transition process from basic to applied research or that the nature of the research is both basic and applied due to the interdisciplinary approach of most projects in this cluster. Moreover, the responses indicated that accompanying research projects in the fields of systems analysis, assessment, and societal framework conditions are classified as either basic or applied research.

In answering the question of how long the scientists have been active in the field of bioeconomy, 43% of the respondents stated that they had already done research on bioeconomy before starting the research project with the Bioeconomy Research Program Baden-Württemberg. Fifty-seven percent entered the field at the start of the project (Figure 3). In the biogas network, more than half of the researchers had previous experience in bioeconomy research, and the majority of respondents in the lignocellulose group (59%) first came into contact with the topic through the current project. In the microalgae network, this share was even higher, at 67%. These differences can be explained by the fact that Baden-Württemberg has been seen as a pioneer in biogas research and implementation since the 1980s, while biotechnology-driven microalgae and lignocellulose research are quite new fields. Overall, the answers reflected the mixed composition of the project teams, which consist of professors or project managers with many years of experience in the field of bioeconomy and doctoral candidates who have recently entered the subject.

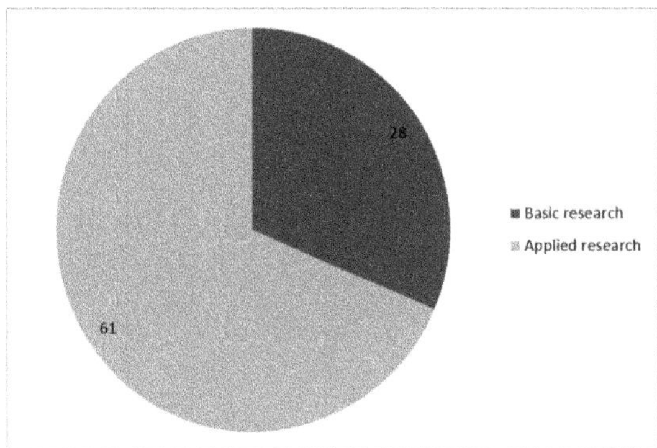

Figure 2. Respondents' assignment of projects to type of research (n = 80, multiple answers possible).

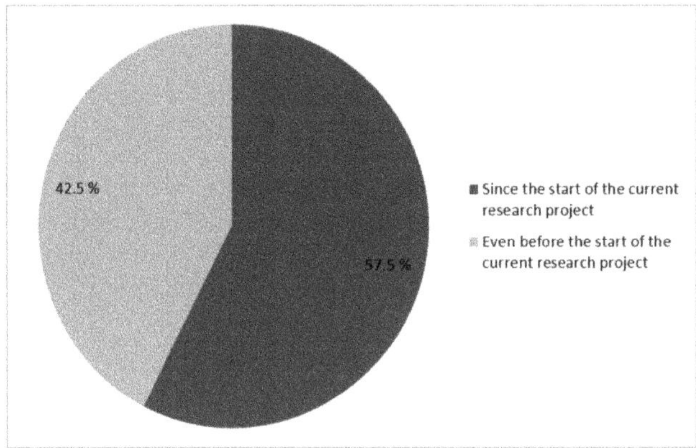

Figure 3. Respondents' familiarity with the bioeconomy topic (n = 80).

3.2. Understanding of Bioeconomy

In a first step, the scientists were asked to define bioeconomy in their own words. The aim of this open question was to get insight into the respondents' understanding of bioeconomy, regardless of already existing definitions.

The 56 definitions given were assigned to the following categories:

- Resource-oriented definition;
- Resource-oriented definition with focus on substitution;
- Resource-oriented definition with focus on sustainability;
- Resource-oriented definition with focus on science and technology;
- Technology-oriented definition.

Almost all of the definitions (89%) saw the production and use of biogenic resources as core elements of the bioeconomy (resource-oriented definition, see Figure 4). Thirty-six percent of the definitions were related to the resource-oriented understanding, and 37% went beyond the pure resource orientation and outlined that biomass use should also meet environmental and social sustainability criteria (sustainability focus). Sixteen percent of the definitions emphasized the substitution of fossil resources with renewable feedstock (substitution focus). Only 7% of the definitions stressed the importance of science and technology in the context of a resource-oriented understanding (science and technology focus). Besides these resource-oriented understandings, 4% of the definitions concentrated on the use of innovative technologies for biomass conversion (technology-oriented definition). While the resource-oriented definition played an important role in all research clusters, the science and technology focus was most supported by respondents from the lignocellulose cluster, and the technology-oriented definition appeared only in the microalgae cluster. Table 1 provides some examples of definitions for the five categories.

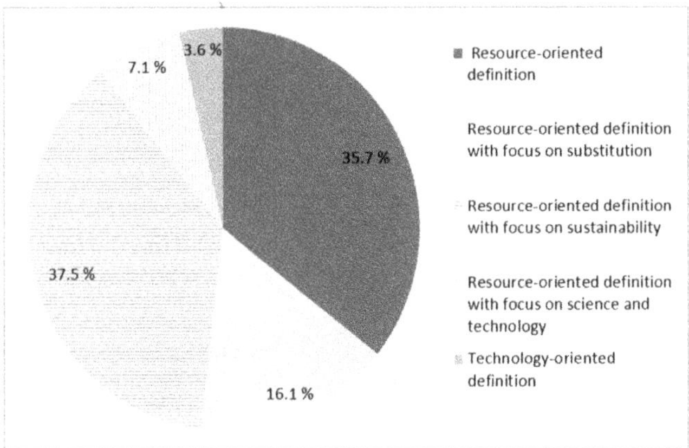

Figure 4. Assignment of the definitions given to categories of bioeconomy understandings ($n = 56$).

Table 1. Examples of bioeconomy definitions formulated by the interviewed scientists.

Type of Definition	Formulation
Resource-oriented definition	"Any steps involving the production, processing, and use of bio-based raw materials." "Use and efficient transformation of biological raw materials (with complex compositions) into defined products for society/humans."
Resource-oriented definition with focus on substitution	"Bioeconomy is the approach to replace the petrochemical-based industry with a bio-based industry, i.e., through renewable raw materials and residues." "Using biomass, instead of fossil fuel, to provide renewable and green energy and food, and establishing a whole industrial and social circulation."
Resource-oriented definition with focus on sustainability	"Bioeconomy describes an economy based on renewable raw materials instead of fossil raw materials. Furthermore, the economy should be as environmentally friendly and sustainable as possible." "Transformation process of an economy that is characterized by fossil fuels, profit, and exploitation (nature and humankind) toward an economy that is based on the careful use of renewable resources, with particular attention to social and ecological effects."
Resource-oriented definition with focus on science and technology	"Bioeconomy means using biological resources. Bioeconomy is based on the latest scientific findings and builds a bridge between technology, ecology, and an efficient economy." "Bioeconomy encompasses the sustainable production of biological resources and their conversion into food, feed, and high-value products/energy via innovative technologies from academic and industrial origins."
Technology-oriented definition	"Bioeconomy is a novel economic pattern that would generate economic effectiveness based on the research and application of biotechnology." "Bioeconomy is the combination of biotechnology and economy. That means it is based on the technology of biology to produce biomass, explore and apply production in order to gain economic benefit."

In a second step, the respondents were asked to indicate to what extent they agree with the definitions formulated in official bioeconomy strategies. This question aimed to find out about support or skepticism toward official bioeconomy definitions by scientists working in the field of bioeconomy.

The OECD definition was selected as an example of a technology-oriented understanding. The European Commission's definition focuses on the sectors involved in the bioeconomy (sector-oriented definition). The German Federal Ministry of Education and Research based its definition on the resources used (resource-oriented definition), and the definition by the German Federal Ministry of Food and Agriculture can be seen as an example of a target-oriented definition. The wording in the questionnaire was as follows:

- **Technology-oriented definition.** Bioeconomy is the development and application of modern biotechnology and life sciences (OECD [23]);
- **Resource-oriented definition.** Bioeconomy covers agriculture and forestry as well as all producing sectors with their respective service sectors that develop, produce, process, and work on biological resources (plants, animals, microorganisms) or use them in any form (Federal Ministry of Education and Research [24]);
- **Sector-oriented definition.** Bioeconomy covers the production of renewable biological resources and their conversion into products that generate additional value, such as food and animal feed, bio-based products, and bioenergy (European Commission [25]);
- **Target-oriented definition.** Bioeconomy combines economy and ecology in an intelligent way and thus enables bio-based and sustainable economic growth (Federal Ministry of Food and Agriculture [26]).

Overall, all types of definitions originating from official bioeconomy strategies were supported by the majority of respondents (Figure 5). However, the highest level of full agreement (over 60%) was recorded for the resource-oriented definition, followed by the target-oriented definition. This was in line with the results of the respondents' own definitions, where the majority of definitions were solely resource-oriented or had an additional focus on sustainability. The sector-oriented definition showed a lower level of full agreement. Nevertheless, summing up the answers "fully agree" and "rather agree", the sector-oriented definition received the highest agreement, together with the resource-oriented definition. The lowest level of full agreement was achieved for the technology-oriented definition. The proportion of people slightly disagreeing with this definition was highest, about 30%. The comparison to the researchers' own definitions showed that the given technology-oriented definition received more support than a similar understanding formulated by the respondents themselves.

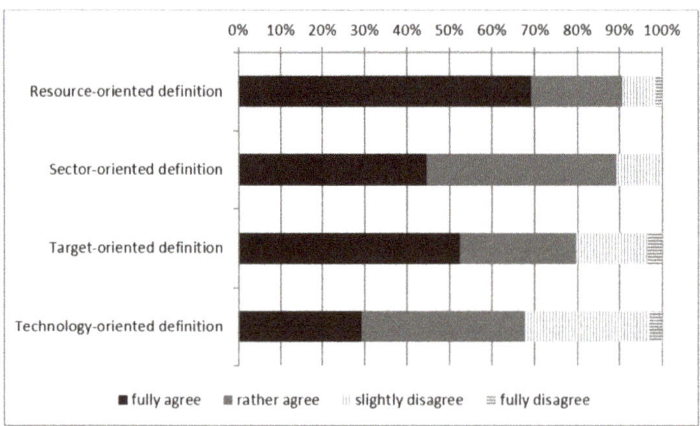

Figure 5. Respondents' agreement with official bioeconomy definitions differing in their thematic focus (*n* = 66).

When looking at the different research clusters, it became obvious that researchers from all clusters strongly agreed with the resource-oriented definition, which was in accordance with the researchers' own understanding. While the majority of the biogas cluster respondents rejected the technology-oriented definition, agreement was rather high among the respondents from the microalgae cluster, where biotechnology plays an important role.

3.3. Visions and Objectives

The bioeconomy strategies of different countries and institutions include visions and far-reaching societal goals that should be achieved through the realization of a bio-based economy. The researchers were asked to rate the goals of the German Policy Strategy Bioeconomy according to their importance. The goals relate to sustainability issues, international competitiveness, specific fields of technology and research, as well as cooperation between science, business, and society [26] (pp. 20 ff.). The results showed that all objectives were predominantly considered essential or important (Figure 6). The majority of the respondents viewed the following two targets as essential: "production of raw materials in line with environmental and climate protection" and "conversion of the economic production base from fossil to biogenic sources in the long run". This result matched well with the respondents' predominant agreement with the resource-oriented definition of the bioeconomy (see Section 3.2). Furthermore, the goals "implementation of cascade use", "sustainable consumption", "networking between science and business", as well as "inter- and transdisciplinary research" were regarded as relevant. The leading role of Germany in solving global challenges was rated as the least important goal. Between 20% and 30% of respondents considered "economic growth", "job creation", "strengthening of industrial biotechnology as an economic field", and "international competitiveness" as less important.

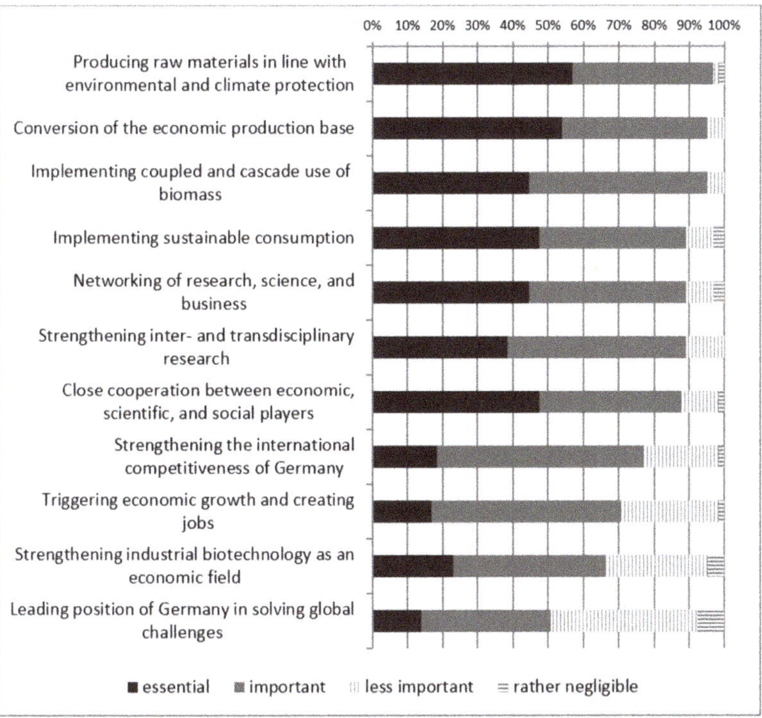

Figure 6. Respondents' evaluations of the objectives of the German Policy Strategy Bioeconomy according to importance (*n* = 65).

The comparison between the clusters showed that the aim of Germany taking a leading role in solving global challenges was considered to be of little importance in the biogas and microalgae clusters, while the lignocellulose network considered this goal essential or important. The production of raw materials in line with the objectives of environmental and climate protection was regarded as less important in the microalgae group than it was in the biogas and lignocellulose consortium, for which the proportion of "essential" mentions was significantly higher. This may have been due to the fact that microalgae production is a technical process that is independent from agricultural production and does not compete for land resources with other biomass utilization pathways such as food production. Furthermore, it is striking that half of the respondents from the microalgae cluster considered a strengthening of industrial biotechnology essential or important, while the other half considered this less or not important. Thus, there were different views on the economic importance of the broad industrial application of biotechnology in the algae network, which could be explained by the early development stage of algae technologies and the resulting uncertain applications. The majority of respondents from the biogas cluster attached importance to strengthening industrial biotechnology as an economic sector with great market and value-added potential, while the majority of this group refused to equate bioeconomy with modern biotechnology (see Section 3.2). This was in line with the prevailing understanding in Germany (and the EU) that biotechnology is a key technology, but also that relevant technology and innovation approaches to bioeconomy comprise more than biotechnology.

In addition, participants were asked to assess the expected contribution of their research to the objectives of the Baden-Württemberg Research Strategy Bioeconomy. The objectives were related to, e.g., the country's resource independence; competitiveness and innovation capacity; new conversion technologies and bio-based products; local value and job creation; and the information flow between business, research, and society [21] (p. 8 f.). Nearly 70% of respondents answered that their project results could contribute significantly or moderately to "using the country's diversity of plant biomass", followed by "developing new technologies for the conversion of biomass", "increasing international visibility and competitiveness of the country", and "networking between the country's research institutes" (Figure 7). The fact that the most significant contributions were expected for "developing new technologies for the conversion of biomass" and "developing new bio-based substances and materials" could be explained by the technical focus of many projects. The proportion of respondents saying that their research could only make a small or even no contribution was over 50% and 60%, respectively, for the objectives "developing cascade systems" and "safeguarding a high standard of nutrition".

The comparison between the clusters showed that the differences were quite small. One major difference was that the respondents from the microalgae cluster indicated that their projects could contribute to ensuring a high standard of nutrition, while this was, of course, not the case in the biogas and lignocellulose teams. It should be noted that, logically, the specific projects cannot contribute to all objectives of the research strategy, but only a few. In particular, answers from the accompanying assessment projects showed that some of them can only make limited contributions to achieving the strategy's objectives, since critical reflection rather than fulfillment of objectives is in the foreground of this type of research.

3.4. Alternative Implementation Pathways

Both in scientific and societal debate on the concept of bioeconomy, different implementation paths are discussed. An analysis of the European discourse, carried out by the authors of this article, revealed that the understandings can be assigned to two main strands of discussion: the technology-based and socio-ecological approaches [8]. Based on a selection of contrasting positions (Table 2), respondents were asked to indicate to what extent they agreed with individual elements of possible implementation pathways. The selected positions covered a wide spectrum of topics relevant to the bioeconomy, from resource use and consumer behavior to innovation focus and production approach to perspectives on nature and participation.

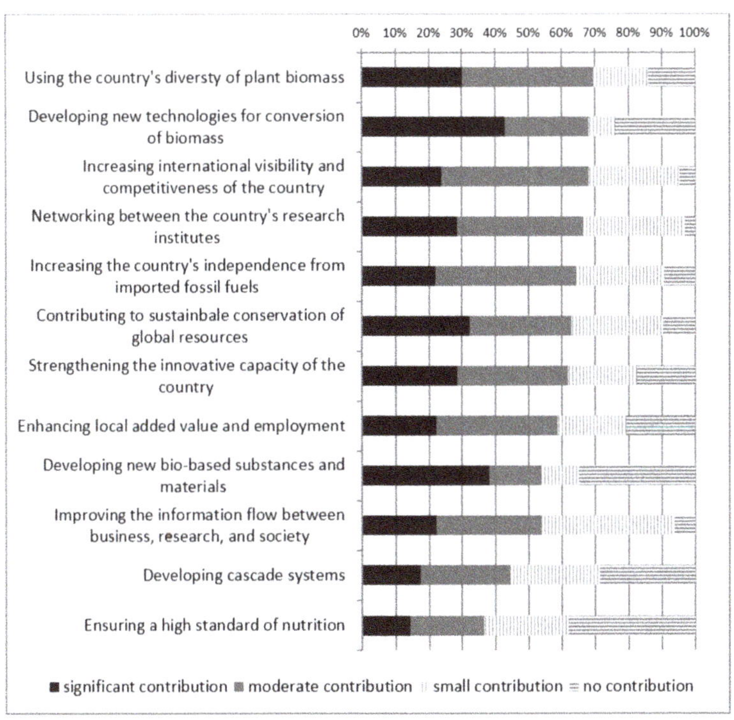

Figure 7. Contribution of the projects to realizing the objectives of the Baden-Württemberg Research Strategy Bioeconomy (*n* = 63).

Table 2. Compilation of contrasting positions on shaping the future bioeconomy.

Elements	Technology-Based Approach	Socio-Ecological Approach
Resource use and consumer behavior	Improving resource efficiency along bio-based value-added chains (less raw material input per product unit, closed cycles, coupled and cascade use)	Promoting sustainable consumption patterns (e.g., reducing meat consumption, consuming regionally produced food, avoiding food waste, reducing fossil energy use)
Innovation focus	Extending technological leadership, intellectual property (e.g., patents), and economic influence of multinational organizations	Promoting social innovations and the use of local knowledge and experience of various stakeholders (farmers, small and medium-sized enterprises, non-governmental organisations)
Production approach	Promoting research and innovation in the field of life sciences as key technologies (genetic engineering, synthetic biology, DNA sequencing, bioinformatics, etc.)	Strengthening multifunctional, agroeconomic land cultivation oriented toward criteria such as diversity, resilience, and self-regulation
Perspectives on nature	Adjusting nature to industrial processes and cycles	Adjusting industrial metabolisms to natural cycles
Participation	Establishing close partnerships between companies, research, and politics to promote investments, skills, and experiences in the fields of key technologies	Involving civil society in shaping and advancing the bioeconomy

Source: Excerpt from Reference [8] (p. 15).

The majority of participants agreed with all statements, except for two (Figure 8). The highest level of full agreement (more than 50% and 60% of the respondents, respectively) was with "improving resource efficiency" and "promoting sustainable consumption". Total agreement was very high (nearly 90% up to 100%) for the elements "improving resource efficiency", "adjusting industrial metabolisms to natural cycles", "establishing close partnerships between companies, research, and politics in the fields of key technologies", "strengthening multifunctional, agroecological land cultivation", and "involving civil society in shaping and advancing the bioeconomy". Agreement with statements on shaping the future bioeconomy was in line with the respondents' evaluations of the objectives of the German National Policy Strategy Bioeconomy (see Figure 6). This was especially the case for improved resource efficiency for environmental and climate protection, sustainable consumption, and close cooperation between companies, research, politics, and social players.

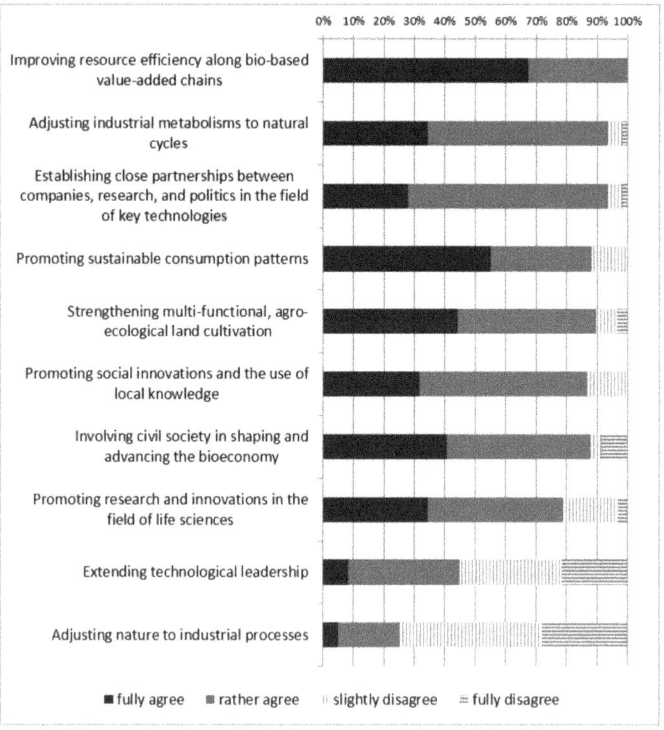

Figure 8. Respondents' agreement with selected positions on shaping the future bioeconomy ($n = 61$).

All in all, the agreement levels were rather similar for all points, which means that the participants considered the positions complementary and not mutually exclusive. Disagreement was expressed by more than 50% and 70% of the respondents, respectively, for the elements "extending technological leadership" and "adjusting nature to industrial processes". The relatively low importance attached to the objective "strengthening the international competitiveness of Germany" corresponded with the limited support for the statement "extending technological leadership".

The research clusters showed some differences. While the biogas and microalgae networks strongly disagreed with "extending technological leadership, intellectual property (e.g., patents), and economic influence of multinational companies", the majority of respondents from the lignocellulose cluster agreed with this approach. One reason for this could be that there is considerable innovation potential in the field of lignocellulose utilization and biorefinery concepts, combined with the ambition to take a pioneering role. Another difference intrinsic to the topics of the three clusters was that full agreement with "promoting research and innovation in the field of life sciences" was highest in the microalgae and lignocellulose networks, while this played a less important role in the biogas network.

3.5. Existing Collaborations

Another part of the questionnaire was about ongoing collaborations the scientists from the research program are involved in (Figure 9). The majority of respondents answered that they were engaged with scientific partners within the cluster and the overall research program. Moreover, collaborations with scientific partners all over Germany and Europe were quite common. A smaller proportion of respondents was involved in collaborations with industry, in particular with companies in the fields of renewable energies, biotechnology, chemistry, and the food industry. Only a small number of respondents cooperate with societal stakeholders and users, with cooperation with users and farmers or agricultural organizations being more common than with nongovernmental groups.

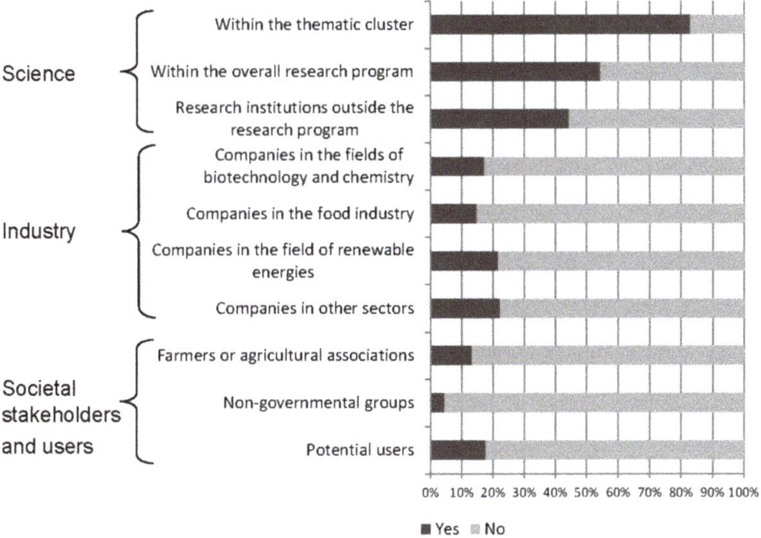

Figure 9. Respondents' collaborations with different types of partners ($n = 48$).

The comparison between the clusters showed a quite similar picture. One difference was that the biogas cluster had more collaboration with farmers and users compared to the other networks. This was due to the fact that manifold contacts have been established during the long history of biogas research in Baden-Württemberg, while the other two topics are quite new and cooperation still needs to be further developed.

Moreover, the scientists were asked to classify their networking activities according to forms of cooperation (Figure 10). The responses showed that traditional forms of scientific cooperation prevail, such as the exchange of information and results, joint definitions of research questions, agreement on methods, joint use of equipment, and agreement on raw materials. The majority of respondents indicated that their collaborations aim to sound out common interests, and only a marginal proportion is involved in collaborations aimed at discussing marketing strategies and developing new business models.

The clusters differed with regard to the discussion and exchange of results. This point was mentioned most frequently in the biogas group. This may have been due to the fact that, in the biogas network, there is a focus on applied research, which takes place in cooperation with partners from practice and involves more exchange, while the lignocellulose and microalgae networks conduct more basic research, which implies increased joint use of equipment and infrastructure and increased coordination of the raw materials to be used. In the microalgae cluster, the different forms of scientific communication play an even more important role than in the other clusters. A possible reason for this could be the specific structure of the research network, where the projects are arranged along the value-added chain from algae cultivation, extraction of ingredients, and conversion up to product design. The project members are therefore dependent on the results of the previous stages, which leads to closer coordination.

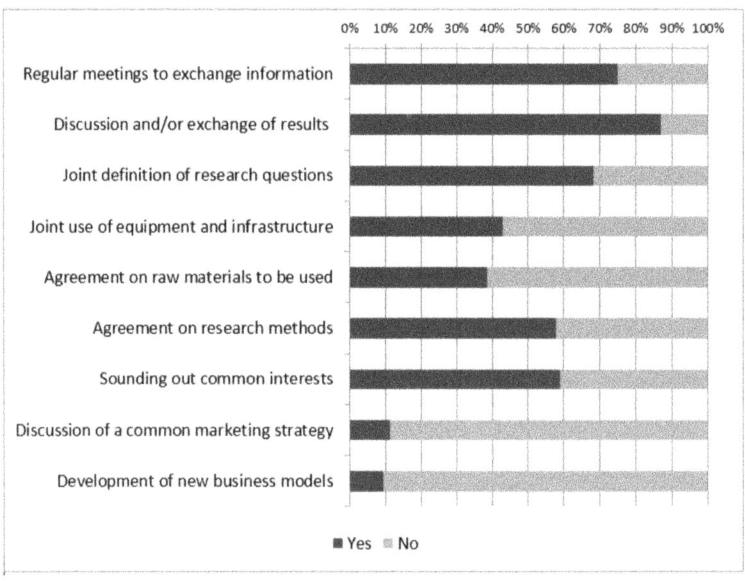

Figure 10. Respondents' forms of cooperation ($n = 48$).

3.6. Obstacles to the Further Development of the Thematic Fields

The participants were asked about possible obstacles to the realization of the biomass value chains they are working on. This included a discussion of competing uses, environmental impacts, and acceptance of the production process or end products by local residents and end users.

3.6.1. Biogas Value Chain

Due to the limited potential of primary biomass and the discourse about adverse effects of energy crop cultivation, there have been attempts to put a stronger focus on the use of residual and waste materials for biogas production in Germany. In our survey, the researchers outlined some challenges for switching to these sources. First, they emphasized that, even with increased mobilization of residues, the demand for energy crops would probably remain the same. Arguments for this assessment were that, on the one hand, the available residual and waste materials would not be sufficient to cover the raw material requirements of the existing biogas plants. On the other hand, it was argued that such feedstocks were technically unsuitable for use in agricultural biogas plants previously operated with energy crops. Second, it was assumed that the expansion of waste fermentation plants would continue to increase. In addition, economic reasons (high transport costs) and legal obstacles (the need to reauthorize agricultural biogas plants) would militate against the widespread use of residual and waste materials. Moreover, it was expected that this could lead to competition between the biogas and lignocellulose paths with regard to certain feedstocks such as landscape conservation material. Third, for this kind of biogas production, the majority of respondents considered sites close to settlements to be useful or necessary, with increased use of alternative substrates such as food waste. This would be useful particularly from an ecological point of view and would also offer possibilities for heat utilization. The respondents mentioned positive examples of near-settlement facilities planned with the involvement of local residents, but also mentioned counterexamples. The main obstacles were seen in the legal requirements for waste fermentation plants and in acceptance by local residents. In addition, respondents expressed doubts about the expected positive environmental impacts related to the use of perennial energy crops, waste, and residual materials.

3.6.2. Lignocellulose Value Chain

The availability of woody biomass was mentioned as a central challenge to the implementation of novel lignocellulose-based value chains. Some respondents assumed that unused wood potential exists, in particular in private forests and in the form of alternative feedstocks such as landscape conservation material, agricultural residues (e.g., straw), and wood waste from the private and commercial sectors. In contrast, the majority of respondents argued that alternative feedstocks are limited and not sufficiently available. Some of the reasons mentioned were competing uses (e.g., the use of agricultural residues in biogas production), conflicts in objectives (e.g., nature conservation opposes the removal of dead wood and cortices from forests), and technical obstacles (e.g., the need for a uniform lignin quality in certain conversion processes, which cannot be ensured when using residues). The majority of respondents expected increasing competition for raw materials, especially in the energy wood market. It was assumed that the innovative value-added chains aspired to by the research cluster would rarely compete with the wood, paper, construction, and furniture industries, but that there might be synergies with these industries in using their residual materials or byproducts.

3.6.3. Microalgae Value Chain

Numerous inhibiting factors were identified for the production and use of microalgae, in particular related to the consumption of resources, acceptance, and competitiveness of algae production. In algae production, there is a considerable need for nutrients, energy, and other inputs. Nitrogen, phosphate, and iron are the main nutrients. If large algae yields are to be achieved, the estimated nutrient requirements are high, but this depends on the type of algae and the cultivation method used. Theoretically, wastewater streams as well as sea and brackish water could be used as nutrient sources. In practice, there are legal barriers because the specifications for nutrient sources used in food production are very high. The use of synthetic fertilizers therefore seems likely. Moreover, further adjuvants and process substances are needed: CO_2, ethanol as an extraction agent, flocculants, disinfectants for the sterilization of reactors, and herbicides. In particular, CO_2 demand is a limiting factor for the cultivation of algae. To ensure the safe use of algae as food, exhaust gases can only be used if no contamination with pollutants can be guaranteed. Therefore, combustion gases cannot be used.

The energy balance of algae production was predominantly classified as negative, though optimization potential was expected from autotrophic cultivation, cultivation in the field, and process improvements. Moreover, industrial algae production has high space requirements. Integration into the city landscape is considered possible. Few respondents expected major impacts on the landscape. However, it was emphasized that there has been no experience with visual impacts and acceptance of larger systems until now. Consumer acceptance is a key factor for realizing the potential of microalgae in nutrition. A clear majority of respondents considered consumer acceptance to be high, especially with regard to use as food ingredients, e.g., in functional foods. Acceptance was expected to be particularly high among younger, health-conscious and environmentally conscious consumers. Acceptance of algae as a substitute for animal-derived food was assumed to be rather low. The following inhibiting factors were mentioned: highly technically processed products do not meet the demand for naturalness; quality losses due to bacterial contamination; difficult imitation of the taste of meat in connection with cultural reservations about the color, taste, and smell of algae; and low popularity in general. Almost all respondents assessed the food industry's interest in algae-based products to be very high, especially with regard to algae-based ingredients and niche products (e.g., in the vegan food sector or fitness industry). However, there was consensus that the demand will hardly shift from animal-based food products to algae-based products. Besides the lack of consumer acceptance, the main reason is the lack of competitiveness of algae products.

4. Discussion

4.1. Predominance of Resource-Oriented Understanding of Bioeconomy

The definitions of bioeconomy, expressed in the responding scientists' own words, with or without specifications, were predominantly resource-oriented. In line with this, the resource-oriented definition presented in the questionnaire achieved the highest agreement of all official definitions. According to this definition, bioeconomy is related to the production, conversion, and use of biological resources. Thus, the scientists' understanding of bioeconomy was strongly in line with the political understanding in Germany and Europe. This was confirmed by recent findings from Austria [12] and Norway [27], which also indicated the dominance of a bioresource-based understanding of bioeconomy among stakeholders. In the same way, the resource-oriented definition by the European Commission was generally supported by respondents (mainly from science and industry) in a public consultation in preparation for the EU Bioeconomy Strategy [28].

The great consistency in the bioeconomy understanding can be explained by the high engagement of scientific communities in the development of official bioeconomy strategies, both at the European and national level [7]. In Baden-Württemberg, for example, scientists were asked to participate in a steering committee and to develop proposals for strategic orientation [21].

Scientific analyses of bioeconomy strategies interpret different definitions as an indication of divergent understandings of bioeconomy (e.g., Reference [10]). In our survey, the technology-oriented definition was less supported than others, but all proposed definitions were agreed upon by the majority of respondents. Agreement with the technology-oriented definition was higher in the lignocellulose and microalgae networks. This preference can be explained by the stronger role of biotechnology in their research agenda. Despite the resource-oriented definition in official bioeconomy strategies, modern biotechnology is also regarded as a key technology in these strategies (e.g., Reference [24]). Overall, the high level of agreement with various definitions indicates that they are not perceived as contrary, but as complementary.

4.2. High Expectance of Contribution of Research Activities to Objectives of the Regional Bioeconomy Strategy

The goals of the German Policy Strategy Bioeconomy were widely supported by the respondents. Only the goal of taking on a leading role in solving global challenges such as climate protection and food security was seen rather critically. Key objectives in all clusters were "the production of raw materials in line with the objectives of environmental and climate protection and nature conservation" and "the long-term conversion of the economic production base from fossil to biogenic sources". Technology visions and objectives of official strategies can fulfill a guiding function for researchers and actors in the respective research area [29]. Our survey results indicated that the official objectives are well integrated into the thinking of the researchers. It can therefore be assumed that the official bioeconomy strategies achieve the objective of providing guidance.

In a similar way, an evaluation of the German Research Strategy Bioeconomy has shown that there is strong support for the overall approach oriented toward societal goals, though there is a lack of agreement on specific objectives among the heterogeneous research communities involved in bioeconomy research [20].

The majority of respondents expected their research results to make a high or medium contribution to achieving the objectives of the Baden-Württemberg Research Strategy Bioeconomy. There were differences between the research clusters regarding the extent of their contributions to the different objectives. This resulted from different research agendas. Of course, an individual research project can only contribute to a few objectives of the research strategy.

An evaluation of the German Research Strategy Bioeconomy [20] and the European Bioeconomy Strategy [30] showed that funded research has significantly contributed to achieving overall goals and that the majority of funded projects have aligned their objectives to the strategies. The EU review outlined that "there is early indication that the projects funded are generating relevant excellent multidisciplinary research (i.e., the quality of publications in terms of citations)" [30] (p. 12).

4.3. Simultaneous Support for the Elements of Different Pathways in the Implementation of the Bioeconomy

Different implementation pathways are discussed for the development of the bioeconomy (e.g., References [11,14]. They are usually presented as alternative trajectories. In our survey, not all elements of the official technology-based approach were supported by the majority of respondents. For example, "adapting nature to industrial processes" and "extending technological leadership" as key elements of the technology-based approach were approved only by a minority. At the same time, some key elements of the socio-ecological implementation pathway met with much approval: for example, "promoting sustainable consumption patterns" and "strengthening multifunctional, agroecological agriculture" were supported by nearly 90% of the respondents. Similarly, stakeholder surveys in Denmark and the United Kingdom, based on interviews, pointed to a potential overlap of the two approaches, which could give rise to a new concept of quality industrial biomass production, understood as using superior biomass quality attributes in place-specific biorefineries to produce different products depending on the region [19].

Obviously, the survey participants considered the discussed implementation pathways of the bioeconomy as compatible or, more precisely, as combinable. This perception might have been due to the fact that projects on agroecology, local food chains, citizens' initiatives, and so on are incorporated in EU and national research programs, even if only marginally in comparison to the dominating technology-centered funding. However, in the background of this allegedly equal coexistence, rival stakeholder networks compete for influence on research priorities, policy changes, and the preferable bioeconomy trajectory [31].

With regard to sustainable development, the current EU bioeconomy policy is seen to lean strongly toward weak sustainability and to favor economic dimensions and concerns over environmental and social dimensions [32]. In addition, current developments in bioeconomy sectors adhere largely to the principles of weak sustainability [33]. The respondents' strong support for some elements of the socio-ecological approach in the survey can be interpreted as the scientists' recognition of the need for strong sustainability in the bioeconomy.

The assumed combinability of elements from different implementation pathways contrasts with the irreconcilability of pathways formulated in scientific and societal discourses on bioeconomy. While the technology-based approach is understood as an enhancement of current practices in the production, processing, and consumption of biomass with new products and more environmentally friendly technologies, the socio-ecological pathway is conceptualized as a structural remodeling of today's production and consumption practices and as fundamental changes in the ways of thinking and living [8]. The divergence between survey statements and discourse viewpoints can be interpreted in two ways: the compatibility between the two perspectives has not been considered sufficiently in the discourse, or the interviewed scientists underestimate the tensions or potential conflicts between the two approaches. This question cannot be answered within the present study, but some indications can be given:

- Since the bioeconomy has not yet been widely implemented, potential conflicts have not yet fully materialized. Therefore, they are probably underestimated. At the moment, different ideas continue to coexist, and the pressure to decide between the given alternatives is not yet high enough;
- Systemic studies on the connection between the bioeconomy and its wider societal and economic implications are still in their infancy. It remains to be seen whether and how the transition to a bioeconomy will contribute to solving key challenges [11]. Addressing environmental and social concerns and limits has been repeatedly demanded, but is not easy. It is, therefore, difficult to appraise the best implementation pathway for the bioeconomy;

- Competing interests, power relations, and politics are inherent in bioeconomy development and make it difficult for stakeholders to negotiate compromises and reach consensus on the most sustainable pathways for the future [34]. Biomass produced from agriculture is subject to ongoing debates about the future of agriculture. Agricultural paradigms carry their own set of ideological and value assumptions and are fiercely contested by critics [19]. Scientists engaged in bioeconomy research are probably not familiar with these debates.

The evaluation of the German Research Strategy Bioeconomy has also shown that the heterogeneous scientific communities interviewed do not share a common understanding of implementation pathways for the bioeconomy up to now [20] (p. 296). Some of them call for more technical and laboratory research in the field of biotechnology; others want to foster more non-technical research in the fields of sustainability and social innovation.

4.4. Prevailing of Scientific Collaborations

The research clusters had in common that most of their collaborations take place within the scientific system, and only a few collaborations exist with societal stakeholders and users. There are certain peculiarities in the biogas network, where cooperation with the latter group of players and industrial companies is more frequent than in the lignocellulose and microalgae networks. With its long-established biogas technologies, this cluster has been able to build extensive networks with industry and other stakeholders, whereas in the fields of lignocellulose and microalgae, partnerships and cooperation are still evolving due to the early stage of development of the technologies. Overall, the forms of cooperation are dominated by conventional scientific formats, such as meetings for exchanges of information, discussions of results, and the development of research questions, while there are only a few cooperation formats for discussing joint marketing strategies and developing new business models. One reason could be that, in the lignocellulose and algae networks, basic research is in the foreground, and therefore the marketing of products lies in the future.

Similarly, the involvement of societal stakeholders has been lagging behind the European level. As highlighted in a review of the European Bioeconomy Strategy from 2012, the Bioeconomy Panel is the central body for implementing "Area of Action 2: Reinforced Policy Interaction and Stakeholder Engagement" [30] (p. 13). The panel was created in 2013 to support interactions between different policy areas, sectors, and stakeholders in the bioeconomy. So far, the main outputs of the panel have been four bioeconomy stakeholder conferences, held between 2012 and 2016, and the 2017 "Stakeholders Manifesto" [35]. The latter represents a marked shift to the European Commission's own strategy paper from 2012 [12]. It should be taken into account that the panel was established on political initiative. The review provided no information on the level of involvement of societal stakeholders in European research projects. A group of experts evaluating the European strategy concluded that initiatives such as the Stakeholder Panel have increased awareness of the bioeconomy concept among a broad range of stakeholders, but that the involvement of public and private stakeholders and civil society organizations should be reinforced with an updated "Bioeconomy Strategy and Action Plan" [36].

The evaluation of the German Research Strategy Bioeconomy showed that the experts surveyed considered it necessary to incorporate a wide range of stakeholders in research, from business and industry to NGOs in the fields of environment and consumption to citizens and the broader public [20] (p. 285). However, until now, funding aimed at fostering dialogue with civil society and the public has been rather rare, and these groups have been only marginally addressed in the existing funding scheme [20] (p. 293, 296). An evaluation of the individual projects revealed that the goal of intensifying networking with partners and establishing new networks was reached less frequently than that of scientific knowledge exchange and interdisciplinary learning [20] (p. 294).

5. Conclusions

The survey results showed that the scientists involved in the Bioeconomy Research Program Baden-Württemberg agreed with and orient themselves toward the official bioeconomy strategies.

The strategies have thus met the objective of becoming an important guiding tool for research activities. The guidance provided by the strategies could lead to a narrowing of technology options and implementation pathways. Then again, the official bioeconomy strategies could be seen as important catalysts for the development of alternatives, which has been part of a critical discussion in the last years. This interplay between closing and opening pathways for implementation of the bioeconomy demands further investigation. On the one hand, the emergence of alternative bioeconomy approaches, the establishment of actor networks supporting these alternatives, and their influence on policy-making require further scientific investigation. On the other hand, official bioeconomy strategies should include reflexive elements, opportunities for new technical ideas, and space for nontechnical ideas and social innovations. In this way, bioeconomy policies could contribute to the development of enhancements and alternatives to existing approaches rather than deal with critical appraisals from the outside.

Our results showed that the surveyed scientists tend to consider the different implementation pathways for the bioeconomy to be combinable, while in the scientific and societal debates, these are seen as contradictory. In conclusion, it can be stated that the majority of respondents support combining alternative and well-established practices and ideas. More research is needed to better understand how realistic such combinations are in practice. The next step after analyzing competing narratives would be to translate them into scenarios and to examine their preconditions and impacts.

Anyway, a critical reflection of goals and implementation pathways, taking into account societal expectations, is of crucial importance in any update process of bioeconomy strategies. Aspects that are not considered at this stage will most probably not be included in the following research funding. The involvement of civil society is still in its infancy and needs to be further developed. The drafting of new or the revision of existing bioeconomy strategies should be set up as a participatory process that includes a broad spectrum of stakeholders and viewpoints.

The development of the bioeconomy as a comprehensive sociotechnical transformation requires interdisciplinary but also transdisciplinary research approaches. The Baden-Württemberg Bioeconomy Research Program, with its interdisciplinary orientation, largely meets this requirement. Consequently, the results of the survey partly reflected these specific research conditions. For instance, the respondents' understanding of the bioeconomy was closely linked to sustainability issues. In addition, they described several challenges for the future development of the three biomass value chains, emphasizing that sustainability is not easy to achieve and that research needs to be intensified in this field. The analysis of collaborations revealed that exchange within scientific communities prevails. Cooperation with industry, other business actors, and end-users is relatively weak, but is commonly called upon as needing to be intensified. However, it remains open how well stronger cooperation with potential users of new bioeconomy technologies for accelerated adoption goes together with intensified examination of sustainability issues. This tension, e.g., between establishing new bio-based processes and reflections on the limits of biomass utilization, should be deliberated on in collaboration. Furthermore, a framework for an integrated assessment of socially responsible research [37] should be introduced in the bioeconomy research area.

Supplementary Materials: Supplement S1: Questionnaire, available online at: http://www.mdpi.com/2071-1050/11/15/4253/s1.

Author Contributions: Conceptualization and methodology, C.P. and R.M.; survey execution and analysis, C.P.; writing—original draft preparation, C.P. and R.M.; writing—review and editing, C.P. and R.M.; visualization, C.P. and R.M.; project administration, R.M.; funding acquisition, C.P. and R.M..

Funding: This research was supported by the Ministry of Science, Research, and the Arts of Baden-Württemberg within the framework of the Bioeconomy Research Program Baden-Württemberg and is based on the project "Bioeconomy in Baden-Württemberg", grant number 33-7533-10-5/145/1.

Acknowledgments: We thank our retired colleague Juliane Jörissen for developing and conducting the survey with us and all of the scientists involved in the Bioeconomy Research Program Baden-Württemberg who filled out the questionnaire. We acknowledge support from the KIT-Publication Fund of the Karlsruhe Institute of Technology.

Conflicts of Interest: The authors declare no conflicts of interest.

References

1. Food and Agriculture Organization of the United Nations (FAO). *How Sustainability Is Addressed in Official Bioeconomy Strategies at International, National and Regional Levels*; FAO: Rome, Italy, 2016.
2. Dieckhoff, P.; El-Chichakli, B.; Patermann, C. *Bioeconomy Policy. Synopsis and Analysis of Strategies in the G7*; German Bioeconomy Council: Berlin, Germany, 2015; Available online: https://biooekonomierat.de/fileadmin/Publikationen/berichte/BOER_Laenderstudie_1_.pdf (accessed on 18 June 2019).
3. Fund, C.; El-Chichakli, B.; Patermann, C.; Dieckhoff, P. *Bioeconomy Policy (Part II)*; Synopsis of National Strategies around the World; German Bioeconomy Council: Berlin, Germany, 2015; Available online: http://biooekonomierat.de/fileadmin/international/Bioeconomy-Policy_Part-II.pdf (accessed on 18 June 2019).
4. Fund, C.; El-Chichakli, B.; Patermann, C. *Bioeconomy Policy (Part III)*; Update Report of National Strategies around the World; German Bioeconomy Council: Berlin, Germany, 2018; Available online: http://biooekonomierat.de/fileadmin/Publikationen/berichte/GBS_2018_Bioeconomy-Strategies-around-the_World_Part-III.pdf (accessed on 18 June 2019).
5. Overbeek, G.; de Bakker, E.; Beekman, V.; Davies, S.; Kresiewa, Z.; Delbrück, S.; Ribeiro, B.; Soyanov, M.; Vale, M. *Review of Bioeconomy Strategies at Regional and National Level*; BioSTEP Project: Berlin, Germany, 2016; Available online: http://bio-step.eu/fileadmin/BioSTEP/Bio_documents/BioSTEP_D2.3_Review_of_strategies.pdf (accessed on 18 June 2019).
6. Mengal, P.; Wubbolts, M.; Zika, E.; Ruiz, A.; Brigitta, D.; Pieniadz, A.; Black, S. Bio-based Industries Joint Undertaking: The catalyst for sustainable bio-based economic growth in Europe. *New Biotechnol.* **2018**, *40*, 31–39. [CrossRef] [PubMed]
7. Meyer, R. Bioeconomy Strategies: Contexts, Visions, Guiding Implementation Principles and Resulting Debates. *Sustainability* **2017**, *9*, 1031. [CrossRef]
8. Priefer, C.; Jörissen, J.; Frör, O. Pathways to Shape the Bioeconomy. *Resources* **2017**, *6*, 10. [CrossRef]
9. Staffas, L.; Gustavsson, M.; McCormick, K. Strategies and Policies for the Bioeconomy and Bio-Based Economy: An Analysis of Official National Approaches. *Sustainability* **2013**, *5*, 2751–2769. [CrossRef]
10. McCormick, K.; Kautto, N. The Bioeconomy in Europe: An Overview. *Sustainability* **2013**, *5*, 2589–2608. [CrossRef]
11. Bugge, M.M.; Hansen, T.; Klitkou, A. What Is the Bioeconomy? A Review of the Literature. *Sustainability* **2016**, *8*, 691. [CrossRef]
12. Hausknost, D.; Schriefl, E.; Lauk, C.; Kalt, G. A Transition to Which Bioeconomy? An Exploration of Diverging Techno-Political Choices. *Sustainability* **2017**, *9*, 669. [CrossRef]
13. Levidow, L.; Birch, K.; Papaioannou, T. Divergent Paradigms of European Agro-Food Innvoation: The Knowledge-Based Bio-Economy (KBBE) as an R&D Agenda. *Sci. Technol. Hum. Values* **2012**, *38*, 94–125.
14. Levidow, L.; Birch, K.; Papaioannou, T. EU agri-innovation policy: two contending visions of the bio-economy. *Crit. Policy Stud.* **2012**, *6*, 40–65. [CrossRef]
15. Schmid, O.; Padel, S.; Levidow, L. The Bio-Economy Concept and Knowledge Base in a Public Goods and Farmer Perspective. *Bio-Based Appl. Econ.* **2012**, *1*, 47–63.
16. Pfau, S.F.; Hagens, J.E.; Dankbaar, B.; Smits, A.J.M. Visions of Sustainability in Bioeconomy Research. *Sustainability* **2014**, *6*, 1222–1249. [CrossRef]
17. De Besi, M.; McCormick, K. Towards a Bioeconomy in Europe: National, Regional and Industrial Strategies. *Sustainability* **2015**, *7*, 10461–10478. [CrossRef]
18. Kircher, M. The transition to a bio-economy: emerging from the oil age. *Biofuels Bioprod. Biorefin.* **2012**, *6*, 369–375. [CrossRef]
19. Shortall, O.K.; Raman, S.; Millar, K. Are plants the new oil? Responsible innovation, biorefining and multipurpose agriculture. *Energy Policy* **2015**, *86*, 360–368. [CrossRef]
20. Hüsing, B.; Kulicke, M.; Wydra, S.; Stahlecker, T.; Aichinger, H.; Meyer, N. *Evaluation der „Nationalen Forschungsstrategie BioÖkonomie 2030". Wirksamkeit der Initiativen des BMBF—Erfolg der geförderten Vorhaben—Empfehlungen zur strategischen Weiterentwicklung*; Abschlussbericht; Fraunhofer ISI: Karlsruhe, Germany, 2017.

21. Ministerium für Wissenschaft, Forschung und Kunst Baden-Württemberg—Ministry for Science, Research, and the Arts of Baden-Württemberg. *Bioökonomie im System aufstellen. Konzept für eine baden-württembergische Forschungsstrategie „Bioökonomie"*; Ministerium für Wissenschaft, Forschung und Kunst Baden-Württemberg: Stuttgart, Germany, 2013; Available online: https://www.baden-wuerttemberg.de/fileadmin/redaktion/dateien/PDF/Broschuere_Konzept-baden-wuerttembergische-Forschungsstrategie-Biooekonomie.pdf (accessed on 18 June 2019).

22. Mayring, P. *Qualitative Inhaltsanalyse. Grundlagen und Techniken*, 7th ed.; Deutscher Studien Verlag: Weinheim, Germany, 2000.

23. Organisation for Economic Cooperation and Development. *The Bioeconomy to 2030: Designing a Policy Agenda*; OECD Publishing: Paris, France, 2009.

24. Bundesministerium für Forschung und Bildung (BMBF)—German Ministry for Research and Education. *Nationale Forschungsstrategie Bioökonomie 2030. Unser Weg zu einer bio-basierten Wirtschaft*; BMBF: Berlin, Germany, 2010; Available online: https://www.bmbf.de/pub/Nationale_Forschungsstrategie_Biooekonomie_2030.pdf (accessed on 18 June 2019).

25. European Commission. *Innovating for Sustainable Growth: A Bioeconomy for Europe*; Communication from the Commission to the European Parliament, the Council, the European Economic and Social Committee and the Committee of the Regions; European Commission: Brussels, Belgium, 2012.

26. Bundesministerium für Ernährung und Landwirtschaft—Federal Ministry of Food and Agriculture. *Nationale Politikstrategie Bioökonomie—Nachwachsende Ressourcen und biotechnologische Verfahren als Basis für Ernährung, Industrie und Energie*; BMEL: Berlin, Germany, 2014; Available online: https://www.bmbf.de/files/BioOekonomiestrategie.pdf (accessed on 18 June 2019).

27. Scordato, L.; Bugge, M.M.; Fevolden, A.M. Directionality across Diversity: Governing Contending Policy Rationales in the Transition towards the Bioeconomy. *Sustainability* **2017**, *9*, 206. [CrossRef]

28. European Commission. *Bio-Based Economy in Europe: State of Play and Future Potential—Part 2; Summary of the Position Papers Received in Response of the European Commission's Public On-Line Consultation*; European Commission: Brussels, Belgium, 2011; Available online: http://ec.europa.eu/research/consultations/bioeconomy/bio-based-economy-for-europe-part2.pdf (accessed on 18 June 2019).

29. Lösch, A. Technikfolgenabschätzung soziotechnischer Zukünfte. Ein Vorschlag zur wissenspolitischen Verortung des Vision Assessments. *TATuP* **2017**, *26*, 60–65. [CrossRef]

30. European Commission. *Review of the 2012 European Bioeconomy Strategy*; Directorate-General for Research and Innovation: Brussels, Belgium, 2017.

31. Levidow, L. European transitions towards a corporate environmental food regime: Agroecological incorportation or contestation? *J. Rural Stud.* **2015**, *40*, 76–89. [CrossRef]

32. Ramcilovic-Suominen, S.; Pülzl, H. Sustainable development—A 'selling point' of the emerging EU bioeconomy policy framework? *J. Clean. Prod.* **2018**, *172*, 4170–4180. [CrossRef]

33. Bennich, T.; Belyazid, S. The Route to Sustainability—Prospects and Challenges of the Bio-Based Economy. *Sustainability* **2017**, *9*, 887. [CrossRef]

34. Devaney, L.; Henchion, M.; Regan, A. Good Governance in the Bioeconomy. *EuroChoices* **2017**, *16*, 41–46. [CrossRef]

35. European Bioeconomy Stakeholders Panel. European Bioeconomy Stakeholders Manifesto. 2017. Available online: https://ec.europa.eu/research/bioeconomy/pdf/european_bioeconomy_stakeholders_manifesto.pdf (accessed on 18 June 2019).

36. Lescai, F.; Newton, A.; Carrez, D.; Carus, M.; Griffon, M.; Jilkova, J.; Juhász, A.; Lange, L.; Mavsar, R.; Pursula, T.; et al. *Expert Group Report—Review of the EU Bioeconomy Strategy and its Action Plan*; European Commission: Brussels, Belgium, 2017.

37. Daedlow, K.; Podhora, A.; Winkelmann, M.; Kopfmüller, J.; Walz, R.; Helming, K. Socially responsible research processes for sustainability transformation: an integrated assessment framework. *Curr. Opin. Environ. Sustain.* **2016**, *23*, 1–11. [CrossRef]

 *sustainability*

Article

Evaluating the Effectiveness of Climate Change Adaptations in the World's Largest Mangrove Ecosystem

Pramod K. Singh [1,*], **Konstantinos Papageorgiou** [2], **Harpalsinh Chudasama** [1] and **Elpiniki I. Papageorgiou** [3]

1   Institute of Rural Management Anand (IRMA), Anand 388001, India; harpalsinh@irma.ac.in
2   Computer Science Department, University of Thessaly, 35100 Lamia, Greece; konpapageorgiou@uth.gr
3   Faculty of Technology, University of Thessaly, 41500 Larisa, Greece; elpinikipapageorgiou@uth.gr
*   Correspondence: pramod@irma.ac.in or pramodirma@gmail.com

Received: 22 September 2019; Accepted: 20 November 2019; Published: 25 November 2019

Abstract: The Sundarbans is the world's largest coastal river delta and the largest uninterrupted mangrove ecosystem. A complex socio-ecological setting, coupled with disproportionately high climate-change exposure and severe ecological and social vulnerabilities, has turned it into a climate hotspot requiring well-designed adaptation interventions. We have used the fuzzy cognitive maps (FCM)-based approach to elicit and integrate stakeholders' perceptions regarding current climate forcing, consequent impacts, and efficacy of the existing adaptation measures. We have also undertaken climate modelling to ascertain long-term future trends of climate forcing. FCM-based simulations reveal that while existing adaptation practices provide resilience to an extent, they are grossly inadequate in the context of providing future resilience. Even well-planned adaptations may not be entirely transformative in such a fragile ecosystem. It was through FCM-based simulations that we realised that a coastal river delta in a developing nation merits special attention for climate-resilient adaptation planning and execution. Measures that are likely to enhance adaptive capabilities of the local communities include those involving gender-responsive and adaptive governance, human resource capacity building, commitments of global communities for adaptation financing, education and awareness programmes, and embedding indigenous and local knowledge into decision making.

Keywords: climate change adaptation; transformative adaptation; limits to adaptation; adaptation barrier; fuzzy cognitive maps; resilience; sustainability; vulnerability; Sundarbans

---

1. Introduction

The Sundarbans, the world's largest coastal river delta and the largest uninterrupted mangrove forest, is critical biodiversity and climate hotspot [1]. It has a complex ecosystem studded with inter-tidal and estuarine zones stretching for about 10,000 km^2 and is located on the borders of the state of West Bengal in India (~40%) and southern Bangladesh (~60%) where the Ganges, Brahmaputra, and Meghna rivers meet the Bay of Bengal. Coastal river deltas are often naturally low-lying areas close to the local mean sea-level. Changes in regional sea-levels affect coastal river delta systems geomorphologically by altering the base level, coastal erosion, and inundation, not to mention inland propagation of tidal and backwater effects [2,3]. The geomorphological response of a coastal river delta system to sea-level rise is determined by the delta system's capacity to adapt. The system is complex owing to the variety of feedbacks and changes in internal and external boundary conditions, including sediment supply, river discharge, ecological system feedbacks, subsidence, and human intervention; it is also mainly associated with hydrodynamic and ecological responses [2,4]. It influences system functioning and determines the ability to either adapt dynamically, mitigate the effects of regional sea-level, or submerge [5,6]. In the coming decades, the effects of human intervention, causing further subsidence, changes in sediment supply, river discharge, and ecosystems will be dominant in determining the impacts of sea-level change in coastal river deltas [5,6].

The coastal regions of the Bay of Bengal, especially the Sundarbans delta, are among the most vulnerable areas of the world in terms of experiencing the rapid sea-level rise, seawater intrusion, and other climate change impacts [7,8]. The Sundarbans landscape is highly fragile and particularly vulnerable to global and regional climate change impacts because of its complex geomorphology and environmental settings attributable to continuing global warming, rising sea-levels, seawater intrusion, land erosion, gradual subsidence, and cyclones [1,9,10]. The Sundarbans' mangroves, which protect more than 10 million people from cyclonic storms, today stand threatened by cyclonic damage. Cyclones and tidal storm surges cause damage to the floral and faunal biodiversity along the sea-land interface [1].

The sea-level has risen approximately three to four times higher than the global mean between 1993 and 2009 in the tropical western Pacific and the Indian Ocean [11]. The coastal region in the southwest of Bangladesh has undergone a relative sea-level rise varying from 2.8 to 8.8 mm per year in the last few decades. In South Asia, the sea-level in the Ganges-Brahmaputra-Meghna delta of Bangladesh is likely to rise by 0.63 to 0.88 m by 2090 [12]. Frequent cyclones, together with increasing sea-levels, have resulted in flooding, coastal erosion, and recession of coastline in the region [13].

Coastal flooding is driven by multiple factors, including local land elevation, regional sea-level rise, heavy precipitation, tidal waves, storm surges, and tropical cyclones [14–18]. All these factors are likely to adversely affect the lives and livelihoods of the local community.

The remaining part of the paper describes the study area and sampling; it discusses the methodology adopted for climate change modelling, construction of fuzzy cognitive maps, and fuzzy cognitive map-based simulations. After this, we present the results of climate change projections. We describe the climate change impacts and adaptations as perceived by the community before presenting the results of fuzzy cognitive maps (FCM)-based simulations. Finally, we discuss the results in detail, drawing appropriate conclusions. The paper ends with guidance for future research.

Study Area and Sampling

The Indian Sundarbans, comprising 54 islands, is home to about 4.5 million people [19]. An analysis of the SRTM data reveals that these islands are naturally flat, mostly comprising moderately elevated areas (0 to 5 m high from the mean sea-level), with few low-lying and elevated regions (0 m and 5 to 10 m respectively).

Agriculture and fishing are the two major livelihood options available to the communities living in this region. However, both sectors have been facing stress due to seawater intrusion, coastal erosion, and the increasing salinity in agricultural fields and river water. All of this has been causing a sizeable male population to emigrate from the villages. Phenomena like global and regional climatic changes coupled with anthropogenic pressures including poaching, human encroachment for agriculture and fishing, and overexploitation of both timber and non-timber forest produce have led to multiple alterations in the mangrove flora, fauna, and ecosystem dynamics and functions posing, as a consequence, severe threats to the Sundarbans' ecosystem [9]. They also introduce many changes in the ecosystem services vital for human health, wellbeing, and livelihoods [1,9,20].

This perception mapping study engaged 46 community groups (male and female groups constituted 35 and 11 respectively) from seven villages in Sagar and Mousini islands of the South 24 Paraganas district of Sunderbans.

2. Methodology

2.1. Climate Change Modelling

We implemented climate modelling in a GIS environment for reference as well as projected climates. Available monthly climate data were read and converted to variables required for subsequent calculations. We used time-series data from the Climate Research Unit (CRU) at the University of East Anglia, the Global Precipitation Climatology Centre (GPCC), and the EU WATCH Integrated Project. We obtained the CRU TS v3.21 (time-series) datasets from the British Atmospheric Data Centre (BADC), highlighting month-by-month climatic variations over the last century covering the period from January 1901 to December 2012. CRU TS v3.21 data were calculated on 0.25×0.25 degree grids. We downloaded the GPCC v6 full re-analysis data product and concluded the spatial interpolation to 5 arc-minutes resolution for the period 1981–2010. We obtained the daily data at 0.5-degree resolution from the WATCH Integrated Project data repository; we compiled within-month precipitation distribution and computed the deviation of daily temperatures from respective monthly means for each month for the period 1981–2010.

We used the results of the IPCC's AR5 climate model for two representative concentration pathways (RCPs: 2.6 and 8.5) to characterise a range of possible future climate distortions for the periods 2041-2070 (2050s) and 2071-2100 (2080s). The RCPs were developed and documented in a special issue of climate change [21]. We implemented climate model simulations based on RCPs as part of the Coupled Model Inter-comparison Project Phase 5 (CMIP5) [22] and extracted monthly mean temperature and precipitation data from the WorldClim 30 arc-second raster databases [23]. We analysed the multi-model ensembles for two climate forcing levels of the RCPs based on spatial data from the IPCC's AR5 CMIP5 process and corrected data bias, downscaling it 0.25-degree as in the Inter-sectoral Impact Model Inter-comparison Project (ISI-MIP) [24]. In order to calculate the ensemble mean, we used the ISI-MIP data at 0.25-degree resolution of five climate models (GFDLESM2M, HadGEM2-ES, IPSL-CM5A-LR, MIROC-ESM-CHEM, and NorESM1-M).

2.2. Fuzzy Cognitive Mapping

We conducted the perception mapping study aided by the fuzzy cognitive maps-based approach introduced by Kosko in 1986 [25] to document communities' perceptions about the direct and indirect impacts of climate variability and change on different livelihood assets. The FCM approach captures the functioning of a complex system based on people's perceptions [26]. The process of data capture in the FCM approach is considered quasi-quantitative because the quantification of concepts and links can be interpreted in relative terms [27–29]. In order to generate data, the participants debated the cause-effect relations between the qualitative concepts and generated quantitative data based on their experiences, knowledge, and perceptions of inter-relationships between the concepts [26–28,30,31].

2.2.1. Main Aspects of Fuzzy Cognitive Maps

FCMs are graph-based structures, describing signed weighted digraphs [25]. They can handle vagueness while being capable of incorporating and adapting human knowledge through fuzzy logic. They form a component of soft computing providing, thereby, a simple but powerful tool for analysing, representing, and simulating dynamic systems [32].

The structure of FCMs consists of concepts (i.e., nodes) $C1, C2 \ldots C_i$, and connections between them. All this is represented by the adjacency matrix W [ ]. The concepts are mapped to the real-valued activation level where C_i takes values in the interval [0,1], which is the degree to which the observation belongs to the concept (i.e., the value of the fuzzy membership function). As a consequence of the dynamic interactions of connected nodes, the concept's state changes over time. The reasoning is performed as the calculation of Equation (1), where f ( ) stands for a hyperbolic transformation function

Equation (2), ensuring that the concept defined value falls within the interval [0,1] and f ( ) is given by Equation (2), where C—parameter, $C > 0$.

$$C_i^{(t+1)} = f\left(C_i^{(t)} + \sum_{\substack{j=1 \\ i \neq j}}^{n} C_j^{(t)} \times W_{ji} \right) \tag{1}$$

$$f(x) = \frac{e^{\lambda x} - e^{-\lambda x}}{e^{\lambda x} + e^{-\lambda x}}, \; a \in \mathrm{R}^+ \tag{2}$$

The edges W_{ij} displayed in the dimensions of the matrices denote the degrees of the causal relationship (i.e., the weight of the edge or influences between the connected nodes) and typically lie between [−1, 1]; whereas $W_{ij} > 0$ implies that C_i increases C_j, $W_{ij} = 0$ means no relation and $W_{ij} < 0$ means C_i decreases C_j. Therefore, the adjacency matrices are not symmetric as per definition. The diagonal entries (e.g., W_{ii}, W_{jj}) in W reflect the effect on it. With increasing uncertainty, fuzzy rules, or fuzzy numbers may be used to describe the weights of the connections [25,32].

2.2.2. Constructing Fuzzy Cognitive Maps

Step 1: Obtaining fuzzy cognitive maps from community groups

The majority of marginal and poor people across the globe rely on climate-sensitive livelihood activities that are highly susceptible to increase in temperature and variability in precipitation patterns, along with extreme climatic events, such as cyclones, floods, droughts, etc., making them highly vulnerable to climate change [33]. Hence, in order to obtain fuzzy cognitive maps, we selected marginal farmers possessing less than two acres of land and some livestock as our stakeholders. The steps of constructing fuzzy cognitive maps from stakeholders/farmers are given in Section 2.2.2.

A consensus of the local community was obtained with regard to the summer and winter temperatures, as well as precipitation variability increase over the last 10 to 15 years. We also sought community opinion in the context of increasing intensity and frequency of climatic extremes. The communities perceived an overall increase in temperatures and precipitation variability. The communities also perceived an increase in climate-related extremes, including cyclonic storms and floods. We divided the local community members, after the group discussion, into groups of four to five individuals. We formed the groups according to simple wealth-ranking allocating individuals with the same landholdings/ number of livestock within a single group while conducting gender-wise segregation. This helped in neutralising power dynamics within each group. We demonstrated the construction of fuzzy cognitive maps to the participants with the help of a disparate context. We asked the following questions from each of the 46 community groups:

i.   What are the changes in summer and winter temperature observed over the past 10 to 15 years?
ii.   What are the changes in rainfall variability observed over the past 10 to 15 years?
iii.   What are the changes in extreme climatic events (cyclone, flood, etc.) observed over the past 10 to 15 years?
iv.   What are the resulting impacts arising from direct effects due to climate variability, sea-level rise, and changes and climatic extremes?
v.   How have your lives and livelihoods been affected due to these changes?
vi.   What adaptation practices have been taken up for enhancing climate resilience?

The participants designed cognitive maps relevant to the central concept: Increased climate variability and change. They laid out concepts pertinent to the central concept, showing impacts of climate variability and change on their lives and livelihoods and adaptation practices adopted; they also assigned cause–effect interconnections between the concepts. Stakeholders assigned individual weights to each connection on a scale of 1–10, with 1 representing the minimum impact and 10 the maximum. Researchers scaled down these weights to a scale of 0.1–1, with 0.1 representing the minimum impact and 1 the maximum [26,27,30].

Step 2: Coding of individual cognitive maps into adjacency matrices

We coded individual cognitive maps into adjacency matrices listing the same concepts on the vertical and horizontal axes. Weights assigned by the stakeholders were coded into the adjacency matrix. A value is coded into the matrix if a connection exists between two concepts [26–28,34].

Step 3: Quantitative aggregation of individual cognitive maps

We aggregated each coded map to construct a social cognitive map (SCM) through matrix addition [26–28,30]. Thus, the SCM we obtained represents the perception of all the 46 community groups. SCM gives a better representation of system dynamics yielding a more accurate, reliable, and comprehensive understanding of a system [26,27,30].

Step 4: Qualitative aggregation of the social cognitive map

In order to organise the data and make it easier to understand, we condensed the concepts obtained from the SCM into broader categories based on their nature [26,27,30]. We followed it up with calculations for an arithmetic mean of the weights of concepts mentioned in the SCM. We did this to identify interconnections between the broader encompassing concepts [26,28,30,35]. The qualitatively aggregated SCM comprises 23 concepts.

2.3. FCM-Based Simulations

The adaptation strategies that are likely to reduce climate risks and increase resilience adequately may be classified as effective adaptations. What cannot be ruled out is the possibility of an adaptation deficit ('a failure to adapt adequately to existing climate risks' [36]). Having implemented all adaptations in the area, climate risk possibilities can arise, presenting limits to adaptation. Therefore, it is crucial to understand the effectiveness of adaptation strategies.

In order to evaluate the effectiveness of current adaptation interventions, we conducted FCM-based simulations with the help of the aggregated SCM. Simulating the FCM model gives a deeper understanding of the concepts' behaviour, their relations, and the extent to which one concept has an impact on the rest. The simulation process was conducted by 'clamping/activating' the initial values of the key concepts (in Equation (1)) until the system reached a stabilisation point (known as the system steady-state). We developed a baseline by 'clamping/activating' the initial values of concepts C1—'climate variability and change' and C2—'climatic extremes' at |1|. This was done by taking into consideration the climate change projections in the region.

In FCMs, each concept varies from 0 to |1| where 0 means 'non-activated' and |1| means 'activated'. When one or more concepts are 'clamped/activated,' the activation spreads through the matrix following the weighted relationships in the FCM matrix. An iteration produces a new state vector with 'activated' concepts and 'non-activated' concepts [31,37]. The resulting concept values are used to interpret the outcomes of a particular scenario [27,28,31,37]. We multiplied the input vector concepts (Table 1) with the adjacency matrix and applied a squashing function (Equation (2)) after every multiplication as a threshold function. We iterated the process until the system (output vector) reached a steady-state. The FCMWizard tool ran the simulations. The FCMWizard can also perform simulations for different possible scenarios, in various scientific domains, using a very intuitive graphical user interface [38].

Table 1. Various future scenarios using fuzzy cognitive maps (FCM)-based simulations.

Scenarios	Input Vector Concepts Used for Simulations
Baseline	C1—Climate variability and change, C2—Climatic extremes
Scenario 1	C14—Dykes and embankments
Scenario 2	C15—Water resource management
Scenario 3	C18—Sustainable agriculture and aquaculture practices
Scenario 4	C22—Strengthening local institutions
Scenario 5	C14—Dykes and embankments, C15—Water resource management, C18—Sustainable agriculture and aquaculture practices, and C22—Strengthening local institutions

The first established approach in scenario planning is the selection of key concepts. Filtering the key concepts helps in linking storylines to the quantitative model while focusing on significant concepts that often have strong direct or indirect effects on the goal. It can, at the same time, significantly change the balance of the whole system. While conducting the FCM-based scenario analysis, recognition of crucial concepts mainly relies upon communities' perceptions, although some characteristics were elicited from the model facilitates the procedure. We identified four key adaptation strategies in the study area for assessing their effectiveness (Table 1). These concepts were selected as they were among the concepts with the highest centrality (see Table S1(a)) and could well influence the dynamics of the system.

Scenario 1 is devoted to increasing the concept of 'dykes and embankments', by 'clamping' it to one. We adopted the same procedure for the following three key concepts: 'water resource management', 'sustainable agriculture and aquaculture practices', and 'strengthening local institutions' where Scenario 2, Scenario 3, and Scenario 4 referred to the increase in the above corresponding concepts ('clamped' to one). We developed the fifth scenario by combining all the four key concepts/adaptation strategies used in the previous scenarios. We also conducted a sensitivity analysis to ascertain the stability of the system.

3. Results

3.1. Projections of Climate Change in the Study Area

Climate vulnerability refers to a system's susceptibility to change as a consequence of variation in climatic parameters. We assessed climate vulnerability, with the help of climate modelling, for key climatic parameters including temperature, precipitation, and accumulated precipitation on consecutive rainy days in the study area (Table 2).

Table 2. Annual mean temperature and precipitation profile of the study area.

Climatic Parameters	Reference Climate * (1981-2010)	** E-Mean of Projections during 2050s (2041–2070)		** E-Mean of Projections during 2080s (2071–2100)	
		RCP 2.6	RCP 8.5	RCP 2.6	RCP 8.5
Mean Temperature (°C)	26.8	27.9	29.2	27.9	30.8
Mean Precipitation (mm)	1744	1832	1872	1912	1872
Accumulated precipitation on consecutive rainy day (above 30 mm)	118	375	328	372	376

* Reference climate (CRUTS32: 1981–2010); ** E-Mean: Ensemble mean-ESM2M of all five models (CRUTS32, HadGEM2-ES, IPSL-CM5A-LR, MIROC-ESM-CHEM, and NorESM1-M) are taken into consideration to avoid extreme variations in the result.

Projections for the 2050s (2041–2070) and the 2080s (2071–2100) with regard to RCPs 2.6 and 8.5 show a rise from the reference climate (1981–2010) in terms of annual mean temperature, annual mean precipitation, and accumulated precipitation on consecutive rainy days in the region. According to the results, there is a definite rise in temperature. The projected increase is 1.1 to 2.4 °C in the 2050s and 1.1 to 4.0 °C in the 2080s as relevant to the RCPs 2.6 and 8.5, in that order. The quantum of precipitation evinces a rise as well. The projected increase is 88 to 128 mm in the 2050s and 128 to 168 mm in the 2080s pertinent to the RCPs 2.6 and 8.5, in that order. There is a rise in accumulated precipitation on consecutive rainy days exceeding 30 mm. The projected increase is 210 to 257 mm in the 2050s and 254 to 258 mm in the 2080s relevant to the RCPs 2.6 and 8.5, in that order. Continuous precipitation on consecutive rainy days exceeding 30 mm is considered a heavy precipitation incident and is likely to cause flooding.

3.2. Climate-Related Impacts as Perceived by the Communities

Community observations relevant to climate change mirror historical trends and projections regarding climate change in the context of climatic parameters mentioned in the previous section. The communities perceived that manifestations of climate variability and change, along with extreme climatic events, have impacted land and water resources, fisheries, agriculture and fodder, forests, human health, and livestock productivity. All this has caused an inexorable decline in the livelihoods and financial reserves of the communities. The groundwater level has plummeted in the region owing to the increased precipitation variability. It has reduced the productivity of agriculture and horticulture crops predominantly. The resultant seawater intrusion has degraded the quality of several freshwater bodies adversely impacting their aquaculture. Rising sea-levels, not to mention cyclones and storm surges, are responsible for saline water ingress causing underground freshwater aquifers and surface water reserves to pollute. Seawater intrusion has also led to soil salinity. Seawater intrusion, coupled with rising temperatures and consequent higher evaporation rates, causes increased salt concentration in surface water bodies. Seawater intrusion is a prominent issue in the region as it primarily impacts agriculture and fish production. It has forced many families to flee from places where they have been old residents. Degrading water quality has adversely affected the productivity of aquaculture, pisciculture, agriculture, and horticulture, all of which are critical to community livelihood. The communities affirm a greater extent of the negative impact on fisheries and agriculture stimulated by rising temperatures, precipitation variability, and climatic extremes. These variabilities and changes in climate have caused pest invasion incidences to rise, leading to considerable loss of agricultural produce. Rising temperatures have led to increasing occurrences of disease within the fish population, depressing fish production thereby. Increasing temperatures, precipitation variability, and increasing water and soil salinity are also responsible for lowering fodder availability, which in turn, has diminished livestock health and productivity. The communities also perceived declining water quality as being responsible for the diminished availability of drinking water, leading to sanitation-related problems causing the deterioration of human health. The proliferation of mosquitoes, owing to increased temperatures, has fomented malarial outbreak in the region. Owing to rising temperatures, rainfall variability, and anthropogenic interference, mangrove forests have degraded, ruining the fragile ecosystem of the Sundarbans and its unique biodiversity. This ultimately causes soil erosion, loss of infrastructure, and overall environmental degradation. Impacts on agriculture, fishery, and livestock production act as significant contributors to declining incomes and increasing the economic poverty of the communities. This not only affects the quality of life of the stakeholders but also decreases opportunities for education for their children.

3.3. Climate Change Adaptations in the Area

The community also informed about several adaptation practices implemented in the study area in order to curtail the impacts of climate change. WWF-India implements the Climate Adaptation Programme. The state government has facilitated the construction of earthen embankments and dykes along the coasts in order to check seawater intrusion. These embankments have been built to reduce the impact of tidal surges while reducing soil erosion to some extent. Several sections along embankments and dykes have been stabilised by mangrove plantations in order to check seawater intrusion, soil erosion, loss of critical infrastructure, and reduce environmental degradation. The state government has also facilitated the digging of hand pumps and tube wells to ensure freshwater access. Furthermore, healthcare facilities are being provided and vaccination drives conducted in order to improve human and livestock health. The Sundarbans Development Board, in partnership with the state government, has initiated several afforestation measures in order to restore the green cover of the region and to conserve the fragile ecosystem of the Sundarbans. The Greening India Programme—implemented by the Tagore Society for Rural Development (TSRD)—engages local communities for mangrove plantation drives through social forestry in a bid to protect the islands from natural calamities. Alongside afforestation activities, TSRD has also implemented a Disaster Management Programme. TSRD has taken the initiative to spread awareness about the importance of maintaining the soil's natural fertility by promoting organic fertilisers and vermicompost. TSRD has been imparting training to communities regarding alternative livelihood options including handicrafts, poultry, duckery, and goatery.

Adaptation measures involving intensifying sustainable agriculture and aquaculture practices such as crop diversification, the introduction of salt-tolerant crops, and traditional techniques of agriculture and fishery are meant to improve soil fertility and production from agriculture and fisheries. Pest control measures have been undertaken to combat pest invasion while increasing agricultural production in the long term. Diversification of livelihoods is another coping strategy adopted by communities. Since traditional natural resource-based livelihoods are unable to sustain the existing population in the Sundarbans of late, many community members are engaged with the tourism sector providing various services to tourists.

The communities acknowledged that earthen embankments and dykes along with mangrove plantations had reduced seawater intrusion, soil erosion, loss of critical infrastructure, and environmental degradation to a certain extent. Planting fruit-bearing and medicinal trees along the coastline have also helped increase the incomes of some households. Such measures have reduced soil erosion and helped improve water and land resources; it has also increased agriculture and fish production and, consequently, fodder availability. Better management of water resources in the form of rainwater harvesting, pond construction, and farm bunding has also helped to increase the availability of freshwater and productivity of agriculture and fisheries while hand pumps and tube wells have helped to reduce the drudgery of women. A combination of improved agricultural inputs with the introduction of climate-resilient agricultural and piscicultural practices coupled with the re-introduction of indigenous salt-tolerant rice varieties and fish species have helped communities diversify their livelihoods while coping with increased water and soil salinity. However, the communities perceived that these adaptation practices were relatively meagre in relation to curbing the current risks involved in climate change. They also believed, notwithstanding the current adaptation practices, the risk from future impacts of climate change cannot be denied. They expressed a requirement for more concerted efforts towards developing climate resilience.

Figure 1 illustrates the condensed social cognitive map showing the perception of communities regarding climate-related impacts and adaptations.

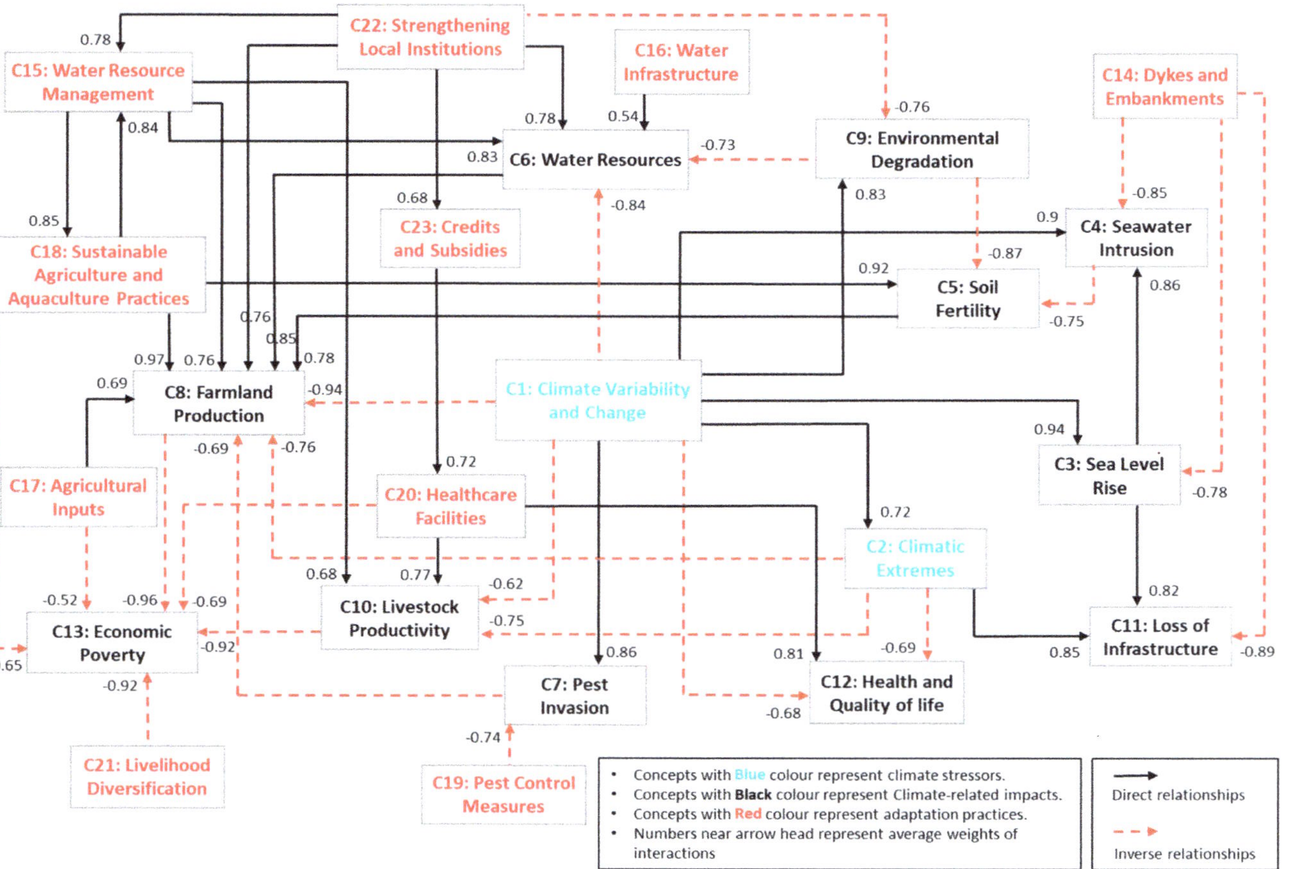

Figure 1. Condensed social cognitive map showing the perception of communities regarding climate-related impacts and adaptations.

3.4. FCM-Based Simulations

The baseline simulates a situation through the existing FCM model in which both 'climate variability and change' and 'climatic extremes' are activated. It illustrates an increase in sea-level rise, seawater intrusion, pest invasion, environmental degradation, loss of infrastructure, and economic poverty. It also suggests a reduction in soil fertility, water resources, agriculture, livestock productivity, and the health and quality of life (Table 3).

After having conducted simulations of all the different scenarios (scenarios 1 to 5, as discussed in Section 2.3), we tabulated the outcome and compared the deviation of each concept against the steady-state of the baseline, as shown in Table 3. Exploring the dynamic change of concepts' values between the baseline steady-state and scenario outcomes enabled a quantitative interpretation of the impact of the key concepts on the system.

The first scenario highlights the effects of 'dykes and embankments'. This scenario does not give relief from sea-level rise while the loss of infrastructure, seawater intrusion, pest invasion, environmental degradation, and economic poverty continue to increase. The second scenario highlights the effects of 'water resource management'. This scenario does not show much deviation from the baseline; it illustrates a substantial increase in sea-level rise, seawater intrusion, pest invasion, environmental degradation, loss of infrastructure, and economic poverty. It also shows a considerable decrease in soil fertility, water resources, agriculture productivity, livestock productivity, and the health and quality of life. The third scenario highlights the effects of 'sustainable agriculture and aquaculture practices'. This scenario also does not show much deviation from the baseline. What it indicates is an increase in sea-level rise, seawater intrusion, pest invasion, environmental degradation, and loss of infrastructure. It also displays a decrease in water resources, health, and quality of life and, consequently, an increase in economic poverty. The fourth scenario highlights the effects of 'strengthening local institutions'. This scenario is unlikely to decrease sea-level rise, seawater intrusion, pest invasion, environmental degradation, and loss of infrastructure. However, water resources, agriculture productivity, and livestock productivity are likely to increase, leading to reduced economic poverty. This scenario also shows an increase in water resource management, sustainable agriculture and aquaculture practices, healthcare facilities, and credits and subsidies because a vibrant local institution is likely to implement interventions in all these areas.

All the concepts deployed for the previous scenarios have been clamped together in the fifth scenario. This integrative scenario displays a marginal reduction in climate change impacts compared to the baseline. However, seawater intrusion, pest invasion, environmental degradation, and economic poverty show a continuous rise. On the other hand, water resources, agriculture production, and livestock productivity show an increase leading to a decrease in the economic poverty. This scenario also indicates an increase in water resource management, sustainable agriculture and aquaculture practices, healthcare facilities, and credits and subsidies (see Table 3 and Figure S1).

Overall, the results of the FCM-based scenario analysis illustrate that the possibilities of climate risk in the region cannot be ruled out in the future even after having implemented all the adaptations. This signifies limits to the ongoing adaptations, meaning the existing adaptations in the area are inadequate in the context of providing resilience to the community against climate stressors.

Table 3. Results of FCM-based simulations.

Concepts	Baseline		Scenario 1		Scenario 2		Scenario 3		Scenario 4		Scenario 5	
	IV	Steady State Value	IV	Steady State Value	IV	Steady State Value	IV	Steady State Value	IV	Steady State Value	IV	Steady State Value
C1: Climate variability and change	1	1	1	1	1	1	1	1	1	1	1	1
C2: Climatic extremes	1	0.9297	1	0.9297	1	0.9297	1	0.9297	1	0.9297	1	0.9297
C3: Sea-level rise	0	0.9564	0	0.6891	0	0.9564	0	0.9564	0	0.9564	0	0.6897
C4: Seawater intrusion	0	0.9986	0	0.9839	0	0.9986	0	0.9986	0	0.9986	0	0.984
C5: Soil fertility	0	−0.9884	0	−0.9881	0	−0.9275	0	−0.9276	0	−0.8334	0	−0.8284
C6: Water resources	0	−0.9921	0	−0.9921	0	−0.9605	0	−0.9605	0	0.7482	0	0.7482
C7: Pest invasion	0	0.9477	0	0.9477	0	0.9477	0	0.9477	0	0.9477	0	0.9477
C8: Agriculture productivity	0	−0.9999	0	−0.9999	0	−0.9961	0	−0.9961	0	0.8096	0	0.8116
C9: Environmental degradation	0	0.9444	0	0.9445	0	0.9445	0	0.9445	0	0.5532	0	0.5532
C10: Livestock productivity	0	−0.9804	0	−0.9804	0	−0.9231	0	−0.9232	0	0.4799	0	0.48
C11: Loss of infrastructure	0	0.9884	0	0.8717	0	0.9884	0	0.9884	0	0.9884	0	0.8721
C12: Health and quality of life	0	−0.9806	0	−0.9806	0	−0.9806	0	−0.9806	0	−0.9031	0	−0.9031
C13: Economic poverty	0	0.9935	0	0.9935	0	0.9746	0	0.9746	0	−0.9981	0	−0.9981
C14: Dykes and embankments	0	0	1	1	0	0	0	0	0	0	1	1
C15: Water resource management	0	0	0	0	1	0.9392	0	0.9391	0	0.9885	1	0.9885
C16: Water infrastructure	0	0	0	0	0	0	0	0	0	0	0	0
C17: Agriculture inputs	0	0	0	0	0	0	0	0	0	0	0	0
C18: Sustainable agriculture and aquaculture practices	0	0	0	0	0	0.9402	1	0.9402	0	0.9455	1	0.9455
C19: Pest control measures	0	0	0	0	0	0	0	0	0	0	0	0
C20: Healthcare facilities	0	0	0	0	0	0	0	0	0	0.9199	0	0.9199
C21: Livelihood diversification	0	0	0	0	0	0	0	0	0	0	0	0
C22: Strengthening local institutions	0	0	0	0	0	0	0	0	1	1	1	1
C23: Credits and subsidies	0	0	0	0	0	0	0	0	0	0.9228	0	0.9228

Note: IV stands for initial value; Concepts shown in Blue are climate stressors, concepts shown in Black are climate-related impacts, and concepts shown in Red are climate change adaptations in the area as perceived by the communities.

4. Discussions

Communities, especially those living in climatically vulnerable regions (such as the Sundarbans) and dependent on climate-sensitive livelihoods, are not only more vulnerable to climate change impacts but also less able to pursue adaptation measures [39]. Adaptation to climate change is often described as a local issue, with specific attention given to context-specific features and factors enabling or constraining adaptations. [40]. Increased climate variability, climatic extremes, rising sea-levels, and coastal flooding can impact natural, human, and financial assets in a coastal river delta like Sundarbans owing to deteriorating water and soil quality attributable to increased seawater intrusion. All of this depresses production pertinent to agriculture, fishery, and livestock while augmenting environmental degradation, worsening human health, and economic poverty in the area. According to the study, the chief climate change adaptation measures in the region are climate change adaptation programme, greening India programme, building dykes and embankments, and the introduction of climate-resilient agriculture and pisciculture. These adaptation measures, to some extent, help reduce seawater intrusion and soil erosion, improve water quality and soil fertility, prevent environmental degradation while increasing the productivity of agriculture, fishery, and livestock in the current climatic conditions. However, the FCM-based scenarios show adverse impacts of climate change in the future in the context of several parameters, including sea-level rise, seawater intrusion, soil fertility, environmental degradation, pest invasion, and loss of infrastructure. The current adaptation interventions in the region are inadequate in terms of reducing ecological and social vulnerabilities and enhancing resilience.

Adaptation interventions, whether planned or autonomous, are not isolated responses and actions. Instead, they are contextual and are known to be influenced by the rate of climate change along with economic, demographic, environmental, social, and technological factors. The lack of knowledge, and level of financial, human, and social capital also limit capabilities of communities for adaptation. The opportunities and constraints regarding adaptations are determined in the contexts of the region, community, or household; which means that they vary across geographies, sectors, communities, and species [41]. Our simulation results reveal the likelihood of limits to adaptations in the region. The adaptation limits are classified as: (a) Hard limits where no adaptation is possible and (b) soft limits where adaption options are currently unavailable [41]. The Intergovernmental Panel on Climate Change [8] refers to adaptation limits as obstacles that tend to be absolute in the real sense while setting up thresholds beyond which adaptation activities cannot be maintained or modified. In some studies [42,43], the hard limits have been referred to as limits to adaptation. Soft adaptation limits, on the other hand, are referred to as adaptation barriers—obstacles that may be overcome through concerted efforts, creative management, prioritisation, and related shifts in resources, institutions, and so on. [43]. Significant barriers to adaptations could include: bio-physical—climate variability and change; socio-economic—lack of climate change awareness among decision-makers; political—asymmetrical governance structures; and other cross-cutting barriers, which carry the potential of evolving into hard limits to adaptations [44–47].

The act of implementing adaptations is riddled with barriers including dated and locally extraneous information along with the paucity of financial resources, appropriate technology, and traditional knowledge pertinent to adaptation strategies, not to mention institutional constraints. Major bio-physical obstacles in the Sundarbans include climate variability and change, leading to rising sea-levels, coastal flooding, and depleting water quality and quantity. The socio-economic barriers in the region include several social, cultural, cognitive, behavioural, economic, and political impediments that influence actions and choices related to adaptations. These include: (a) Inadequate community awareness regarding manifestations of climate change; (b) lack of information, knowledge, infrastructure, and technology to battle with the impacts of climate variability and change; (c) lack of sufficient financial capital; and (d) other hidden social, economic, and political norms that indirectly control adoption and non-adoption of climate change adaptations and translate into the marginalisation of communities. These barriers to climate change adaptations not only shape the current and future climatic vulnerability in the region, but also the implementation and effectiveness of adaptations. Therefore, it is important to overcome these barriers in order to ensure the successful and effective implementation of adaptations. Researchers argue that many limits and barriers, especially social ones, can be overcome with sufficient financial resources, social efforts and support, and political will [48]. Gender-responsive and inclusive leadership, strategic and creative thinking, handholding support, resourcefulness, a collaboration between various players, and effective communication will be required for overcoming these barriers. However, merely overcoming barriers is not as straightforward as building adaptive capacity and does not necessarily ensure a successful adaptation intervention [43]. The need of the hour, hence, is the adoption of community-based adaptation (CBA) and ecosystem-based adaptation (EbA) supporting greater prioritisation of adaptation requirements at the local level while bringing co-benefits to the ecosystems and communities [40,49].

5. Conclusions and Research Directions

This paper sheds light on the diverse impacts of climate variability and change with regard to the Sundarbans ecosystem besides the lives and livelihoods of resident communities. Increased risks attributable to climatic variability and change, in addition to a rapidly growing population in the region, exert pressure on the Sundarbans ecosystem, making it extremely vulnerable to climate-related impacts. According to our findings, the current adaptations in the study area do not emerge as effective strategies against climate change. This calls for immediate action regarding implementing area-specific robust adaptation interventions. After all, adaptation is not just about choosing between technical options; it is also about 'social and political change' [50]. Some of the relatively effective adaptations in the context of river deltas are: Climate-resilient lifestyle and employment, climate-resilient farming systems, and better planning of housing and other infrastructure. Mainstreaming climate change adaptation requires targeted strategies and actions that are beyond mere aspirations in order to be effective while overcoming current and potential adaptation barriers [51].

Innumerable scholars have grappled with the concept of maladaptation [52–55], yet no clear metric has emerged that can identify the threshold between potentially successful adaptation and maladaptation. A holistic approach with a full range of CBA and EbA interventions is likely to enhance resilience against climate variability and change in order to protect the fragile Sundarbans ecosystem. These include profound systemic change requiring the reconfiguration of social and ecological systems [56]. While designing and implementing the CBA and EbA interventions, the role of indigenous and local knowledge cannot be overlooked. Therefore, the success of CBA and EbA relies on gaining a good understanding of the socio-cultural and socio-political context within which the communities operate on the ground. This includes gender, caste, land ownerships and tenure arrangements, local governance, decision-making processes, etc. [47,49,57,58]. Such adaptations are likely to support needs at the local level while bringing co-benefits to the ecosystems and communities [40,49]. Globally, small islands have been focusing increasingly on CBA that seeks to enable community-level ownership of adaptations. They are also looking to EbA approaches that benefit the ecosystems as well as communities [49,59]. Ample evidence is available with regard to the implementation of both CBA and EbA approaches across the small islands. However, one must still work to find robust ways of quantifying their benefits [60].

Transformative adaptation, which describes the need for changes in major systems in order to deal with impacts related to climate change, is usually contrasted with incremental adaptation. This refers to small adjustments made for climate-proofing. However, there is no clear metric available for planning a transformative adaptation. Climate-resilient lifestyle and employment, climate-resilient farming systems, and superior planning of housing and other infrastructure are likely to overcome several barriers while being transformative. However, some of these transformative adaptations may necessitate behavioural change on the part of communities. The Sundarbans needs more than just physical infrastructure for long-term sustainability. It requires a holistic development plan that includes ecosystem management in terms of land management and sustainability of the natural resource base along with community development in terms of creating awareness, capacity building, livelihood enhancement, adaptive governance, and disaster preparedness and management in order to mainstream climate change adaptations.

While CBA and EbA have gained significant traction lately, they too require more in-depth research in the context of coastal river deltas. The importance of locating local and timely adaptation strategies and development goals independently or through mutual co-benefits cannot be undermined. The situation merits an urgency regarding introducing synergy between the natural and social sciences while involving the latter more intensively in climate change adaptation research and application while mainstreaming adaptation interventions into development policies. It is vital to keep innovating and refining climate-related economic studies in order to support areas such as loss and damage along with adaptation costs.

Of late, international research and implementation policies have been focusing on climate change mitigation, adaptations, and development on a global basis to a greater extent. Attention to local problems and development potential have been somewhat overshadowed. Further research is needed to assess the restorative and productive abilities of the fragile ecosystems under changing climatic conditions. Scholars also need to conduct new scientific research with regard to both the effectiveness of adaptations, transformative adaptation, adaptation deficit, and limit to adaptation. This necessitates immediate and adequate financial commitment from global communities in the context of climate change adaptation research and implementation. Such efforts could help maintain climate change adaptations pertinent to the policy agenda while increasing political stakes [51]. Research also needs to be more thorough in terms of identifying the amount of funding required to implement an effective adaptation activity.

Supplementary Materials: The following are available online at http://www.mdpi.com/2071-1050/11/23/6655/s1, Table S1(a): Network statistics of the FCM model; Table S1(b): Structural analysis of the FCM model; Figure S1: Results of various FCM-based simulations.

Author Contributions: All authors contributed equally to this paper.

Funding: This research was funded by the Department of Science and Technology, Government of India.

Acknowledgments: We thank the Department of Science and Technology, Government of India, for providing financial support for the study. We thank Indrani Talukdar for language editing. We acknowledge the community members for participating in the research. We sincerely thank the two anonymous reviewers for raising pertinent questions and providing constructive suggestions.

Conflicts of Interest: The authors declare no conflict of interest.

References

1. Hossain, M.S.; Dearing, J.A.; Rahman, M.M.; Salehin, M. Recent changes in ecosystem services and human well-being in the Bangladesh coastal zone. *Reg. Environ. Chang.* **2016**, *16*, 429–443. [CrossRef]
2. Phillips, J.D. Environmental gradients and complexity in coastal landscape response to sea-level rise. *Catena* **2018**, *169*, 107–118. [CrossRef]
3. IPCC. *Climate Change 2013: The Physical Science Basis. Contribution of Working Group I to the Fifth Assessment Report of the Intergovernmental Panel on Climate Change*; Stocker, T.F., Qin, D., Plattner, G.K., Tignor, M., Allen, S.K., Boschung, J., Nauels, A., Xia, Y., Bex, V., Midgley, P.M., Eds.; Cambridge University Press: Cambridge, UK; New York, NY, USA, 2013; p. 1535.
4. Temmerman, S.; Kirwan, M.L. Building land with a rising sea. *Science* **2015**, *349*, 588–589. [CrossRef]
5. Van De Lageweg, W.I.; Slangen, A.B.A. Predicting Dynamic Coastal Delta Change in Response to Sea-Level Rise. *J. Mar. Sci. Eng.* **2017**, *5*, 24. [CrossRef]
6. Lentz, E.E.; Thieler, E.R.; Plant, N.G.; Stippa, S.R.; Horton, R.M.; Gesch, D.B. Evaluation of dynamic coastal response to sea-level rise modifies inundation likelihood. *Nat. Clim. Chang.* **2016**, *6*, 696–700. [CrossRef]

7. IPCC. *Climate Change 2014: Impacts, Adaptation, and Vulnerability. Part. B: Regional Aspects. Contribution of Working Group II to the Fifth Assessment Report of the Intergovernmental Panel on Climate Change*; Barros, V.R., Field, C.B., Dokken, D.J., Mastrandrea, M.D., Mach, K.J., Bilir, T.E., Chatterjee, M., Ebi, K.L., Estrada, Y.O., Genova, R.C., et al., Eds.; Cambridge University Press: Cambridge, UK; New York, NY, USA, 2014; p. 688.

8. IPCC. *Climate Change 2007: Impacts, Adaptation and Vulnerability. Contribution of Working Group II to the Fourth Assessment Report of the Intergovernmental Panel on Climate Change*; Parry, M.L., Canziani, O.F., Palutikof, J.P., van der Linden, P.J., Hanson, C.E., Eds.; Cambridge University Press: Cambridge, UK; New York, NY, USA, 2007; p. 976.

9. Neogi, S.B.; Dey, M.; Kabir, S.M.L.; Masum, S.J.H.; Kopprio, G.; Yamasaki, S.; Lara, R. Sundarban mangroves: Diversity, ecosystem services and climate change impacts. *Asian J. Med. Biol. Res.* **2016**, *2*, 488–507. [CrossRef]

10. Lara, R.J.; Neogi, S.B.; Islam, M.S.; Mahmud, Z.H.; Islam, S.; Paul, D.; Demoz, B.B.; Yamasaki, S.; Nair, G.B.; Kattner, G. *Vibrio cholerae* in waters of the Sunderban mangrove: Relationship with biogeochemical parameters and chitin in seston size fractions. *Wetl. Ecol. Manag.* **2011**, *19*, 109–119. [CrossRef]

11. Nurse, L.A.; McLean, R.F.; Agard, J.; Briguglio, L.P.; Duvat-Magnan, V.; Pelesikoti, N.; Tompkins, E.; Webb, A. Small Islands. In *Climate Change 2014: Impacts, Adaptation, and Vulnerability. Part B: Regional Aspects. Contribution of Working Group II to the Fifth Assessment Report of the Intergovernmental Panel on Climate Change*; Barros, V.R., Field, C.B., Dokken, D.J., Mastrandrea, M.D., Mach, K.J., Bilir, T.E., Chatterjee, M., Ebi, K.L., Estrada, Y.O., Genova, R.C., et al., Eds.; Cambridge University Press: Cambridge, UK; New York, NY, USA, 2014; pp. 1613–1654. [CrossRef]

12. Kay, J.E.; Deser, C.; Phillips, A.; Mai, A.; Hannay, C.; Strand, G.; Arblaster, J.M.; Bates, S.C.; Danabsoglu, G.; Edwards, J.; et al. The Community Earth System Model (CESM) Large Ensemble Project: A Community Resource for Studying Climate Change in the Presence of Internal Climate Variability. *Bull. Am. Meteorol. Soc.* **2015**, *96*, 1333–1349. [CrossRef]

13. Behera, R.; Kar, A.; Das, M.R.; Panda, P.P. GIS-based vulnerability mapping of the coastal stretch from Puri to Konark in Odisha using analytical hierarchy process. *Nat. Hazards* **2019**, *96*, 731–751. [CrossRef]

14. Vitousek, S.; Barnard, P.L.; Fletcher, C.H.; Frazer, N.; Erikson, L.; Storlazzi, C.D. Doubling of coastal flooding frequency within decades due to sea-level rise. *Sci. Rep.* **2017**, *7*, 1–9. [CrossRef]

15. Wasko, C.; Sharma, A. Global assessment of flood and storm extremes with increased temperatures. *Sci. Rep.* **2017**, *7*, 1–8. [CrossRef] [PubMed]

16. Muis, S.; Verlaan, M.; Winsemius, H.C.; Aerts, J.C.J.H.; Ward, P.J. A global reanalysis of storm surges and extreme sea-levels. *Nat. Commun.* **2016**, *7*, 1–11. [CrossRef] [PubMed]

17. Little, C.M.; Horton, R.M.; Kopp, R.E.; Oppenheimer, M.; Vecchi, G.A.; Villarini, G. Joint projections of US East Coast sea-level and storm surge. *Nat. Clim. Chang.* **2015**, *5*, 1114–1120. [CrossRef]

18. Wahl, T.; Jain, S.; Bender, J.; Meyers, S.D.; Luther, M.E. Increasing risk of compound flooding from storm surge and rainfall for major US cities. *Nat. Clim. Chang.* **2015**, *5*, 1093–1097. [CrossRef]

19. Government of India. Census 2011. Available online: http://www.censusindia.gov.in/2011-common/census_2011.html (accessed on 20 August 2019).

20. Uddin, M.S.; Shah, M.A.R.; Khanom, S.; Nesha, M.K. Climate change impacts on the Sundarbans mangrove ecosystem services and dependent livelihoods in Bangladesh. *Asian J. Conserv. Biol.* **2013**, *2*, 152–156.

21. Van Vuuren, D.P.; Stehfest, E.; den Elzen, M.G.J.; Kram, T.; van Vliet, J.; Deetman, S.; Isaac, M.; Goldewijk, K.K.; Hof, A.; Beltran, A.M.; et al. RCP2.6: Exploring the possibility to keep global mean temperature change below 2°C. *Clim. Chang.* **2011**, *109*, 95–116. [CrossRef]

22. Taylor, K.E.; Stouffer, R.J.; Meehl, G.A. An overview of CMIP5 and the experiment design. *Bull. Am. Meteorol. Soc.* **2011**, *93*, 485–498. [CrossRef]

23. Hijmans, R.E.; Cameron, S.E.; Parra, J.L.; Jones, P.G.; Jarvis, A. Very high resolution interpolated climate surfaces of global land areas. *Int. J. Clim.* **2005**, *25*, 1965–1978. [CrossRef]

24. Hempel, S.; Frieler, K.; Warszawski, L.; Schewe, J.; Piontek, F. A trend-preserving bias correction—The ISI-MIP approach. *Earth Syst. Dyn.* **2013**, *4*, 219–236. [CrossRef]

25. Kosko, B. Fuzzy Cognitive Maps. *Int. J. Man Mach. Stud.* **1986**, *24*, 65–75. [CrossRef]

26. Özesmi, U.; Özesmi, L.S. Ecological models based on people's knowledge: A multi-step fuzzy cognitive mapping approach. *Ecol. Model.* **2004**, *176*, 43–64. [CrossRef]

27. Singh, P.K.; Chudasama, H. Pathways for drought resilient livelihoods based on people's perception. *Clim. Chang.* **2017**, *140*, 179–193. [CrossRef]

28. Singh, P.K.; Chudasama, H. Assessing impacts and community preparedness to cyclones: A fuzzy cognitive mapping approach. *Clim. Chang.* **2017**, *143*, 337–354. [CrossRef]

29. Solana-Gutiérrez, J.; Rincón, G.; Alonso, C.; García-de-Jalón, D. Using fuzzy cognitive maps for predicting river management responses: A case study of the Esla River basin, Spain. *Ecol. Model.* **2017**, *360*, 260–269. [CrossRef]

30. Singh, P.K.; Nair, A. Livelihood vulnerability assessment to climate variability and change using fuzzy cognitive mapping approach. *Clim. Chang.* **2014**, *127*, 475–491. [CrossRef]

31. Papageorgiou, E.; Kontogianni, A. Using Fuzzy Cognitive Mapping in Environmental Decision Making and Management: A Methodological Primer and an Application. In *International Perspectives on Global Environmental Change*; Young, S., Ed.; InTech: Lonodn, UK, 2012; pp. 427–450. ISBN 978-953-307-815-1.

32. Papageorgiou, E. (Ed.) *Fuzzy Cognitive Maps for Applied Sciences and Engineering—From Fundamentals to Extensions and Learning Algorithms, Intelligent Systems Reference Library, 54*; Springer (Springer-Verlag): Heidelberg, Germanay, 2014; ISBN 978-3-642-39738-7.

33. Allen, M.R.; Dube, O.P.; Solecki, W.; Aragón–Durand, F.; Cramer, W.; Humphreys, S.; Kainuma, M.; Kala, J.; Mahowald, N.; Mulugetta, Y.; et al. Framing and Context. In *Global warming of 1.5 °C. An IPCC Special Report on the impacts of global warming of 1.5 °C above Pre-Industrial Levels and Related Global Greenhouse Gas Emission Pathways, in the Context of Strengthening the Global rEsponse to the Threat of Climate Change, Sustainable Development, and Efforts to Eradicate Poverty*; Masson-Delmotte, V., Zhai, P., Pörtner, H.O., Roberts, D., Skea, J., Shukla, P.R., Pirani, A., Moufouma-Okia, W., Péan, C., Pidcock, R., et al., Eds.; IPCC: Geneva, Switzerland, 2018; pp. 49–91.

34. Ziv, G.; Watson, E.; Young, D.; Howard, D.C.; Larcom, S.T.; Tanentzap, A.J. The potential impact of Brexit on the energy, water and food nexus in the UK: A fuzzy cognitive mapping approach. *Appl. Energy* **2018**, *210*, 487–498. [CrossRef]

35. Jetter, A.J.; Kok, K. Fuzzy cognitive maps for futures studies—A methodological assessment of concepts and methods. *Futures* **2014**, *61*, 45–57. [CrossRef]

36. UNDP. *Adaptation Policy Frameworks for Climate Change: Developing Strategies, Policies and Measures*; Lim, B., Spanger-Siegfried, E., Burton, I., Malone, E., Huq, S., Eds.; Cambridge University Press: Cambridge, UK; New York, NY, USA, 2004; p. 248.

37. Nápoles, G.; Papageorgiou, E.; Bello, R.; Vanhoof, K. On the convergence of sigmoid Fuzzy Cognitive Maps. *Inf. Sci.* **2016**, *349*, 154–171. [CrossRef]

38. Papageorgiou, E.; Papageorgiou, K.; Dikopoulou, Z.; Mouhrir, A. A web-based tool for Fuzzy Cognitive Map Modeling. In Proceedings of the 9th International Congress on Environmental Modelling and Software (iEMSs), Fort Collins, CO, USA, 24–29 June 2018.

39. Asfaw, S.; Pallante, G.; Palma, A. Diversification Strategies and Adaptation Deficit: Evidence from Rural Communities in Niger. *World Dev.* **2018**, *101*, 219–234. [CrossRef]

40. Nalau, J.; Becken, S.; Schliephack, J.; Parsons, M.; Brown, C.; Mackey, B. The Role of Indigenous and Traditional Knowledge in Ecosystem-Based Adaptation: A Review of the Literature and Case Studies from the Pacific Islands. *Weather Clim. Soc.* **2018**, *10*, 851–865. [CrossRef]

41. Klein, R.J.T.; Midgley, G.E.; Preston, B.L.; Alam, M.; Berkhout, F.G.H.; Dow, K.; Shaw, M.R. Adaptation Opportunities, Constraints, and Limits. In *Climate Change 2014: Impacts, Adaptation, and Vulnerability. Part A: Global and Sectoral Aspects. Contribution of Working Group II to the Fifth Assessment Report of the Intergovernmental Panel on Climate Change*; Field, C.B., Barros, V.R., Dokken, D.J., Mach, K.J., Mastrandrea, M.D., Bilir, T.E., Chatterjee, M., Ebi, K.L., Estrada, Y.O., Genova, R.C., et al., Eds.; Cambridge University Press: Cambridge, UK; New York, NY, USA, 2014; pp. 899–943.

42. Barnett, J.; Evans, L.S.; Gross, C.; Kiem, A.S.; Kingsford, R.T.; Palutikof, J.P.; Pickering, C.M.; Smithers, S.G. From barriers to limits to climate change adaptation: Path dependency and the speed of change. *Ecol. Soc.* **2015**, *20*, 5. [CrossRef]

43. Moser, S.C.; Ekstrom, J.A. A framework to diagnose barriers to climate change adaptation. *PNAS* **2010**, *107*, 22026–22031. [CrossRef] [PubMed]

44. Petzold, J.; Magnan, A.K. Climate change: Thinking small islands beyond Small Island Developing States (SIDS). *Clim. Chang.* **2019**, *152*, 145–165. [CrossRef]

45. Shackleton, S.; Ziervogel, G.; Sallu, S.; Gill, T.; Tschakert, P. Why is socially-just climate change adaptation in sub-Saharan Africa so challenging? A review of barriers identified from empirical cases. *WIREs Clim. Chang.* **2015**, *6*, 321–344. [CrossRef]
46. Antwi-Agyei, P.; Dougill, A.J.; Stringer, L.C. Barriers to climate change adaptation: Evidence from northeast Ghana in the context of a systematic literature review. *Clim. Dev.* **2014**, *7*, 297–309. [CrossRef]
47. Nunn, P.; Aalbersberg, W.; Lata, S.; Gwilliam, M. Beyond the core: Community governance for climate-change adaptation in peripheral parts of Pacific Island Countries. *Reg. Environ. Chang.* **2013**, *14*, 221–235. [CrossRef]
48. Adger, W.N.; Dessai, S.; Goulden, M.; Hulme, M.; Lorenzoni, I.; Nelson, D.R.; Naess, L.O.; Wolf, J.; Wreford, A. Are there social limits to adaptation to climate change? *Clim. Chang.* **2009**, *93*, 335–354. [CrossRef]
49. Nalau, J.; Handmer, J. Improving development outcomes and reducing disaster risk through planned community relocation. *Sustainability* **2018**, *10*, 3545. [CrossRef]
50. Petzold, J.; Ratter, B.M.W. Climate change adaptation under a social capital approach—An analytical framework for small islands. *Ocean. Coast. Manag.* **2015**, *112*, 36–43. [CrossRef]
51. Runhaar, H.; Wilk, B.; Persson, Å.; Uittenbroek, C.; Wamsler, C. Mainstreaming climate adaptation: Taking stock about "what works" from empirical research worldwide. *Reg. Environ. Chang.* **2018**, *18*, 1201–1210. [CrossRef]
52. Neset, T.; Asplund, T.; Käyhkö, J.; Juhola, S. Making sense of maladaptation: Nordic agriculture stakeholders' perspectives. *Clim. Chang.* **2019**, *153*, 107–121. [CrossRef]
53. Antwi-Agyei, P.; Dougill, A.J.; Stringer, L.C.; Codjoe, S.N.A. Adaptation opportunities and maladaptive outcomes in climate vulnerability hotspots of northern Ghana. *Clim. Risk Manag.* **2018**, *19*, 83–93. [CrossRef]
54. Juhola, S.; Glaas, E.; Linnér, B.; Neset, T. Redefining maladaptation. *Environ. Sci. Policy* **2016**, *55*, 135–140. [CrossRef]
55. Barnett, J.; O'Neil, S. Maladaptation. *Glob. Environ. Chang.* **2010**, *20*, 211–213. [CrossRef]
56. De Coninck, H.; Revi, A.; Babiker, M.; Bertoldi, P.; Buckeridge, M.; Cartwright, A.; Dong, W.; Ford, J.; Fuss, S.; Hourcade, J.C.; et al. Strengthening and implementing the global response. In *Global Warming of 1.5 °C. An IPCC Special Report on the Impacts of Global Warming of 1.5 °C above Pre-Industrial Levels and Related Global Greenhouse Gas Emission Pathways, in the Context of Strengthening the Global Response to the Threat of Climate Change, Sustainable Development, and Efforts to Eradicate Poverty*; Masson Delmotte, V., Zhai, P., Pörtner, H.O., Roberts, D., Skea, J., Shukla, P.R., Pirani, A., Moufouma-Okia, W., Péan, C., Pidcock, R., et al., Eds.; IPCC: Geneva, Switzerland, 2018; pp. 313–443.
57. Parsons, M.; Brown, C.; Nalau, J.; Fisher, K. Assessing adaptive capacity and adaptation: Insights from Samoan tourism operators. *Clim. Dev.* **2018**, *10*, 644–663. [CrossRef]
58. McNamara, K.E.; Buggy, L. Community-based climate change adaptation: A review of academic literature. *Local Environ.* **2016**, *22*, 1–18. [CrossRef]
59. Remling, E.; Veitayaki, J. Community-based action in Fiji's Gau Island: A model for the Pacific? *Int. J. Clim. Chang. Strateg. Manag.* **2016**, *8*, 375–398. [CrossRef]
60. Doswald, N.; Munroe, R.; Roe, D.; Giuliani, A.; Castelli, I.; Stephens, J.; Möller, I.; Spencer, T.; Vira, B.; Reid, H. Effectiveness of ecosystem-based approaches for adaptation: Review of the evidence-base. *Clim. Dev.* **2014**, *6*, 185–201. [CrossRef]

Article

A Systemic Design Approach Applied to Rice and Wine Value Chains. The Case of the InnovaEcoFood Project in Piedmont (Italy)

Eleonora Fiore *, Barbara Stabellini and Paolo Tamborrini

Department of Architecture and Design, Politecnico di Torino, 10126 Torino, Italy;
barbara.stabellini@polito.it (B.S.); paolo.tamborrini@polito.it (P.T.)
* Correspondence: eleonora.fiore@polito.it; Tel.: +39-3287303068

Received: 14 October 2020; Accepted: 6 November 2020; Published: 8 November 2020

Abstract: Attention to food waste is an increasingly growing phenomenon today, especially in the context of a circular economy. The InnovaEcoFood project investigates the use of by-products of the Piedmontese rice and wine production chains to valorize their untapped potential in the food sector by applying the Systemic Design approach. We collected, systematized, and visualized a range of solutions for exploiting these by-products, starting from an in-depth literature review on the two value chains. With the support of a consortium of partners from both multidisciplinary industrial and academic sectors, it was possible to validate the links that have been generated. Eventually, the project created food products that integrated these outputs as ingredients (like flour and butter) because they have antioxidant properties and are rich in proteins. InnovaEcoFood has successfully tested how value could be created from waste. Moreover, using rice hull, marc flour, and bran lipid (butter) is of immediate technical and economic feasibility. It could be considered a viable way that deserves further experimentation.

Keywords: systemic design; rice; wine; value chains; by-products; circular economy

---

1. Introduction

Attention to food waste is an increasingly growing phenomenon today, especially in the context of a circular economy. Considering the entire supply chain, food waste can occur at every stage of the process: during production and processing, distribution and storage, and eventually during preparation and final consumption [1,2]. Europe's agro-food industry has an important share in the economy [3]. However, around 88 million tons of food waste is generated yearly [4].

On this premise, the InnovaEcoFood project explores the use of the outputs of the Piedmontese rice and wine production chains, which are two centuries-old production activities with a cultural and gastronomic tradition recognized and appreciated at an international level. The project aims to demonstrate that the competitiveness of companies can be improved using Systemic Design (SD) as an approach, going beyond the concept of recycling waste, by promoting the creation of value from by-products considering them as value-added raw materials. This project also applies the latest technologies to enhance the unexploited qualities of agricultural waste, production processes and transformation by-products of the two supply chains to obtain environmental, economic and social benefits.

Technological integration applied to an important sector such as grape growing and rice cultivation is an objective to pursue to increase regional and national economic resilience. InnovaEcoFood promotes industrial development and aims to create zero waste supply chains according to the principles of the circular economy, with the design purpose of aligning with EU policy, following Global Goal 12 of Agenda 2030 (responsible consumption and production), including the creation of value from waste deriving from the production processes. The project brings together multidisciplinary skills ranging from SD to chemical engineering, from food science and technology to communication, intending to develop guidelines that can become the basis for sustainable and efficient production. Making a production chain that is historically and culturally recognized, as part of a system is essential. On the one hand, it aims to minimize the impact due to the disposal of outputs, on the other hand, it develops resilient, sustainable and competitive innovations with strong implications on the economy, society and culture by activating relationships with the territory itself.

The development of new activities and products in the food (additives and functional foods) and pharmaceutical (formulation of supplements or natural products with antioxidant activity) sectors, increase regional competitiveness, generating new revenues from waste products and the exploitation of local know-how.

Section 2 includes a literature review related to the circular economy (CE) and its connection with the SD approach to establish similarities and differences between the two approaches. The importance of using SD as an approach to address the two value chains is defined at the end of that section. In this research, SD approach is used to investigate the two traditional Piedmontese supply chains, comparing the current state of exploitation of the outputs or by-products of the production processes with the value creation obtained with the systemic approach. For this reason, Sections 3 and 4 provide an in-depth analysis of the two value chains, explaining their importance at the regional level, describing their by-products and the potential arising from their use in an SD perspective. Section 5 includes the results of the InnovaEcoFood project, which experiments the production of flours for human nutrition and the extraction of high value-added oils and molecules from the vegetable matrices of the two agricultural activities. It uses both mechanical and chemical processes, involving the fractionation and micronzation of by-products and the extraction of the active substances contained in them. Section 6 is dedicated to discussing the results, the impact at the European level, limits, and further research. In Section 7, we draw conclusions.

2. Circular Economy Strategies and Systemic Design Approach

2.1. Circular Economy

In the last few years, CE is receiving increasing attention worldwide as a way to overcome the current production and consumption model, the so-called 'take, make and dispose' [5] or linear model, based on continuous growth and increasing resource throughput. By promoting the adoption of closing-the-loop production patterns within an economic system, CE aims to increase the efficiency of resource use, to achieve a better balance and harmony between economy, environment and society [6]. Many studies have been conducted on this topic [7–9] mainly rooted in environmental and political aspects [10] as well as economic and business ones [6,9].

Generally known as the 'Reduce, Reuse, Recycle' (3R) strategy, now these strategies have extended to nine, from refuse to recover [11]: the so-called R-strategies or R-list (Figure 1).

Results evidence that CE origins are mainly rooted in ecological and environmental economics and industrial ecology (IE) [6,12–18]. Some authors attribute the origins of the CE in General System theory [19,20]. Nevertheless, more often the origins of the CE are attributed to more recent theories such as regenerative design, performance economy, cradle to cradle, biomimicry and blue economy, that contribute to the further refinement and development of the concept of CE [6,21]. In Europe, CE primarily emerged in Germany in 1976 with the Waste Disposal Act, while at European Community level CE was promoted much later, through the Waste Directive 2008/98/EC [22] and more specifically with the Circular Economy Package [6,23,24].

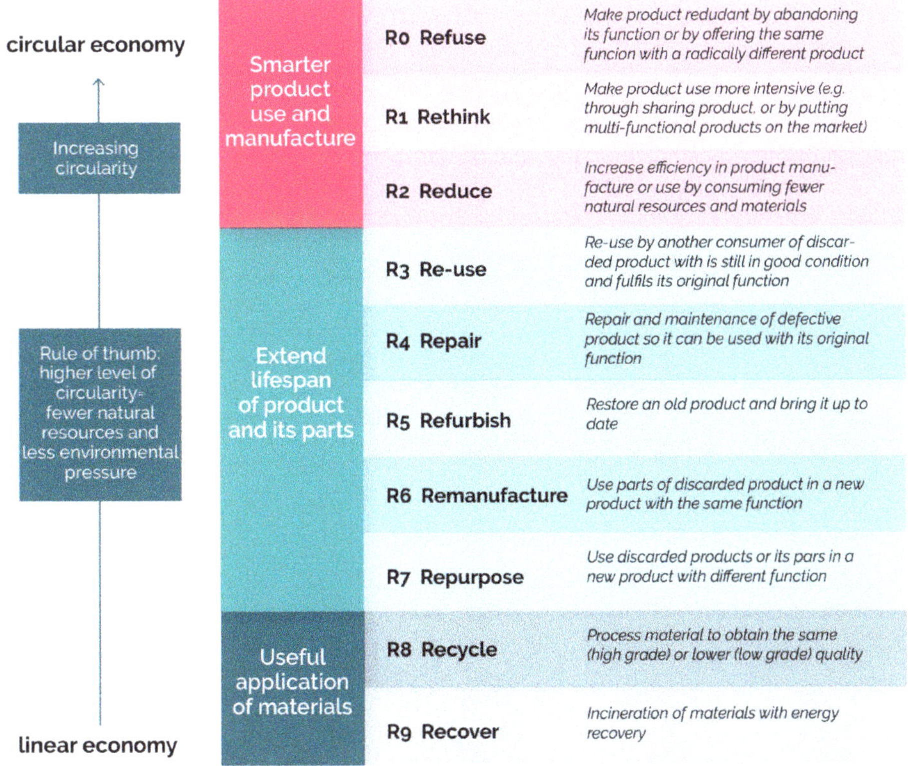

Figure 1. Circularity strategies within the production chain in order of priority (credit).

'Reduce, Reuse and Recycle' are three principles that, with some modifications, are also included in the waste hierarchy of European Waste Directive 2008/98/EC [25] since 1989 as well as in United States solid waste Agenda [6,26–28]. It must be pointed out that the CE in the European Union is a tool to design bottom-up environmental and waste management policies [6]. According to Ghisellini and colleagues [6] "CE implies the adoption of cleaner production patterns at company level, an increase of producers' and consumers' responsibility and awareness, the use of renewable technologies and materials (wherever possible) as well as the adoption of suitable, clear and stable policies and tools".

CE principles and limits have been widely discussed [5–8,10,22,26–49], and this is not the space to give further evidence. This short preamble is, instead, instrumental in pointing out similarities and differences between the principles and aims of the CE and the SD.

2.2. Systemic Design: Similarity and Differences with CE

The growing interest of the EU in CE has renewed interest in the SD as an approach that can lead to new business models. SD approach applies mainly to the industrial sector and is particularly suitable and declinable on value chains in the agri-food field. A systemic vision requires designing radically alternative solutions, as well as growing attention towards the interaction between the processes involved and the environment and the actor of a specific area. This way, the regeneration does not consist solely to material or energy recovery but, instead, it becomes an improvement of the entire living and economic model compared to previous business-as-usual economy and resource management [6]. Moreover, treating the productive systems as complex systems of complementary and symbiotic activities rather than disconnected entities is fundamental to share resources, know-how and technologies. It means there is no longer any reason for the growth of the single reality to the detriment of another. The relationships generated within the system make it becomes autopoietic [50–52], i.e., a system that produces itself and tends to evolve autonomously on the onset of change.

Regarding productive organizations as complex adaptive systems allows a new management model to generate economic, social and environmental benefits [53]. It is interesting to notice how complex entities interact openly with their environments and evolve continually by acquiring new, 'emergent' properties [54]. Complex systems are generally dynamic, nonlinear and capable of self-organization to sustain their existence. This approach is patterned after the self-organizing behavior of living systems. These systems show inherent 'resilience' by taking advantage of fundamental properties such diversity (existence of multiple forms and behaviors), efficiency (performance with modest resources consumption), adaptability (flexibility to change in response to new pressures) and cohesion (existence of unifying forces or linkages) [55,56].

While CE proved to be rather rigidly linked to product manufacturing and the concept of the life cycle of industrial processes, SD seems to accommodate the concept of value chains better, also introducing the idea of material and energy flows.

Therefore, the five principles for the application of SD are the following [57]:

(1)   The output (waste) of one system becomes the input (resource) for another, which creates an increase in cash flow and new job opportunities;
(2)   relationships generate the system itself: each relationship contributes to the system, and it can be within or outside of the system;
(3)   self-producing systems support and reproduce themselves, thus allowing them to define their own paths of action and jointly co-evolve; it means that industries connected each other in a systemic approach are in dynamic balance and will change their sets easily to adapt themselves to the continuous changes of the environmental conditions (market, supply chains, …);
(4)   act locally: acting locally values local resources (human, culture and material) and helps to solve local problems by creating new opportunities. The innovative solutions can come from all over the world, but they should be appropriate for the local context, and the real values should come from the expertise, resources, knowledge of a specific area;
(5)   people at the center of the project to be connected to their own environmental, social and cultural context. The real needs of people are the focus of the design process and not generate false longing to satisfy the companies' wishes.

Rather than focusing on waste, the SD treats by-products as outputs that become input for other processes, mainly new processes, so rather than closing the loop like the CE, the SD connects different loops and creates open systems made of relationships and connections between local realities.

Although biological nutrients, that in general are nontoxic, "can return safely to the biosphere or in a cascade of consecutive uses" [6] SD intends to generate as much value as possible from these by-products, before returning them to the ground and closing the loop. InnovaEcoFood project, indeed, will go further from the current CE methodology as it opens a collaboration between knowledge from diverse sources. Moreover, different flows will be considered in a proximity environment, making a profit from local partners and easing materials and information exchange. The SD is also responsible for granting the sustainability of the circular value chain model around Piedmont. Furthermore, this will boost bio-economy in the Piedmontese territory.

SD theory is rooted in the General System Theory, the generative science [56] and cybernetics [56,58], sharing a similar multidisciplinary approach. SD also derives from other eco-management theories, such as the open living systems [56], Cluster Theory [14,59], IE [60] and Industrial Symbiosis [14,61–63]. Among the pioneers of this approach, we counted the Austrian biologist Karl Ludwig von Bertalanffy and the physicist Fritjof Capra. The theories about complexity help the management of the entire food systems and the design approaches help the planning of different divergent elements. Those theories are the lens that SD research team at Politecnico di Torino applies to value chains' analysis, including them in complex systems made of people, resources and relationship among the activities. It is interesting to note that the CE and SD have common roots [21,33,39,64–70] like IE and according to some authors [6] also the General Systems Theory [19,20].

To summarize, CE and SD share several principles and goals, despite the existence of some differences:

- SD opens the use of output in different sectors, not only the one from which it comes;
- SD aims territorial valorization and not just a geographical concentration of industries;
- SD goes over the competition among enterprises, in favor of real collaboration.

Furthermore, SD is very close to the IE [60] and industrial symbiosis [61], even if SD considers some more flows (like information) and acts in open systems instead of just in a closed loop. SD aims to redesign human production systems to imitate natural ones, efficient par excellence [57]. The geographic proximity is neither necessary nor sufficient; turn waste in business opportunities reduces demands on the earth's resources and provides a stepping-stone towards creating a CE [64].

Preferable CE strategies such as reuse, repair, refurbish, remanufacturing, repurpose, find no agri-food application. In the project, we will detail later on, by-product valorization seems to fall along with the guideline 'recycle' rather than 'reuse', which is considered one of the less preferable strategies for products [11,71] according to the R-list (Figure 1), in which it is the penultimate strategy listed.

Unlike other sectors, in agri-food, recycling is not meant to be a downcycling, as waste generates economic value. In the systemic processes implemented, recycling is not comparable to recycling materials that either remain unchanged or lose performance. Indeed, it implies the creation of value from agricultural or process waste that cannot be framed within the rigid CE grid. It is a matter of transforming value chains and building systems able to consolidate relationships between companies in the reuse of outputs. Therefore, it is not correct to consider this kind of strategy as mere recycling. For this reason, for the InnovaEcoFood project, we proposed the SD theoretical framework, which is more complex because it follows the logic of complex systems, deals with open loop, and leaves room for creating relationships between companies and the dynamics managed at a local level.

3. Case Study: The Rice Supply Chain

Italy, with an annual rice consumption per capita of about 5.5 kg, is the largest rice producer in Europe. In Italy, rice cultivation covers about 220,000 hectares. It is mainly located in the lower Po Valley and the narrow strip reaching as far as the Pre-Alps between Lombardy and Piedmont. It that areas, large quantities of water are available for irrigation. The provinces of Vercelli, Pavia, Novara, Milano, which alone account for 90% of the total Italian area invested in rice, are the most rice-growing provinces. Sporadic traces of rice cultivation are also present in central Italy (Siena, Grosseto) and on the islands (mainly Sardinia), which means that rice can be grown anywhere, provided there is plenty of water. However, Piedmont is the first rice-producer in Italy with a wide selection of rice varieties due to its wide range of paddy fields covering about 217,195 hectares in the territory (all data on production and cultivation of rice, grapes and wine come from the Istat database (http://dati.istat.it), regarding the year 2018) and with nearly 2,000 companies involved in its production, which amounts to several million quintals (1 quintal = 100 kilos) per year. The Piedmontese province of Novara and Vercelli alone cover 52.7% of Italian production. Vercelli is the capital of rice production and alone produces more than a third of the national total and the entire province accounts for three million quintals of cultivated, produced and processed rice. The history of rice in Piedmont is linked to two main factors. One is the need to exploit clayey soils, which are sterile. The other is the possibility of using the water resources of the large glacial snowy rivers that flow down from the Aosta Valley, Monte Rosa and several smaller rivers of resurgent or mountain origin.

3.1. Analysis of Rice Supply Chain Waste

Rice production generates large quantities of waste, co-products and by-products (about 40%) compared to the total raw material entering the supply chain, which are not currently valorized.

The cultivation of rice and its processing thus results in a series of by-products that are considered waste for the food industry.

This analysis, performed in the Piedmont region (Italy), shows in detail the process of rice transformation and the characteristics of every output. In detail:

- Straw: it is obtained even before harvesting, during threshing in the field; it is composed of cellulose, lignin, waxes, minerals and silicates, and is used as compost, fertilizer [72], feed, fuel [73] or is used in the construction sector, or for the creation of new packaging material or cloths [74–76].
- Husk: the woody part of the seed is almost totally used in combustion plants to produce electricity, thanks to its adequate calorific power [77]. From combustion, 16% ash residue, rich in carbon and silica, is obtained and used in the production of refractory bricks, tyres and steelworks as thermal insulation and antioxidant in castings [78–80]. The adsorption potential of rice husk allows its use for the treatment and purification of drinking water [81–83].
- Broken rice: the size of the broken rice varies depending on the variety of rice from which it comes: relatively small breaks are obtained from round rice, much larger from long rice, but in any case, the size of the piece is variable because the grains of rice can break at any point or even in multiple parts. Broken rice is largely used to obtain rice flour [84], but it is also used as it is for example in the production of beer, puffed products and animal feed, and a large part is added to rice destined for the poorest countries, where it reaches even 50% of the product.
- Green grain and stained grains: they are used exclusively in animal feed.
- Gem: it has a low ash content and a high content of proteins, vitamins and fats, and it is possible to obtain an oil that is used in niche foods; its primary use remains, however, animal feed.
- Hull or bran: it is an abundant and underutilized by-product of rice milling and polishing (Figure 2a). It is rich of bioactive components and emerging evidence reveals rice bran (and its extraction products) as a functional food supplement with broad health benefits. Bran is used primarily for animal feed and in the pharmaceutical industry, by extracting calcium, magnesium oesophosphates, gamma-oryzanols, phenolics, flavonoids, tocopherols and inositols [85].
- Flour (rice middlings): a by-product of the first polishing of de-husked rice (Figure 2b). It mainly consists of particles of the aleuron layer, endosperm and germ. It is obtained after hulling, and it is rich in protein, potassium, iron and zinc. It is intended for animal feed.

(a)   (b)

Figure 2. (a) Hull (right) (b) Rice middlings. Source: Agrindustria.

It should be noted that, in the literature, the terms hull and husk are often confused and used one in place of the other.

Applications and Critical Issues

In this contribution, we will focus on post-threshing by-products from what is called paddy rice (or rice in the husk). Therefore, from the data obtained during the field visit to the Vercelli farm 'Gli Aironi' (http://www.gliaironi.it/) and supported by literature evidence [86–88], we derive that the production of 1 kg of rice generates 200 g of husk, 50–70 g of hulls, and up to 60 g of broken rice that can be valorized. The inadequate disposal of these by-products generates negative environmental impacts, as well as an important cost for the companies, even though some of these by-products have a minimum economic recognition when used as feed.

If we consider 100% of paddy rice, the percentages of products and by-products are detailed as follows:

- rice 62%;
- husk 20%;
- broken rice 6%;
- green grain 4%;
- gem, hulls and flour 7%;
- stained grains 1%.

The main critical points of the process can be summarized as follows:

- the wastewater generated by the food processing of rice and the cleaning of machinery is rich in organic compounds characterized by high BOD levels as a pollutant load.
- the husk produced is thrown into a landfill or waste-to-energy plant to produce silica, with high air emissions that are filtered but produce particles and ashes that are pressed and delivered to the landfill.
- hulls and flour are used only as animal feed while they have characteristics suitable for human consumption.

In this paper, we will deal with the last point in detail.

3.2. Potential and Scalability

Currently, the outputs deriving from the rice transformation are resold and reused, but not fully exploited. Indeed, they are very rich in nutrients and chemical-physical characteristics that could be valorized for other uses in sectors where they would acquire greater value (food, bioplastics, building materials, ...). Considering that Italian rice production is about 14.7 million quintals, the total amount of residues can be estimated at 0.7 million quintals of hulls and 2.94 million quintals of husks. Hull is a seasonal by-product obtained in the post-harvest period of rice, which goes from October to June of the following year, through various processes that divide the rice into various parts (husks, hull and rice). Hull is listed on the feed market with a price of 170.00 €/ton (2018). This residue, however, represents a significant source of organic matter, gamma-oryzanol and lipid substances.

4. Case Study: Wine Supply Chain

In Italy, the cultivation of wine grapes is widespread in almost all regions of the country, but is mainly concentrated in the following regions: Puglia, Tuscany, Piedmont. With about 44 thousand hectares of vineyards, Piedmont is the sixth largest region by extension, and annually about 2.6 million hectoliters of wine are produced from the roughly 20,000 wineries in the area. It has a wide selection of varieties (more than 60) and it produces 17 DOCG and 42 types of DOC wines recognized as Piedmontese products [89]. As regards the value of wine production, Piedmont is the fourth region with almost 460 million euros in 2018, after Veneto (900 million euros), Puglia (600 million) and Tuscany (500 million) [90].

The geography of the Piedmontese territory, e.g., orography and morphological feautures, are fundamental factors for the identity of the wine and the production of numerous types of wine, thanks in particular to the creation of microclimates in the different production areas (hills, Alpine and pre-alpine regions). Piedmontese wines, with few exceptions, are monovarietal, i.e., produced with a single grape. In Piedmont, the first examples of zoning of wine-growing areas began, defining concepts such as terroir and cru: a specific wine is produced exclusively with grapes from a single vineyard or parcel whose name appears on the label.

It is convenient to state a difference between the process of making red wine and the process of white and rosé wines. There are little differences among these processes, which generate different by-products in the process. In white and rosé winemaking process, the pressing operation is taken before the alcoholic fermentation. However, in red winemaking process the pressing is taken after alcoholic fermentation, thus the marc is fermented and alcoholic, while for white and rosè processes, marcs are non-alcoholic and sugar-rich. Through this process, wineries generate a large amount of solid waste, estimated around 30% of the material used. Wine production entails the generation of large amounts of by-products mainly consisting of organic matter (grape pomace containing seeds, pulp and skins, stalks, vine pruning and grape leaves) and wastewater.

The implementation of waste management and its subsequent by-products valorization is a pending task in the Wine Industry. Disposal of such amount of waste induces significant environmental effects similar to all food-processing waste.

4.1. Analysis of the Wine Supply Chain Waste

Composition of grapes is variable depending on its variety and climatic or vinicultural factors. Still, it is important to consider in nature and composition on the organic by-products and waste generated during the winery process, to understand the potential added value of innovative technologies in the process. The outputs of the wine production chain can be divided into:

- crop residues (vine shoots, pruning and stalks);
- organic residues from the winery (grape seeds, grapes skins, pomaces, lees and distillation residues);
- wastewater that contains solid processing residues (seeds, skins etc.), traces of products used in wine treatment (fining agents) and residues of cleaning and disinfection products.

Wooden residues from vineyards are often burnt to avoid the transmission of diseases from year to year. Waste from the winery is usually destined for distillation or pressed and disposed of in landfills. The components of wastewater are easily biodegradable elements, except for polyphenols [91].

Applications and Critical Issues

Organic matter is generated from the vineyard until the end of the process when the wine is bottled. First, when grapes are collected, approximately 2–3% of the total weight is lost in branches, stems and stalks. In detail:

- Vine shoots and pruning: branches and stems (i.e., woody parts) are usually disposed of generating environmental problems because they are burned in the field. This operation is nowadays more and more sporadic because it is considered ecologically incorrect. Indeed, it causes the emission of fumes and the mineralization of the organic substance, precluding the formation of humus. As a replacement for this practice, the vine shoots are chopped and buried, thus constituting a source of organic matter in the soil (with an annual replenishment equal to about a quarter of the required quota) and of natural nutrients coming from slow mineralization. An alternative is their use for energy production as biomass. Vineyard pruning residues can amount to a few tons per hectare, with production varying according to the vigor of the vine and the form of training adopted; the energy yield also varies according to numerous factors. The annual biomass per hectare is between 1.5 and 3 tons and provides energy equivalent to 0.5–0.9 tons of diesel.
- Stalks: the stems of the white and rosé grapes come from the destemming phase. They have high fiber content, mainly lignin and cellulose, as well as a high percentage of nutritive mineral

elements such as nitrogen and potassium. They are primarily used for composting [92–94] and are subsequently spread in the soil. The resulting compost can also be used as a substrate for the cultivation of Agaricus Bisporus, the most widely used species of mushroom in traditional cooking [95]. Another similar waste is obtained through thinning, i.e., the pruning of some ripe bunches that are abandoned in the field to reduce fruit production in favor of a higher quality finished product, wine.

- Marcs: After pressing, in the white and rosé wines, the remaining solid parts are not fermented marc (grapes skins, pomaces). This marc (Figure 3a) is a high moisture content mixture, which is also rich in sugar (around 14%). In red wine processing, indeed, are produced fermented marc, reverting sugars into alcohol This output has a variable chemical composition depending on various factors, such as the seasonal trend, the place of origin, the variety of grape, the time of harvest and the different techniques used in winemaking. Among the main uses are: direct spreading on the land for agronomic use; composting and subsequent agronomic use [96]; energy use as biomass through biogas or combustion plant, pharmaceutical and/or cosmetic use [97,98], food use, enocyanin extraction, zootechnical use, in the preparation of animal diets, animal feed [99], production of tannin-based materials and textile dye [100–105].

- The marc also contains the seeds (grape pips or 'vinaccioli' in Italian) which may be separated later by drying and centrifuging. By cold pressing grape seeds, an oil can be obtained without the use of chemical solvents, used both in food and cosmetics [106]. Grape seeds are rich in calcium, phosphorus and flavonoids and organic acids with high lightening properties. More precisely, they contain a good quantity of linoleic acid, an essential fatty acid, rich in omega 6, a well-known antioxidant and anti-cholesterol. Rich in polyphenols, it is modest in vitamin E content, especially when compared to other vegetable oils, such as corn, soybean, wheat germ or sunflower oil.

- The lees are the solid waste that remains after the fermentation. They are a mixture of dead yeasts and other solid wastes like tartaric acid and pigments (Figure 3b). Lees are traditionally an essential raw material to produce ethanol and tartaric acid [107,108]. The latter has many applications in the food industry, as it is an excellent stabilizer, replacing citric acid [109]. Lees can be used for the recovery of high value-added products [110], among which phenolic compounds stand out. It consists of yeasts, potassium salts, calcium and tartaric acid.

<table>
<tr><td>(a)</td><td>(b)</td></tr>
</table>

Figure 3. (**a**) Grape marc from crushing (right) (**b**) Solid by-products. Source: Deta webpage [111].

The composition of organic waste from wine is shown in Table 1.

Table 1. Composition of organic waste from wine.

Marcs	Skins and Seeds	Lees
Sugar (14%)	Lipids (15%)	Tartaric acid (12%)
High moisture (65%)	Proteins (10%)	Proteins (20%)
	Fiber, as cellulose, pectin, hemicellulose, lignin, polyphenols (65%)	Fiber, as cellulose, pectin, hemicellulose, lignin, polyphenols (25%)
		Sugar and pigments (10%)
		Lipids (4%)

Moreover, as mentioned before, water is used in several steps of the winery process: in cleaning operations (of harvesting tools, trucks, hoppers, boxes and destemmers, presses, deposits, boots and barrels), but also in the clarification process. Wastewater in winery processes is also rich in organic and inorganic matter. This is a problem when organic matter's natural decomposition process takes place, as it consumes the dissolved oxygen in water affecting the aquatic biota. The large volume of wastewater generated, and its seasonality is also a problem for its management.

Sewage sludge from wastewater can contain many nutrients, including nitrogen and phosphorus. As with civil sewage sludge, composting in combination with other substrates of wine origin is the most common use [92]. The Italian legislation also allows its direct use in agriculture if it complies with the limits [112]. The sludge has also been used as a co-substrate in anaerobic co-digestion, together with wine lees, under both mesophilic and thermophilic conditions, to produce biogas and bio-stabilized effluent [113]. The list of products that can be obtained from the by-products of winemaking is very rich:

- from the lees, it is possible to obtain alcohol for food and industrial use, grappa (in association with marc), calcium tartrate, natural tartaric acid, colorants, ethanol, beta-glucans, food;
- from the marc it is possible to obtain grappa or alcohols (for food and/or industrial use), natural tartaric acid, lactic acid, proteins, bioemulsifiers, biotensives, tannins, polyphenols, antiallergens, hydrolytic enzymes, bioethanol, fertilizers, soil improvers, compost (in association with pruning residues) absorbents for decontamination of heavy metals, substrates for human food or micro-organisms (in association with pruning residues), biocontrol agents, electricity (in association with pruning residues), resveratrol, anthocyanins;
- from grape pips, it is possible to obtain tannins, antioxidants, antimicrobials, flour, edible oil, cosmetics, biodiesel, lubricants.

In this paper, we will focus on the by-products of winemaking, without considering the residues from the field. Therefore, from the analysis carried out through field visits at the Asti winery Bocchino (www.vinibocchino.it) and supported by some evidence in the literature [86–88] the production of 1 kg of wine grapes generates about 100 g of marc and up to 60 g of grape seeds that can be exploited.

As reported above, the inadequate disposal of these by-products generates negative environmental impacts, as well as a relevant cost for companies. From 100% fresh grapes, the percentages of products and by-products are detailed as follows:

- 80% wine and must;
- 10% of grape marc;
- 6% of grape pips (seeds);
- 4% of stalks.

The main critical points of the process can be summarized as follows:

- the vine pruning currently has no specific use, but are managed by each farm to reduce the damage (both environmental and economic);
- the marc, skins, stalks and seeds of the grapes are currently destined for distillation, where they lose the organoleptic qualities that would allow them to be reused in the food sector;

- the lees, a residue deposited after the fermentation of the wine, is currently disposed of.

  During the InnovaEcoFood project, we dealt with the second point, focusing on the marc.

4.2. Potential and Scalability

These residues represent a significant source of organic matter, polyphenols, nitrogen, macro- and trace elements. According to current legislation, the by-products of winemaking are subjected to management methods that, with defined timeframes, provide for the obligation of total or partial delivery to the distillery, or their controlled reuse for alternative uses, mainly in the feed sector. In fact, at present, grape marc is usually sold to the large distilleries that store the product. Then they perform classifications, separations and distillations. After distillation, the remaining (grape skin) has a market for feed use of about 200 €/ton, while the grape seeds can reach 500.00 €/ton. The grape marc or pomace is a seasonal by-product. It is produced from the end of August to the end of September, after pressing. The pomace is conferred to distillation companies from September onwards; it is stored and processed throughout the year. The pomace obtained from winery can be 'fermented' or 'virgin' (the latter also called 'sweet'). In fermented grape pomace, yeasts have transformed sugars into alcohol. Usually, fermented grape pomace is obtained during the production of red wine, as the grape pomace remains in contact with the must for at least 5/6 days. The virgin (or sweet) grape pomace has not yet undergone fermentation. It comes from white/rosè wine processing, in which the skins and grape seeds are separated from the must before alcoholic fermentation. A completely different type of marc is obtained after distillation, which varies considerably, especially in organoleptic properties.

Innovative and sustainable systems for the management of this kind of waste as valuable resources are an opportunity for Europe in terms of reduction of environmental impacts and the creation of new jobs. Moreover, the reduction of waste production and the management of it as resources is part of the path towards sustainability defined in the Europe 2020 Resource-efficient Europe Flagship.

Italian production is approximately 75 million quintals of wine grape or 54 million hectoliters of wine. Based on an estimation performed in the Piedmont region, 5% of Italian wine comes from wineries that produce annually no more than 25 hL. In this case, they are not required to deliver the marc to distillery nor alternative use. The national quantity of residues can be estimated about 7.5 million quintals of marc and 2.5 million quintals of grape seeds. The Piedmont Region potential of marc and lees available every year, instead, can be estimated at 0.4 million quintals of marc and 0.2 million quintals of grape seeds. The EU is a leading global producer of wine, producing 167 million hectolitres every year (https://ec.europa.eu/info/food-farming-fisheries/plants-and-plant-products/plant-products/wine_en). Waste, coproducts and by-products of these productions are rarely valued, and their improper disposal generates negative environmental impacts, as well as an important cost for EU. Besides, the application of this strategy in this specific agri-food system offers a high potential of replicability. Moreover, other value chains (e.g., olive oil industry) could be suitable for applying this approach, but it can also be applied in a wide variety of territories. Waste and by-products represent about 20% in weight of the produced wine.

5. InnovaEcoFood Project

The project took place in Piedmont, an important wine and rice producer region. Rural areas have great potential to include new and interesting business models that create new opportunities and quality jobs, including social and economic aspects, and tackle Europe's limited resources. InnovaEcoFood project will upgrade two current value chains to a more circular value chain that expands to other sectors revalorizing agriculture and process by-products. This multi-actor approach can reach other sectors like food, textile, polymers, or bioactive compounds, diversifying and revitalizing the economy, considering the reality of local needs. In general, the agri-food sector and agri-communities are integrated systems requiring a holistic approach to face major current challenges and avoid economic and social decadence. At the same time, Italian producers are still not benefitting from the untapped potential of agri-food by-products valorization.

5.1. Objectives

InnovaEcoFood aims to create value from rice and vine waste, typical of the Piedmontese culture, by obtaining food products. This project's overall goal is to boost rural development by demonstrating and rolling out innovative 'small-scale' 'bioeconomy-based' 'systemic design-enabled' sustainable business models.

InnovaEcoFood aims to demonstrate that linear winery and rice value chains can be analyzed and transformed into progressively more systemic value chains. Subsequently, these can be integrated into circular networks of value-generating rural business cases considering other bio-based processes and expanding toward other sectors, such as construction, textile, polymers.

Within the InnovaEcoFood project, the research team contributes to experimenting with the residues of Piedmontese rice and wine-winery production value chain using the SD approach. The project aims to show that companies' competitiveness can be improved by applying SD and integrating the latest technology to valorize unused agricultural residues. For that reason, technological partners of this project own technologies capable of processing by-products or extracting bioactive compounds that could be reintroduced in the process being applied to food.

5.2. Technological Partners

A sample of this new player in the value chain is one of the partners of InnovaEcoFood project: Agrindustria di Tecco (Cuneo, Italy; www.agrind.it), an SME operating since 1985 in the field of the valorization of secondary plant products deriving from regional value chains. It produces vegetable granules and micronized products for successive chemical extraction; it processes vegetable matrices and biomaterials for different production sectors (Figure 4).

Figure 4. Illustration of Agrindustria processes (cryogenic mill and steam auger). Source: Agrindustria.

In this project, Agrindustria dealt with mechanical treatment (mainly drying, grinding and micronization) of grape marc and rice hull to obtain flours that can act as an alternative to wheat flour, with high protein content (Figure 5). On the other hand, a new actor like Exenia Group (www.exeniagroup.it)from the province of Turin (Italy), deals with tech applications in biotechnology. Exenia was founded in November 1995 with the specific entrepreneurial project to become one of the first examples, in Italy, of a company able to combine high-tech technologies and new products for highly dynamic markets. The constant investment in R&D responds to a model widely spread in the USA (e.g., Dedicated Biotechnology Companies), Japan and Northern Europe. Exenia Group is involved in research and experimental development: extraction of active ingredients from plants, extracts and production of essential oils, research and development on 'clean' technologies for the extraction of raw materials for food, cosmetics, nutraceutical and pharmaceutical products, research and development on photosynthetic microorganisms, cultivation, study, research, development and industrial use, also on behalf of third parties, of new products in the field of biotechnology and natural products. Exenia uses supercritical fluids (CO_2) for food, pharmaceutical and cosmetic applications. Extraction conducted with fluids under supercritical conditions is a valid alternative to traditional separation systems, such as fractional distillation, steam current extraction, solvent extraction. This type of extraction offers the possibility to continuously vary the fluid's solvent power to conduct the extraction with little pressure and/or temperature variations.

Figure 5. Mechanical treatment of wine and rice by-products from Agrindustria to obtain flours. Source: DISAT.

During the project, Exenia obtained the physical-chemical refining of high value-added molecules from rice hull (or bran), cooperating with the Department of Applied Science and Technology (DISAT) of the Politecnico of Torino and the production of a butter (rice bran lipid) rich in gamma-oryzanol for food or nutraceutical use.

5.3. Systemic Design Method and Schemes

As can be found in the Appendix A, a desk research highlighted several possible applications of by-products. The applications were collected, summarized and schematized in Figure 6 for the rice supply chain and Figure 7 for the wine sector, combined with field data collected from Aironi and Bocchino companies in 2018. These are two systemic visualizations that have been subsequently submitted to different actors for validation.

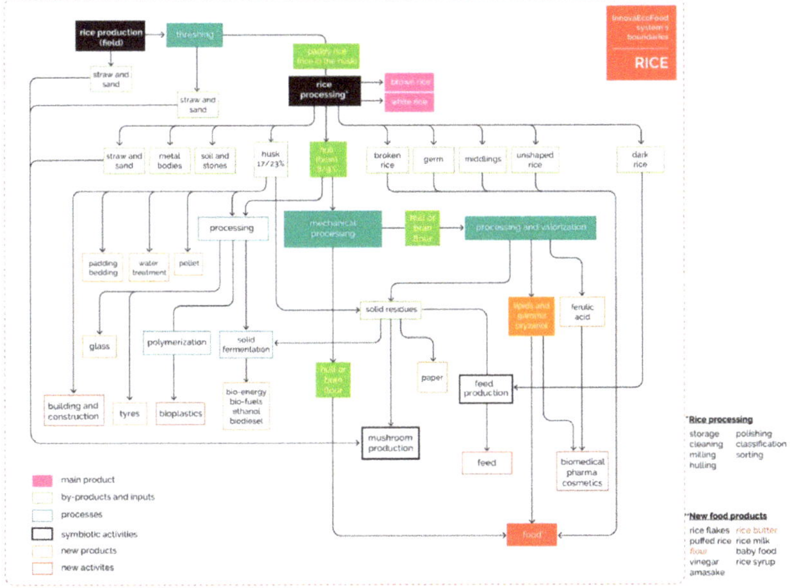

Figure 6. A systemic value chain proposed for rice.

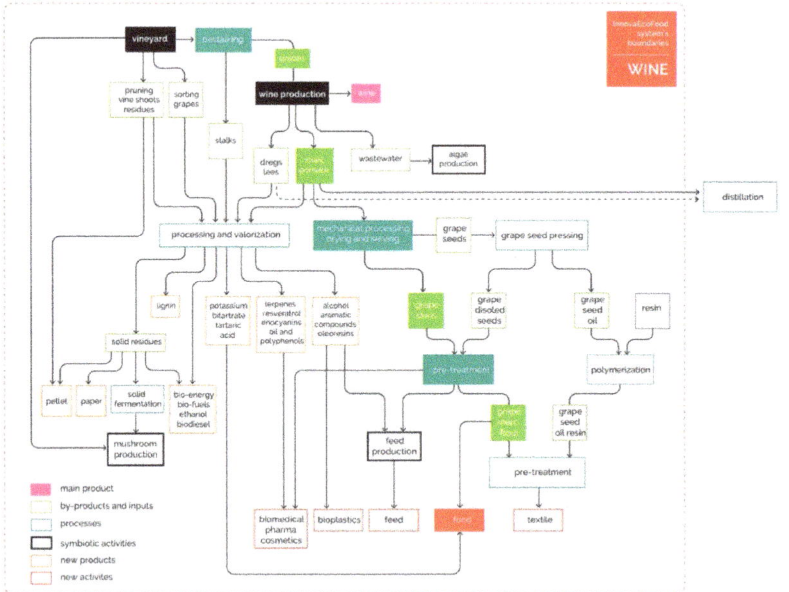

Figure 7. A systemic value chain proposed for wine.

The diagrams shown in Figures 6 and 7 are intended to provide a general overview of all the Piedmont applications for the by-products (output) of these specific sectors. By-products are generated at various levels in the cultivation and processing of rice (Figure 6), as well as during viticulture and winemaking (Figure 7).

To provide readers with a shared legend, we attribute some color codes. The two supply chains share one phase of harvesting and processing, indicated in the figures as full black boxes. The main products obtained are highlighted in pink, i.e., refined and whole-grain rice in one case, wine in the other. Green is for the outputs treated as new inputs (e.g., husks, bran, straw, etc. for rice; marc, lees, grape seeds, pruning waste, etc. for wine) and the products circulating as flows through the processing (paddy rice, bran, and bran flour, but also bunches, marc and so on). In blue, instead, are reported the processing. The orange boxes distinguish the new products (finished products, or molecules with high added value). The red boxes are the sectors and other production activities with which the project could relate. Instead, the black frame indicates possible symbiotic activities, intending to achieve an industrial symbiosis.

Finally, the full boxes distinguish the processes that have been activated during the InnovaEcoFood project. In contrast, the framed or outlined box indicates the processes and products that could be obtained, which have not been thoroughly investigated yet during this project but are reported in the scientific literature as viable solutions. We indicate the boundaries of the InnovaEcoFood project system with a red dotted line. For the rice value chain, we can see that all the activities fall within the system's boundaries activated by the project. In the case of the wine value chain, instead, distillation (the current linear process for marcs) has not been included within the system boundaries, as its outputs that are difficult to exploit on a systemic level. For this reason, it is considered outside the project. The production of the food products resulting from the project is detailed in the following paragraph.

5.4. Production of Food from Waste

During the InnovaEcoFood project, we decided to operate in human nutrition, in particular for the study and development of food products with high added value such as creams and bakery products. Indeed, the project investigates the production of new ingredients to produce new food products. The two companies mentioned above have provided know-how and technologies to enable the food company Quasani - Fattoria della Mandorla (www.fattoriadellamandorla.it/progetto-quasani) to develop foods with high nutritional, organic and health value supported by scientific literature and primary prevention guidelines. Within the project, they have been involved in integrating these outputs as ingredients, creating ad hoc recipes containing hull flour, marc flour and rice butter rich in gamma-oryzanol. The products are all processed without lactose, animal milk protein, hydrogenated fats, GMOs, cholesterol and gluten. The company has integrated the by-products supplied by Agrindustria and Exenia into foods such as food creams (core business), but also crackers and taralli, thus inserting the rice bran resulting from hulling, flour derived from post-winery grape skins, and rice butter from supercritical CO_2 extraction from rice hull. The company evaluates the expansion of its creams and other products, integrating processed waste from rice and wine production chains and high value-added molecules. Rice is a product widespread in the territory and has high organoleptic qualities but is scarcely used in the cream sector due to the greater difficulty of processing compared to other vegetable products such as soy. However, the use of rice and its by-products offers interesting potential for food innovation, as well as the possibility to trace the raw material. Bran is a by-product containing vitamins, minerals, essential fatty acids, dietary fiber, and other sterols that make it suitable for human nutrition. As already mentioned, numerous studies already tested its healthy properties, such as rebalancing thyroid hormones, improving muscle endurance, regulating cholesterol levels, preventing heart disease, and the formation of kidney stones. It has anti-carcinogenic properties, regulates the intestine, stabilizes blood sugar.

5.5. The Relationships Implemented by the Project

Regarding the processes activated by the InnovaEcoFood project and the relationships established, the project has experimented with flour from grape marc and rice hull, and rice butter produced through supercritical CO_2 extraction. The project partner Agrindustria Tecco Srl supplied the grape pomace and rice hulls by activating new relationships with unusual suppliers to find the material to be tested. The dry and moist grape pomace and hulls have been found in the Piedmont area. InnovaEcoFood aims to create new products for the food sector and no longer feed, as it is not very profitable. The cooperation activated during the project suggests that it is possible to create clusters of companies dedicated to valorizing specific by-products, providing for symbiotic activities and cascade approaches. Agrindustria partner provided the connection between the producers and the high-tech partner (Exenia), effectively collecting, storing and preparing the secondary raw materials for subsequent applications. Indeed, mechanical processing was carried out by Agrindustria, including cryogenic grinding, bacterial load reduction, and final drying. In the scheme, we refer to these activities as 'mechanical processing'. The rice hull was then transferred to Exenia's site (60 km), where the supercritical CO_2 extraction process took place. In the scheme, we refer to these processes by indicating them as 'processing and valorization'. The hull and grape pomace flour obtained from the project were sent to Quasani—Fattoria della Mandorla (1000 km away) to produce crackers, 'taralli', and creams. The rice butter obtained from the experimentation was also sent to Quasani. Evaluating these interactions with a circular economy view supports the hypothesis that we cannot refer to this process as recycling. We like the idea of being part of the CE strategy that rethink supply chains to be more circular. Systemizing activities and possible solutions allow anyone to visualize the connections between them and identify the gaps for connecting activities at the national or even EU levels, as discussed in the next section. One strength of the SD approach is the ease of changing scale, from micro to macro and vice versa.

6. Discussions

6.1. Study Limitations & Recommendations for Future Research

The project aimed to valorize some materials destined for zootechnical use, investigating whether hull and pre-distillation marc, correctly processed, could be valorized in a food market production. Post-distillation marc has lower organoleptic characteristics comparing to winemaking by-products, thus its use in the food market is not advisable. Indeed, to create a production process for the food sector, distillation should be avoided, and pomace should be processed as soon as possible, after pressing and winemaking. The fundamental process to be done immediately after pressing should be the stabilization (dehydration/drying) of the byproduct, to stabilize and process it over time.

During the project, an economic analysis was carried out by the Department of Management and Production of Politecnico di Torino based on data provided by partners through a survey. It highlighted the creation of value in the various processes on the material, which suggests that using the by-product in the food sector is of immediate technical and economic feasibility and could be considered a viable way that deserves further experimentation. In this regard, the project investigates also obtaining high added value molecules from marc flour by performing extraction with supercritical CO_2 to obtain polyphenols, anthocyanins, and resveratrol. However, the flour turned out not to be as rich in these molecules as it was forecasted. This mismatch could be attributed to the grape variety. More likely, it can be attributed to the volatility and sensitivity of these compounds that degrade in contact with light or heat. The extracts, therefore, prove to be even more deficient in these components compared to literature. Anthocyanins, polyphenols and trans-resveratrol are, in any case, present in the flour, which can be used as it is, and without prior extraction, for the preparation of food. On the other hand, as far as hull flour is concerned, it has also been successfully tested to obtain rice oil/rice butter, rich in gamma-oryzanol, a molecule with antioxidant properties. In particular, 20 g of butter is extracted from 1 kg of rice flour. The concentration of gamma-oryzanol is 0.6% of the extracted oil component according to the analysis carried out by the Department of Applied Science and Technology of Politecnico di Torino. The presence of gamma-oryzanol inside the rice butter gives unique properties to the final product, compared to typical milk-derived butter. Several experimental studies demonstrated that gamma-oryzanol, introduced in the diet, even if in minimal quantities, has powerful antioxidant and anti-inflammatory effects and has a positive effect on lipid metabolism and cholesterol levels regulation. Butter derived from rice hull was used in all three final products. In the final composition of 'taralli', crackers and cream produced, gamma-oryzanol content varies between 336 and 360 mg every 500 g of product. There are no real exact recommended doses at the time of writing, but the daily amounts vary in the wide range between 50 mg per day and 800 mg because experimental studies on these molecules are still in progress. The final products fit perfectly within this range, making them excellent allies for health. However, the project would require further analysis to understand how to optimize the supercritical CO_2 extraction process to reduce costs and make it economically feasible in the food sector. There are no real impediments for its use except its cost: the final rice butter price is about 250 €/kg, already considering an industrial process at full production capacity. At this stage, the use of rice butter in pharmaceuticals could be investigated in more detail, given the relatively high gamma-oryzanol concentration.

6.2. Impacts

In Europe, the food and drink industry is one of the most significant sectors with a turnover of around 1100 billion euros and 4.24 million people employed [114]. The reduction of food waste is a problem that the sector faces for ethical, economical and limitation of natural resources [1]. Around 88 million tons of food waste are generated annually within the EU, with an estimated cost of 143 billion euros [4]. The United Nations has settled its Sustainable Development Goals (SDGs) in which they firmly bet for food waste and loses reduction [2,3]. Therefore, the InnovaEcoFood project is part of the Sustainable Development Goal n.12-responsible consumption and production.

Grape is a product rich in sugars and antioxidants of high nutritious interest. The partners' technological capability made it possible to diversify the sector, introducing new actors that will find an added value to by-products. On the other hand, rice is rich in lipids and gamma-oryzanol, with antioxidant, anti-inflammatory effects, acting positively on lipid metabolism and cholesterol regulation.

In this project, the partners developed a new methodology to incorporate new players in the value chain of rice and wine processing to jump from a linear system to a circular one using the SD as a basis. Another EU's priority is diversifying and revitalizing the economy to create quality jobs in a rural environment. In Europe, 19.8% of the population lives in rural areas, but the reality shows a negative trend since decades ago with the examples of the aging of the population, low birth rate, and most companies in cities [115]. The InnovaEcoFood project presents a new approach to transforming the rice and wine sector's linear value chains into a systemic approach that can generate value in rural environments, expanding to other sectors like food, biopolymers, and textiles. On the one hand, focusing on the wine sector is interesting to maximize its impact and replicability among the European territory. On the other hand, focusing on the rice sector helps give importance to territorial/local cultivation. Beyond the rice and wine sectors, the InnovaEcoFood project will also reach other value chains, enhancing its impact.

The project could have an impact at the European level, especially considering wine waste. Indeed, grape crops are one of the main extended agro-economic activities globally. Currently, up to 210 million tons of grapes (*Vitis vinifera* L.) are produced annually, and 15% are addressed to the wine industry. Figure 8 shows the distribution of crops. Within the agro-food industry, grapes and wine productions generate one of the most valued products worldwide. As we already mentioned in Section 4.2, waste and by-products represent approximately 20% of the produced wine. The wine sector has specific importance in Europe's food and drinks sector as the region is one of the leaders in production. The sector has a revenue of more than 40 billion euros [116] and for some European areas revenues from wine represent a high share compared to other agricultural activities as indicated in Figure 9. Currently, the European wine industry's supply chain is a linear one in which by-products are considered waste, or they either meet minimum economic value and value is created by maximizing the number of end products produced/sold only. InnovaEcoFood has developed a pilot plant, including three different small enterprises. It is considered as a first milestone to boost the development of a bio-based economy in Europe.

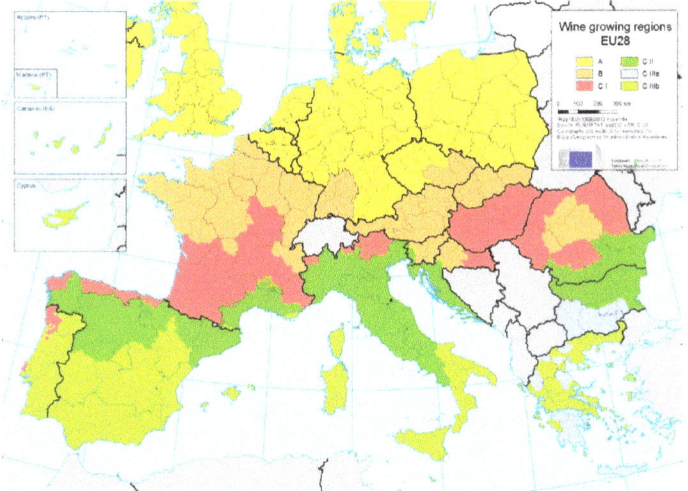

Figure 8. Winegrowing regions. Source: EU Wine Market Data Portal [117].

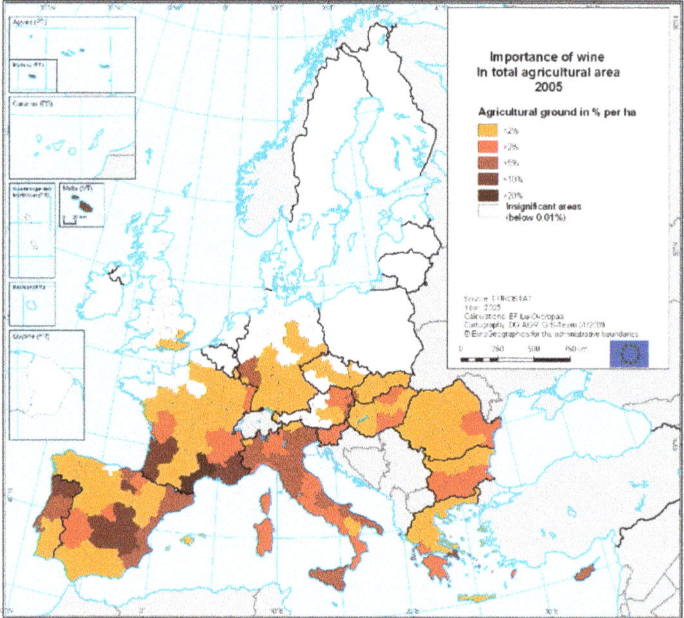

Figure 9. The importance of wine in total agricultural area (2005). Source: EU Wine Market Data Portal [117].

On the other hand, rice cultivation is not so diffused in Europe. As can be noticed from the map in Figure 10, Italy is the largest producer of rice in Europe. Italy alone accounts about 49% of all EU production [118], and the two regions mentioned above (Piedmont and Lombardy) are two regions that produce between 400 and 800 MT/year each. However, this value chain gives us an example of how it is possible to enhance regional peculiarities and their by-products with the use of SD.

Therefore, this paper presents two very different value chains, which have in common their relevance for the Piedmont region:

- in the case of the wine value chain, the vine is a widely-extended crop in the whole Europe territory, as a grant for replicability of the generated model, (it has European relevance and offers the possibility to think about its scalability and the replicability of the approach);
- in the case of the rice value chain, on the other hand, it is a local/regional value chain. Re-integrating them into local food and diet provides an example of how the approach allows the valorization of regional by-products.

Nowadays, agricultural by-products' valorization is still a great challenge, especially in a small-scale setting in rural areas. These by-products represent great potential for the generation of high added-value products. In this context, it is necessary to create and develop new business models adapted to rural areas that diversify the income sources of these small producers, creating jobs and ultimately revitalizing rural areas thanks to the bio-economy.

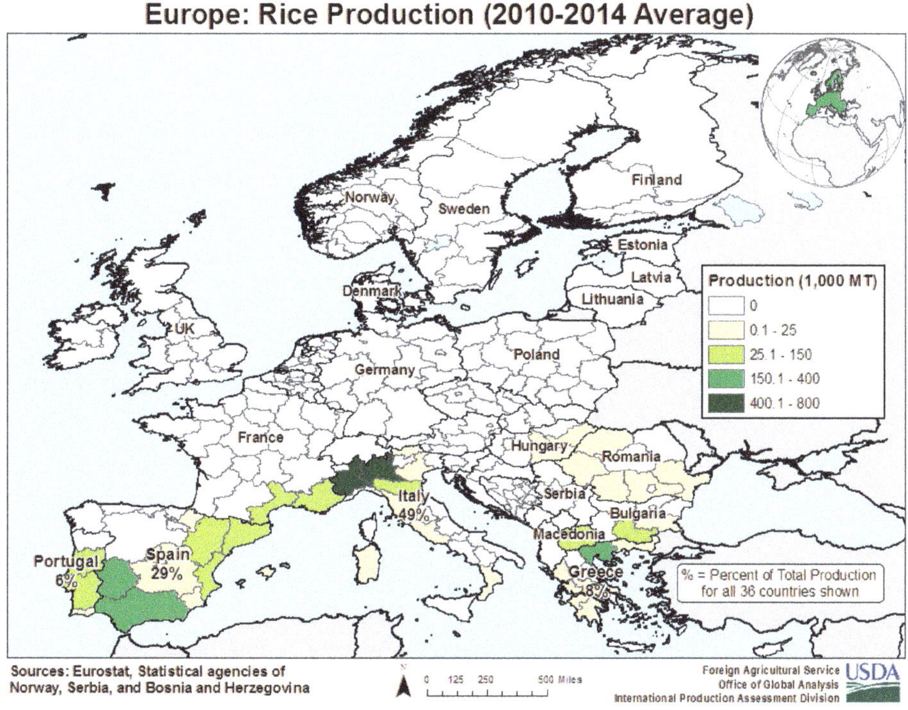

Figure 10. Rice production (2010–2014 average). Source: Eurostat [118].

6.3. Limitations of the SD Approach

The diffusion of SD is often limited by national/regional policies that do not support its development. Another limit is the difficulty of connecting companies and promoting dialogue for a common goal. The difficulty of stimulating systemic changes emerges when economic benefits are not immediately measurable, but positive effects will be measurable in a few years. The experience shows how many CE opportunities can be found. However, we are aware that each connection of the scheme and consequent valorization should face regulatory, technical, cultural, and financial barriers, so policymakers can play a crucial role in helping businesses overcome these barriers. Regulatory barriers include, for example, the definitions of waste that hinder trade and transport of by-products, the tight division in sectors, and so on. Technical barriers are related to the innovations that allow giving higher quality to the waste (for example, the quality of recycled materials requires R&D actions), and the ability to scale them up at the industrial level. In other cases, the value chains are not complete, especially from the supply of agricultural by-products. How to guarantee constant quality and quantity of supply is another key issue related to seasonality and other variables. At a cultural level, the is the need to increase the awareness of the potential use of secondary materials, especially in the food sector. The Complex Systems approach defines:

- Non-linear problems-the whole is greater than the sum of its parts
- Adaptive behavior-both the system and its constituent parts adjust over time to the changes in the environment, within the system, and within the components
- Self-organizing capacity-components self-organize without central direction
- Emergent properties-it is hard to anticipate the system outcome of interventions carried out at the component level [119].

These properties set the difference with 'closing the famous circles of production', or finding alternative waste destinations to landfills. The goal is to shift the focus from the product to the territory [119]. However, there is a lack of successful case studies history in each sector able to ensure that the application of SD will bring assured economic benefits since each project is tailored to a specific value-chain. The complex dynamics involved have often prevented the comparison of different business models.

6.4. Future Research

The project focused only on a tiny portion of the overall system since the regional funds from which it drew were limited. However, the project can start from this first part and extend suddenly to fill some current experimentation gaps. For the wine supply chain, the continuation of the project should include experimentation in the direction of:

- using different types of grapes, possibly of biological origin, to evaluate the differences in the extraction phase, but also the organoleptic differences when flour is used directly in food formulation;
- introducing variations in the storage phase and the transport of the output (immediate, refrigerated, and without exposure to light), at the same time of bottling for red wine or destemming for white wine. In fact, in the white winemaking process, stalks, grape seeds, and skins are removed before fermentation. Evaluating the differences between the output of the red and white wines would be of great interest, as they have radically different characteristics;
- introducing small machinery for the treatment of the output on-site (at winemakers and social cellars) to preserve the characteristics as much as possible and optimize storage while reducing costs and impacts. It would be possible to transport the dehydrated product without transporting the liquid component, which is unused.
- modifying the processes that use heat (such as dehydration) preferring processes at low temperatures to avoid incurring the degradation of compounds, which deteriorate in contact with heat.

Another step forward could be to expand and diversify the sector by mobilizing a more comprehensive range of players in the consortium, including small businesses, farmers, and their associations. Indeed, the scalability discussed in Sections 5.5 and 6.2 would deserve further investigation, proposing pilot cases in other EU wine-producing regions. It would also be interesting to see how different value chains can be connected in other regions.

7. Conclusions

The project took place in Piedmont, an important wine and rice producer region. Rural areas have great potential to include new and interesting business models that create new opportunities and quality jobs. The InnovaEcoFood project has addressed more circular value chains expanding to other sectors (food, pharma, textile, polymers), valorizing agriculture and process by-products by connecting the actors on a territory, diversifying and revitalizing the economy, considering the reality of local needs. The project starts from the current value chains highlighting the by-products generated in each step, either in the field or during transformation processes. For each output, it highlighted different scenarios that lead to their valorization and the creation of connections between the different activities, finally focusing on the valorization of marc and hull for food purposes. Thanks to the project consortium, the use of these by-products has been successfully experimented—after mechanical processing and extraction with supercritical CO_2-ingredients inside bakery products and creams. At present, their use in the food sector is limited due, among the other things, to legislative limitations regarding the classification of by-products as waste and the impossibility of handling them as raw materials. However, there is room to open different scenarios. As Europe is facing a problem of finite resources available, the introduction of added-value by-products as new ingredients (products with high functionality) in the food sector is advisable in the next future. It should also be considered primary importance to reduce food waste. The project highlights the approach's scalability at the EU level to other production realities and value chains.

Author Contributions: All authors equally contributed to the article (conceptualization, methodology, validation, analysis). All authors have read and agreed to the published version of the manuscript.

Funding: This research was co-funded by the Piedmont Region-POR FESR 2007/13 -I.1.3 Innovazione e PMI.

Acknowledgments: We thank the project partners for this valuable experimentation and for giving us access to their specific experimentation data. We thank Silvia Barbero, who contributed to setting the methodological framework thanks to her fundamental expertise in the field of SD. We want to thank also the Department of Applied Science and Technology (DISAT) of the Politecnico of Torino, in particular Debora Fino, Tonia Tommasi, and Silvia Fraterrigo for the chemical part, and the Department of Management and Production (DIGEP) of the Politecnico of Torino and in particular Claudio Castiglione for the economic analysis of those processes.

Conflicts of Interest: The authors declare no conflict of interest.

Appendix A

This appendix aims to collect research and case studies to validate and verify the schemes of the two supply chains shown in Figures 6 and 7. The literature search has been carried out in the Scopus database, and therefore may not consider different indexing.

Specifically, the research carried out skimming of the phenomenon by investigating the keywords 'circular economy' connected first to 'rice' and then to 'wine'. Among titles, abstracts, and keywords, the extractions counted 462 and 161 publications, respectively. These were then further filtered with new keywords such as 'waste' (421 results), 'husk' (165 results), 'hull' (41 results) for the database concerning the rice supply chain, and 'waste' (142 results), 'wastewater' (45 results), 'lees' (83 results), 'pomace' (43 results), 'marc' (12 results), 'stalks' (10 results), for the wine supply chain.

Further manual research has evidenced a small number of publications compared to those extracted from the system, which addresses chain waste reuse in new applications. The research highlights how often attention is not paid to the single product, but rather to the whole output set. Moreover, difficulties emerge in differentiating between hull and husk, often confused in terminology. The two tables below summarize the main results of interest.

Table A1. Application of by-products from wine production.

By-Product	Application
Lees and pomace	Improving the quality of fish feeds in terms of organoleptic characteristics and health benefits [110]
Pomace/marc	Valorization of grape agro-waste to produce bioactive molecules and new polymeric materials [100]. Extraction of molecules, fractions and biologically active biomolecules with a possible use in the nutraceutical and cosmeceutical industry from vinery marcs [97,98]. Production of compost from marc [96]. Improving animal fleshes with different pomace powder preparations [99]. Production of tannin-based adhesives for wood industry, tannin-based materials such as biocomposites and rigid foams [101]. Production of textile dye from grape pomace [102–105]
Seeds	Extraction of bioactive compounds with high added value before using biomass for energy purposes (e.g., in food, cosmetics, and pharmaceuticals sectors, biopolymers and energy sector to produce biohydrogen and biomethane [106].
Wastewater	Analysis of water consumption in wine production to identify wastewater treatment and management improvements towards water reuse [120].
Winery waste	Biorefinery opportunities from winery waste (biomass production) [121]. Employing winery waste to re-balance soil fertility; valorization of these in the agricultural sector or different industrial chains (e.g., cosmetics, nutraceuticals, etc.) [122,123].
	Production of compost and biofertilizer from viticulture waste [124].

Table A2. Application of by-products from rice production.

By-Product	Application
Straw	Natural fertilizer is used to remove phosphorus loading in water [72]. Extraction of water-soluble phenolic compounds to incorporate into bioactive starch-based films, producing bioactive food packaging [74]. Creation of innovative cloths from straw rice [75]. Transformation of rice straw in glucose for bio-carburant production [73].
Husk	Rice husk to purify colored wastewater [81]. Production of jet fuel through fluidized-bed fast pyrolysis, hydro-processing and hydro-cracking/isomerization [125]. Utilization of rice husk for a potential waste-water treatment due to their adsorption potential across a variety of common drinking water contaminants [82,85]. Production of energy and fertilizer by using rice waste [77].
Husk ash	Production of sustainable plastic composites from ashes (including rice husk ash) [78]. Production of eco-friendly concretes from rice husk ash [79].
Husk and straw	Rice husks and rice straw used as substrates for solid-state fermentation with dikaryotic and monokaryotic strains of Pleurotus sapidus [126]. Production of insulating materials for green building from rice straw mixed with waste wool. Production of biofillers from husk for polymer composites, mono- and di-glyceride mixtures. Extraction of high-added-value molecules for the food industry from bran [80].
Husk and bran (hull)	Development of a novel bio-fertilizer using rice bran and husks [127,128].
Straw, husk and bran (hull)	Development of new products as biofuels, enzymes, biodegradable material food contact, single cell protein, bio-adsorbent, nanoparticles, bio alcohol, bioactive compounds like fibers, phytochemicals, minerals, so on [76].
Bran (hull)	Oil extraction from defatted rice bran for bioethanol, lactic acid, and biobutanol production [129]. Extraction of fatty acid profile and bioactive compounds such as phenolics, flavonoids, gamma-oryzanols, and tocopherols, from bran rice [85].
Broken rice	Development of gluten-free products that require pre-gelatinized starch, such as pasta, from broken rice flour [84].

References

1. Audet, R.; Brisebois, É. The Social Production of Food Waste at the Retail-Consumption Interface. *Sustainability* **2019**, *11*, 3834. [CrossRef]
2. Martin-Rios, C.; Demen-Meier, C.; Gössling, S.; Cornuz, C. Food waste management innovations in the foodservice industry. *Waste Manag.* **2018**, *79*, 196–206. [CrossRef] [PubMed]
3. Lemaire, A.; Limbourg, S. How can food loss and waste management achieve sustainable development goals? *J. Clean. Prod.* **2019**, *234*, 1221–1234. [CrossRef]
4. Food Waste. Available online: https://ec.europa.eu/food/safety/food_waste_en (accessed on 30 January 2020).
5. Ness, D. Sustainable urban infrastructure in China: Towards a factor 10 improvement in resource productivity through integrated infrastructure system. *Int. J. Sustain. Dev. World Ecol.* **2008**, *15*, 288–301.
6. Ghisellini, P.; Cialani, C.; Ulgiati, S. A review on circular economy: The expected transition to a balanced interplay of environmental and economic systems. *J. Clean. Prod.* **2016**, *114*, 11–32. [CrossRef]
7. Ellen Macarthur Foundation. Towards the Circular Economy. 2012. Available online: http://www.ellenmacarthurfoundation.org/business/reports (accessed on 14 September 2019).
8. Prendeville, S.; Sanders, C.; Sherry, J.; Costa, F. Circular Economy: Is it Enough? 2014. Available online: http://www.edcw.org/sites/default/files/resources/Circular%20Ecomomy-%20Is%20it%20enough.pdf (accessed on 5 June 2020).
9. Club of Rome. The Circular Economy and Benefits for Society, Swedish Case Study Shows Jobs and Climate as Clear Winners. 2015. Available online: http://www.clubofrome.org/cms/wp-content/uploads/2015/04/Final-version-Swedish-Study-13-04-15-till-tryck-ny.pdf (accessed on 15 April 2018).
10. Birat, J.-P. Life-cycle assessment, resource efficiency and recycling. *Met. Res. Technol.* **2015**, *112*, 206. [CrossRef]
11. Potting, J.; Hekkert, M.; Worrell, E.; Hanemaaijer, A. *Circular Economy—Measuring Innovation in the Product Chain*; Policy Report; PBL Netherlands Environmental Assessment Agency: The Hague, the Netherlands, 2017. Available online: www.pbl.nl/sites/default/files/cms/publicaties/pbl-2016-circular-economy-measuring-innovation-in-product-chains-2544.pdf (accessed on 15 December 2019).
12. Clift, R. Clean technology and industrial ecology. In *Pollution—Causes, Effects and Control*; Harrison, R.M., Ed.; Royal Society of Chemistry: Cambridge, UK, 2001; pp. 411–444.

13. Graedel, T.E.; Allenby, B.R. *Industrial Ecology*, 2nd ed.; Prentice Hall: Upper Saddle River, NJ, USA, 2003.
14. Lanzavecchia, B.; Tamborrini, S.P. *Il Fare Ecologico. Il Prodotto Industriale e i Suoi Requisiti Ambientali*; Edizione Ambiente: Milan, Italy, 2012.
15. Ayres, R.U. Industrial metabolism: Theory and policy. In *Industrial Metabolism: Restructuring for Sustainable Development*; Ayres, R.U., Simonis, U.K., Eds.; United Nations University Press: Tokyo, Japan, 1994; pp. 3–20.
16. Frosch, R.A. Industrial ecology: A philosophical introduction. *Proc. Natl. Acad. Sci. USA* **1992**, *89*, 800–803. [CrossRef]
17. Erkman, S. Industrial ecology: An historical view. *J. Clean. Prod.* **1997**, *5*, 1–10. [CrossRef]
18. Chiu, A.S.F.; Geng, Y. On the industrial ecology potential in Asian developing countries. *J. Clean. Prod.* **2004**, *12*, 1037–1045. [CrossRef]
19. Von Bertalanffy, L. An outline of general system theory. *Br. J. Phil. Sci.* **1950**, *1*, 134–165. [CrossRef]
20. Von Bertalanffy, L. *General System Theory*; George Braziller: New York, NY, USA, 1968; p. 295.
21. Ellen Macarthur Foundation. The Circular Model and Brief History and School of Thought. 2013. Available online: http://www.ellenmacarthurfoundation.org/circulareconomy/circular-economy/the-circular-model-brief-history-and-schools-ofthought (accessed on 19 September 2019).
22. He, P.; Lü, F.; Zhang, H.; Shao, L. Recent developments in the area of waste as a resource, with particular reference to the circular economy as a guiding principle, in waste as a resource 2013. In *Environmental Science and Technology*; Hester, R.E., Harrison, R.M., Eds.; The Royal Society of Chemistry: Cambridge, UK, 2013; Volume 37.
23. EC, European Commission. MEMO, Questions and Answers on the Commission Communication towards a Circular Economy and the Waste Targets Review. 2014. Available online: http://europa.eu/rapid/press-release_MEMO-14-450_en.htm (accessed on 24 June 2020).
24. EC, European Commission. *Communication from the Commission to the European Parliament, the Council, the European Economic and Social Committee and the Committee of the Regions. Towards a Circular Economy: A Zero Waste Programme for Europe*; European Commission: Bruxelles, Belgium, 2014. Available online: https://eur-lex.europa.eu/resource.html?uri=cellar:50edd1fd-01ec-11e4-831f-01aa75ed71a1.0001.01/DOC_1&format=PDF (accessed on 29 September 2020).
25. EU. Directive 2008/98/EC of the European Parliament and of the Council of 19 November 2008 on waste and repealing certain directives. *Official Journal of EU L* **2008**, *312*. Available online: http://eur-lex.europa.eu/LexUriServ/LexUriServ.do?uri\protect$\relax\protect{\begingroup1\endgroup\@@over4}$OJ:L:2008:312:0003:0030:en:PDF (accessed on 10 June 2020).
26. Thomas, J.S.; Birat, J.P. Methodologies to measure the sustainability of materials e focus on recycling aspects. *Rev. Metall.* **2013**, *110*, 3–16. [CrossRef]
27. Bakker, C.; Wang, F.; Huisman, J.; Hollander, M.D. Products that go round: Exploring product life extension through design. *J. Clean. Prod.* **2014**, *69*, 10–16. [CrossRef]
28. Park, J.Y.; Chertow, M.R. Establishing and testing the "reuse potential" indicator for managing wastes as resources. *J. Environ. Manag.* **2014**, *137*, 45–53. [CrossRef] [PubMed]
29. Gwehenberger, G.; Erler, B.; Schnitzer, H. A Multi e Strategy Approach to Zero Emissions. Presented at the Technology Foresight Summit, Budapest, Hungary, 27–29 March 2003.
30. Cagno, E.; Trucco, P.; Tardini, L. Cleaner production and profitability: Analysis of 134 industrial pollution prevention (P2) project reports. *J. Clean. Prod.* **2005**, *13*, 593–605. [CrossRef]
31. Davis, G.G.; Hall, J.A. In: Circular Economy Legislation, the International Experience. Executive Summary. 2006. Available online: https://www.reusablepackaging.org/insights/circular-economy-legislation-the-international-experience/ (accessed on 23 March 2019).
32. Schnitzer, H.; Ulgiati, S. Less bad is not good enough. *J. Clean. Prod.* **2007**, *15*, 1185–1189. [CrossRef]
33. Ren, Y. The circular economy in China. *J. Mater. Cycles Waste Manag.* **2007**, *9*, 121–129.
34. Feng, Z.; Yan, N. Putting a circular economy into practice in China. *Sustain. Sci.* **2007**, *2*, 95–101.
35. Zhu, D. Background, pattern and policy of China for developing circular economy. *Chin. J. Popul. Resour. Environ.* **2008**, *6*, 3–8.
36. Sakai, S.-I.; Yoshida, H.; Hirai, Y.; Asari, M.; Takigami, H.; Takahashi, S.; Tomoda, K.; Peeler, M.V.; Wejchert, J.; Schmid-Unterseh, T.; et al. International comparative study of 3R and waste management policy developments. *J. Mater. Cycles Waste Manag.* **2011**, *13*, 86–102. [CrossRef]
37. Bilitewsky, B. The circular economy and its risks. *Waste Manag.* **2012**, *32*, 1–2. [CrossRef]

38. Lazarevic, D.; Aoustin, E.; Buclet, N.; Brandt, N. Plastic waste management in the context of a European recycling society: Comparing results and uncertainties in a life cycle perspective. *Resour. Conserv. Recycl.* **2010**, *55*, 246–259. [CrossRef]

39. Preston, F. A Global Redesign? Shaping the Circular Economy. Briefing Paper. 2012. Available online: http://www.chathamhouse.org/sites/default/files/public/Research/Energy%20Environment%20and%20Development/bp0312_preston.pdf (accessed on 22 September 2019).

40. Lett, L.A. Las amenazas globales, el reciclaje de residuos y el concepto de economía circular. *Revista Argentina de Microbiología* **2014**, *46*, 1–2. [CrossRef]

41. Su, B.; Heshmati, A.; Geng, Y.; Yu, X. A review of the circular economy in China: Moving from rhetoric to implementation. *J. Clean. Prod.* **2013**, *42*, 215–227. [CrossRef]

42. Reh, L. Process engineering in circular economy. *Particuology* **2013**, *11*, 119–133. [CrossRef]

43. Stahel, W.R. Policy for material efficiency e sustainable taxation as a departure from a throwaway society. *Phys. Eng. Sci.* **2013**, *371*, 20110567. [CrossRef] [PubMed]

44. Stahel, W.R. Reuse Is the Key to the Circular Economy. Available online: http://ec.europa.eu/environment/ecoap/about-eco-innovation/experts-interviews/reuse-is-the-key-to-the-circular-economy_en.htm (accessed on 12 March 2020).

45. Figge, F.; Young, W.; Barkemeyer, R. Sufficiency or efficiency to achieve lower resource consumption and emissions? The role of the rebound effect. *J. Clean. Prod.* **2014**, *69*, 216–224. [CrossRef]

46. Manomivibool, P.; Hong, J.H. Two decades, three WEEE systems: How far did EPR evolve in Korea's resource circulation policy? *Resour. Conserv. Recycl.* **2014**, *83*, 202–212. [CrossRef]

47. Mirabella, N.; Castellani, V.; Sala, S. Current options for the valorization of food manufacturing waste: A review. *J. Clean. Prod.* **2014**, *65*, 28–41. [CrossRef]

48. Sevigné-Itoiz, E.; Gasol, C.M.; Rieradevall, J.; Gabarrell, X. Environmental consequences of recycling aluminum old scrap in a global market. *Resour. Conserv. Recycl.* **2014**, *89*, 94–103. [CrossRef]

49. Castellani, V.; Sala, S.; Mirabella, N. Beyond the throwaway society: A life cycle-based assessment of the environmental benefit of reuse. *Integr. Environ. Assess. Manag.* **2015**, *11*, 373–382. [CrossRef]

50. Prigogine, I.; Stengers, I. *Order out of Chaos: Man's New Dialogue with Nature*; Heinemann: London, UK, 1984.

51. Maturana, H.R.; Varela, F.J. Autopoiesis and cognition. The realization of the living. In *Boston Studies in the Philosophy of Science*; Cohen, R., Wartofsky, M., Eds.; Reidel Publishing: Boston, MA, USA, 1980.

52. Kauffman, S.A. *The Origins of Order. Self-Organization and Selection in Evolution*; Oxford University Press: Oxford, UK, 1993.

53. Pisek, P.E.; Wilson, T. Complexity, leadership, and management in healthcare organizations. *Br. Med. J.* **2001**, *323*, 746–749.

54. Heylighen, F.; Joslyn, C. Cybernetics and second order cybernetics. In *Encyclopedia of Physical Science & Technology*, 3rd ed.; Meyers, R.A., Ed.; Academic Press: New York, NY, USA, 2001; Volume 4, pp. 155–170.

55. Fiksel, J. Designing resilient, sustainable systems. *Environ. Sci. Technol.* **2003**, *37*, 5330–5339. [CrossRef]

56. Barbero, S. *Systemic Energy Networks. The Theory of Systemic Design applied to Energy Sector*; Lulu Press: Morrisville, NC, USA, 2012; Volume 1.

57. Bistagnino, L. *Systemic Design, Designing the Productive and Environmental Sustainability*, 2nd ed.; Slow Food: Bra, Italy, 2011.

58. Shannon, C.E.; Weaver, W. The mathematical theory of communication. *Bell Syst. Tech. J.* **1948**, *27*, 379–423, 623–656. [CrossRef]

59. Porter, M. *The Competitive Advantage of Nations*; Macmillan: London, UK, 1990.

60. Frosch, R.A.; Gallopoulos, N.E. Strategies for manufacturing. *Sci. Am.* **1989**, *261*, 144–152. [CrossRef]

61. Chertow, M. Industrial symbiosis: Literature and taxonomy. *Annu. Rev. Energy Environ.* **2000**, *25*, 313–337. [CrossRef]

62. Ellen McArthur. Rethinking the Economy. Available online: https://www.ellenmacarthurfoundation.org/news/rethinking-the-economy (accessed on 20 April 2020).

63. Ellen McArthur. Kalundborg Symbiosis. Effective Industrial Symbiosis. Available online: https://www.ellenmacarthurfoundation.org/case-studies/effective-industrial-symbiosis (accessed on 20 April 2020).

64. Pearce, D.W.; Turner, R.K. *Economics of Natural Resources and the Environment*; Harvester Wheatsheaf: Hemel Hempstead, UK, 1989.

65. Ehrenfeld, J.; Gertler, N. Industrial ecology in practice: The evolution of interdependence at Kalundborg. *J. Ind. Ecol.* **1997**, *1*, 67–79. [CrossRef]
66. Van Berkel, R.; Willems, E.; Lafleur, M. The relationship between cleaner production and industrial ecology. *J. Ind. Ecol.* **1997**, *1*, 51–65. [CrossRef]
67. Andersen, M.S. An introductory note on the environmental economics of the circular economy. *Sustain. Sci.* **2007**, *2*, 133–140. [CrossRef]
68. Zhu, D.; Wu, Y. Plan C: China's development under the scarcity of natural capital. *Int. Eco. Res. Eff.* **2011**, *5*, 175–186. [CrossRef]
69. Mathews, J.A.; Tan, H. Progress towards a circular economy: The drivers and inhibitors of eco-industrial initiative. *J. Ind. Ecol.* **2011**, *15*, 435–457. [CrossRef]
70. Iung, B.; Levrat, E. Advanced maintenance services for promoting sustainability. *Procedia CIRP* **2014**, *22*, 15–22. [CrossRef]
71. den Hollander, M.C.D.; Bakker, C.A.; Hultink, E.J. Product Design in a Circular Economy: Development of a Typology of Key Concepts and Terms. *J. Ind. Ecol.* **2017**, *21*, 517–525. [CrossRef]
72. Carricondo Anton, J.M.; Gonzalez Romero, J.A.; Mengual Cuquerella, J.; Turegano Pastor, J.V.; Oliver Villanueva, J.V. Alternative use of rice straw ash as natural fertilizer to reduce phosphorus pollution in protected wetland ecosystems. *Int. J. Recycl. Org. Waste Agric.* **2020**, *9*, 61–74.
73. Belaud, J.P.; Prioux, N.; Vialle, C.; Buche, P.; Destercke, S.; Sablayrolles, C. Framework for sustainable management of agricultural by-product valorization. *Chem. Eng. Trans.* **2019**, *74*, 1255–1260.
74. Menzel, C.; González-Martínez, C.; Vilaplana, F.; Diretto, G.; Chiralt, A. Incorporation of natural antioxidants from rice straw into renewable starch films. *Int. J. Biol. Macromol.* **2020**, *146*, 976–986. [CrossRef]
75. Heffernan, S. Innovative Natural Yarn Manufactured from Waste. In Proceedings of the International Scientific-Technical Conference MANUFACTURING, Poznan, Polan, 19–22 May 2019; Springer: Cham, Switzerland, 2019; pp. 485–494.
76. Belc, N.; Mustatea, G.; Apostol, L.; Iorga, S.; Vlăduţ, V.N.; Mosoiu, C. Cereal supply chain waste in the context of circular economy. In Proceedings of the E3S Web of Conferences, Târgovişte, Romania, 6–8 June 2019; EDP Sciences: Les Ulis, France, 2019; Volume 112.
77. Vaskalis, I.; Skoulou, V.; Stavropoulos, G.; Zabaniotou, A. Towards Circular Economy Solutions for The Management of Rice Processing Residues to Bioenergy via Gasification. *Sustainability* **2019**, *11*, 6433. [CrossRef]
78. Assi, A.; Bilo, F.; Zanoletti, A.; Ponti, J.; Valsesia, A.; La Spina, R.; Bontempi, E. Review of the reuse possibilities concerning ash residues from thermal process in a medium-sized urban system in Northern Italy. *Sustainability* **2020**, *12*, 4193. [CrossRef]
79. Nicoara, A.I.; Grumezescu, A.M.; Vrabec, M.; Šmuc, N.R.; Sturm, S.; Ow-Yang, C.; Gulgun, M.A.; Bundur, Z.B.; Ciuca, I.; Vasile, B.S. End-of-Life materials used as supplementary cementitious materials in the concrete industry. *Materials* **2020**, *13*, 1954. [CrossRef]
80. Overturf, E.; Ravasio, N.; Zaccheria, F.; Tonin, C.; Patrucco, A.; Bertini, F.; Canetti, M.; Avramidou, K.; Speranza, G.; Bavaro, T.; et al. Towards a more sustainable circular bioeconomy. Innovative approaches to rice residue valorization: The RiceRes case study. *Bioresour. Technol. Rep.* **2020**, *11*, 100427. [CrossRef]
81. Mladenovic, N.; Makreski, P.; Tarbuk, A.; Grgic, K.; Boev, B.; Mirakovski, D.; Jordanov, I. Improved dye removal ability of modified rice husk with effluent from Alkaline scouring based on the circular economy concept. *Processes* **2020**, *8*, 653. [CrossRef]
82. Sharifikolouei, E.; Baino, F.; Galletti, C.; Fino, D.; Ferraris, M. Adsorption of Pb and Cd in rice husk and their immobilization in porous glass-ceramic structures. *Int. J. Appl. Ceram. Technol.* **2019**, *17*, 105–112. [CrossRef]
83. Grace, M.A.; Clifford, E.; Healy, M.G. The potential for the use of waste products from a variety of sectors in water treatment processes. *J. Clean. Prod.* **2016**, *137*, 788–802. [CrossRef]
84. Fradinho, P.; Sousa, I.; Raymundo, A. Functional and thermorheological properties of rice flour gels for gluten-free pasta applications. *Int. J. Food Sci. Technol.* **2018**, *54*, 1109–1120. [CrossRef]
85. Benito-Román, O.; Varona, S.; Sanz, M.T.; Beltrán, S. Valorization of rice bran: Modified supercritical CO_2 extraction of bioactive compounds. *J. Ind. Eng. Chem.* **2019**, *80*, 273–282. [CrossRef]
86. Moongngarm, A.; Daomukda, N.; Khumpika, S. Chemical compositions, phytochemicals, and antioxidant capacity of rice bran, rice bran layer, and rice germ. *Apcbee Procedia* **2012**, *2*, 73–79. [CrossRef]

87. Limtrakul, P.; Semmarath, W.; Mapoung, S. Anthocyanins and Proanthocyanidins in natural pigmented rice and their bioactivities. In *Phytochemicals in Human Health*; Rao, V., Mans, D., Rao, L., Eds.; IntechOpen: London, UK, 2019.

88. Shafie, N.H.; Esa, N.M. The healing components of rice bran. In *Functional Foods: Wonder of the World. Evidence-Based Functional Foods in Health & Disease*; Azlan, A., Ismail, A., Eds.; UPM Press: Serdang, Malaysia, 2017; pp. 341–368.

89. Vini a Denominazione di Origine, DOCG e DOC. Available online: https://www.regione.piemonte.it/web/temi/agricoltura/viticoltura-enologia/vini-denominazione-origine-docg-doc (accessed on 23 July 2020).

90. Direzione Agricoltura. Linee Guida per la Tenuta Degli albi dei Vigneti e per la Conduzione Delle Superfici Vitate Iscritte. 2008. Available online: https://areadocumentale.vignaioli.it/_literature_108439/LINEE_GUIDA_PER_LA_TENUTA_DEGLI_ALBI_DEI_VIGNETI_E_PER_LA_CONDUZIONE_DELLE_SUPERFICI_VITATE_ISCRITTE (accessed on 15 February 2020).

91. Brito, A.F.; Broderick, G.A.; Reynal, S.M. Effects of different protein supplements on omasal nutrient flow and microbial protein synthesis in lactating dairy cows. *J. Dairy Sci.* **2007**, *90*, 1828–1841. [CrossRef]

92. Bertran, E.; Sort, X.; Soliva, M.; Trillas, M.I. Composting winery waste: Sludges and grape stalks. *Bioresour. Technol.* **2004**, *95*, 203–208. [CrossRef]

93. Diaz, M.J.; Madejon, E.; Lopez, F.; Lopez, R.; Cabrera, F. Optimization of the rate vinasse/grape marc for co-composting process. *Process Biochem.* **2002**, *37*, 1143–1150. [CrossRef]

94. Mustin, M. *Le Compost: Gestion de la Matière Organiqu*; Éditions François Dubusc: Paris, France, 1987; p. 954.

95. Pardo, A.; Pardo, J.E.; de Juan, J.A.; Zied, D.C. Modelling the effect of the physical and chemical characteristics of the materials used as casing layers on the production parameters of Agaricus bisporus. *Arch. Microbiol.* **2010**, *192*, 1023–1030. [CrossRef]

96. Olejar, K.J.; Vandermeer, C.; Fedrizzi, B.; Kilmartin, P.A. A Horticultural medium established from the rapid removal of Phytotoxins from Winery Grape Marc. *Horticulturae* **2019**, *5*, 69. [CrossRef]

97. Tacchini, M.; Burlini, I.; Bernardi, T.; De Risi, C.; Massi, A.; Guerrini, A.; Sacchetti, G. Chemical characterisation, antioxidant and antimicrobial screening for the revaluation of wine supply chain by-products oriented to circular economy. *Plant Biosyst. Int. J. Deal. All Asp. Plant Biol.* **2018**, *153*, 809–816. [CrossRef]

98. Fidelis, M.; de Moura, C.; Kabbas Junior, T.; Pap, N.; Mattila, P.; Mäkinen, S.; Putnik, P.; Kovačević, D.B.; Tian, Y.; Yang, B. Fruit seeds as sources of bioactive compounds: Sustainable production of high value-added ingredients from by-products within circular economy. *Molecules* **2019**, *24*, 3854. [CrossRef] [PubMed]

99. Mainente, F.; Menin, A.; Alberton, A.; Zoccatelli, G.; Rizzi, C. Evaluation of the sensory and physical properties of meat and fish derivatives containing grape pomace powders. *Int. J. Food Sci. Technol.* **2019**, *54*, 952–958. [CrossRef]

100. Ferri, M.; Vannini, M.; Ehrnell, M.; Eliasson, L.; Xanthakis, E.; Monari, S.; Tassoni, A. From winery waste to bioactive compounds and new polymeric biocomposites: A contribution to the circular economy concept. *J. Adv. Res.* **2020**, *24*, 1–11. [CrossRef]

101. Brosse, N.; Pizzi, A. *Tannins for Wood Adhesives, Foams and Composites*; CRC Press: Boca Raton, FL, USA, 2017; pp. 197–220.

102. Lagoa, B.; Campos, L.; Silva, T.; Moreira, M.J.; Castro, L.M.; Veloso, A.C.; Pinheiro, M.C. Winery wastes: A potential source of natural dyes for textiles. In *Wastes: Solutions, Treatments and Opportunities III, Proceedings of the Selected Papers from the 5th International Conference Wastes 2019, Lisbon, Portugal, 4–6 September 2019*; CRC Press: Boca Raton, FL, USA, 2019; p. 69.

103. Baaka, N.; Haddar, W.; Ben Ticha, M.; Mhenni, M.F. Eco-friendly dyeing of modified cotton fabrics with grape pomace colorant: Optimization using full factorial design approach. *J. Nat. Fibers* **2019**, *16*, 652–661. [CrossRef]

104. Baaka, N.; Ticha, M.B.; Haddar, W.; Amorim, M.T.P.; Mhenni, M.F. Upgrading of UV protection properties of several textile fabrics by their dyeing with grape pomace colorants. *Fibers Polym.* **2018**, *19*, 307–312. [CrossRef]

105. Bechtold, T.; Mahmud-Ali, A.; Mussak, R. Anthocyanin dyes extracted from grape pomace for the purpose of textile dyeing. *J. Sci. Food Agric.* **2007**, *87*, 2589–2595. [CrossRef]

106. Lucarini, M.; Durazzo, A.; Romani, A.; Campo, M.; Lombardi-Boccia, G.; Cecchini, F. Bio-based compounds from grape seeds: A biorefinery approach. *Molecules* **2018**, *23*, 1888. [CrossRef]

107. Versari, A.; Ferrarini, R.; Parpinello, G.P.; Galassi, S. Concentration of grape must by nanofiltration membranes. *Food Bioprod. Process.* **2003**, *81*, 275–278. [CrossRef]

108. Braga, F.G.; Lencart e Silva, F.A.; Alves, A. Recovery of winery by-products in the Douro demarcated region: Production of calcium tartrate and grape pigments. *Am. J. Enol. Vitic.* **2002**, *53*, 42–45.

109. Boulton, R.B.; Singleton, V.L.; Bisson, L.F.; Kunkee, R.E. *Principles and Practices of Winemaking*; Springer: Boston, MA, USA, 1999.

110. Câmara, J.S.; Lourenço, S.; Silva, C.; Lopes, A.; Andrade, C.; Perestrelo, R. Exploring the potential of wine industry by-products as source of additives to improve the quality of aquafeed. *Microchem. J.* **2020**, *155*, 104758. [CrossRef]

111. Solid by-products to produce calcium tartrate. Available online: https://detadistilleria.it (accessed on 29 September 2020).

112. Saviozzi, A.; Riffaldi, R.; Levi-Minzi, R.; Panichi, A. Properties of soil particle size separates after 40 years of continuous corn. *Commun. Soil Sci. Plant Anal.* **1997**, *28*, 427–440. [CrossRef]

113. Da Ros, C.; Cavinato, C.; Pavan, P.; Bolzonella, D. Winery waste recycling through anaerobic co- digestion with waste activated sludge. *Waste Manag.* **2014**, *34*, 2028–2035. [CrossRef] [PubMed]

114. Data & Trends of the European Food and Drink Industry. 2018. Available online: https://www.fooddrinkeurope.eu/publication/data-trends-of-the-european-food-and-drink-industry-2018/ (accessed on 23 February 2019).

115. Statistical Factsheet. Available online: https://ec.europa.eu/agriculture/sites/agriculture/files/statistics/factsheets/pdf/eu_en.pdf (accessed on 23 February 2019).

116. Statista. Available online: https://www.statista.com/outlook/10030000/102/wine/europe (accessed on 1 February 2019).

117. European Commission. Wine Growing Regions. Source: EU Wine Market Data Portal. Available online: https://ec.europa.eu/info/food-farming-fisheries/farming/facts-and-figures/markets/overviews/market-observatories/wine_en (accessed on 1 February 2019).

118. Crop Production Maps. Available online: https://ipad.fas.usda.gov/rssiws/al/europe_cropprod.aspx (accessed on 10 March 2020).

119. Barbero, S. *Systemic Design Method Guide for Policymaking: A Circular Europe on the Way*; Umberto Allemandi: Torino, Italy, 2017.

120. Oliveira, M.; Costa, J.M.; Fragoso, R.; Duarte, E. Challenges for modern wine production in dry areas: Dedicated indicators to preview wastewater flows. *Water Supply* **2019**, *19*, 653–661. [CrossRef]

121. Ahmad, B.; Yadav, V.; Yadav, A.; Rahman, M.U.; Yuan, W.Z.; Li, Z.; Wang, X. Integrated biorefinery approach to valorize winery waste: A review from waste to energy perspectives. *Sci. Total Environ.* **2020**, *719*, 137315. [CrossRef]

122. Manniello, C.; Statuto, D.; Di Pasquale, A.; Giuratrabocchetti, G.; Picuno, P. Planning the flows of residual biomass produced by wineries for the preservation of the rural landscape. *Sustainability* **2020**, *12*, 847. [CrossRef]

123. Manniello, C.; Statuto, D.; Di Pasquale, A.; Picuno, P. Planning the flows of residual biomass produced by wineries for their valorization in the framework of a circular bioeconomy. In *International Mid-Term Conference of the Italian Association of Agricultural Engineering*; Springer: Cham, Switzerland, 2019; pp. 295–303.

124. Cortés, A.; Silva, L.F.O.; Ferrari, V.; Taffarel, S.R.; Feijoo, G.; Moreira, M.T. Environmental assessment of viticulture waste valorisation through composting as a biofertilisation strategy for cereal and fruit crops. *Environ. Pollut.* **2020**, *264*, 114794. [CrossRef]

125. Chen, Y.K.; Lin, C.H.; Wang, W.C. The conversion of biomass into renewable jet fuel. *Energy* **2020**, *201*, 117655. [CrossRef]

126. Pinela, J.; Omarini, A.B.; Stojković, D.; Barros, L.; Postemsky, P.D.; Calhelha, R.C.; Ferreira, I.C. Biotransformation of rice and sunflower side-streams by dikaryotic and monokaryotic strains of Pleurotus sapidus: Impact on phenolic profiles and bioactive properties. *Food Res. Int.* **2020**, *132*, 109094. [CrossRef]

127. Kadoglidou, K.; Kalaitzidis, A.; Stavrakoudis, D.; Mygdalia, A.; Katsantonis, D. A Novel Compost for Rice Cultivation Developed by Rice Industrial By-Products to Serve Circular Economy. *Agronomy* **2019**, *9*, 553. [CrossRef]

128. Chen, W.; Oldfield, T.L.; Katsantonis, D.; Kadoglidou, K.; Wood, R.; Holden, N.M. The socio-economic impacts of introducing circular economy into Mediterranean rice production. *J. Clean. Prod.* **2019**, *218*, 273–283. [CrossRef]
129. Alexandri, M.; López-Gómez, J.P.; Olszewska-Widdrat, A.; Venus, J. Valorising agro-industrial wastes within the circular bioeconomy concept: The Case of Defatted Rice Bran with Emphasis on Bioconversion Strategies. *Fermentation* **2020**, *6*, 42. [CrossRef]

 *sustainability* 

Article

Impact of Ethanol Plant Location on Corn Revenues for U.S. Farmers

Ani L. Katchova [1,*] and Ana Claudia Sant'Anna [2]

[1] Department of Agricultural, Environmental, and Development Economics, The Ohio State University, Columbus, OH 43210, USA
[2] Division of Resource Economics and Management, West Virginia University, Morgantown, WV 26506, USA; ana.santanna@mail.wvu.edu
* Correspondence: katchova.1@osu.edu

Received: 22 October 2019; Accepted: 15 November 2019; Published: 19 November 2019

Abstract: Ethanol production has rapidly expanded over the past few years. The opening of an ethanol plant can increase local demand for corn, pressuring increases in local corn basis. But how does this affect corn contract prices and revenues? At the farm level, the impact of an ethanol plant on local corn contract revenues is still unknown. Data from the USDA Agricultural Resource Management Survey suggests that corn contract revenues in counties with ethanol plants are higher than corn contract revenues in counties without ethanol plants at similar prices. We estimate the impact of ethanol plants on local corn contract revenues by running non-spatial and spatial difference-in-difference models. A statistically significant effect of ethanol plant location on corn contract revenues within the same county was not found, but rather a statistically significant effect of ethanol plants on corn contract revenues for farmers located in adjacent counties. Local competitive advantage, not the presence of an ethanol plant, may be the reason for observed higher revenues in counties with an ethanol plant. Therefore, policymakers should focus their resources in promoting greater efficiency in corn production to boost farmers' revenues.

Keywords: biofuels; spatial difference-in-difference; corn markets

---

1. Introduction

Biofuel markets and policies are often related to matters of national security, the environment, and food security. In the U.S., ethanol production has rapidly expanded over the last few years. From 1999 to 2017, the number of biofuel refineries increased from 50 to 213 [1,2]. Along with the increase in the number of biorefineries came an increase in the share of corn supplied to ethanol production and of corn prices (Figure 1).

Increases in ethanol production may benefit local corn producers. The opening of an ethanol plant, for instance, can increase local demand for corn, pressuring increases in local corn basis [3]. This impact, though, may not be lasting [4,5]. In fact, in some areas, the presence of an ethanol plant has been linked to lower corn prices [6,7], providing mixed evidence for the effect of ethanol plants on corn prices in previous studies. Furthermore, analysis of the data from the USDA Agricultural Resource Management Survey (ARMS), shows that, on average, farmers in counties with ethanol plants receive the same prices for corn as those located in counties without plants (Figure 2a). Therefore, can the opening of an ethanol plant really benefit the local farmer? The same data from the USDA show that farmers in counties with ethanol plants do, however, receive higher corn revenues (Figure 2b), a phenomenon that is not considered in previous studies. Our study fills this gap in the literature by estimating the impact on farmers' corn revenues from having an ethanol plant in their county.

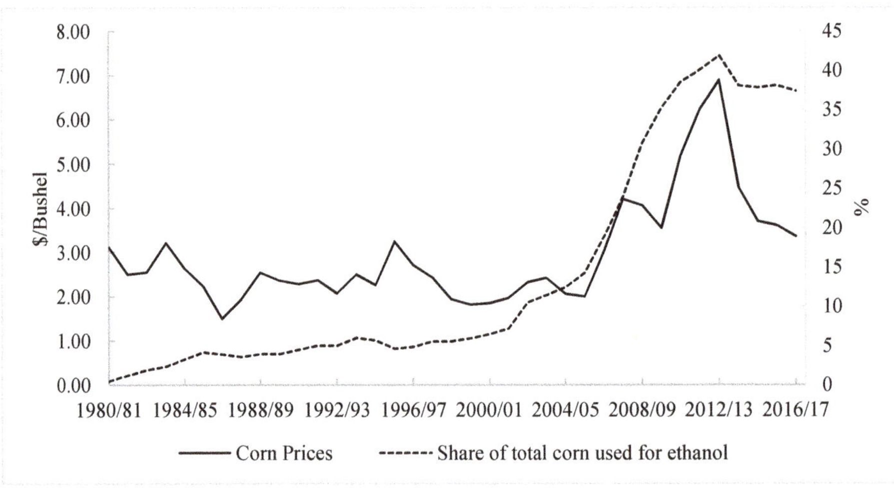

Figure 1. Corn prices against the share of total corn used for ethanol, from 1980 to 2017. Source: USDA 2017.

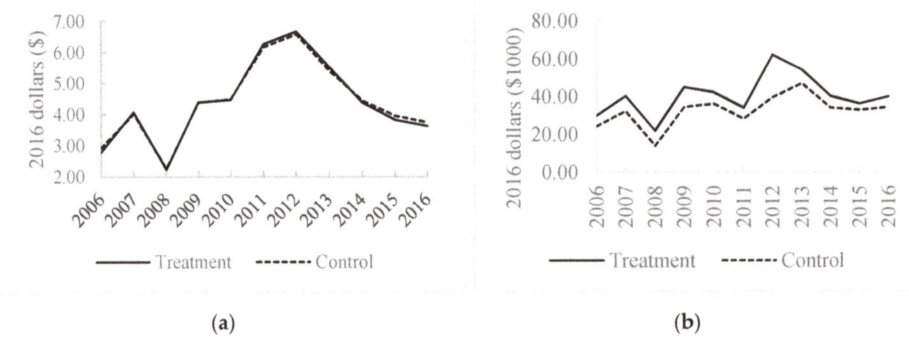

(a)   (b)

Figure 2. (**a**) Median contract corn prices for farmers located in counties with (treated group) and without (control group) an ethanol plant from 2006 to 2016. Source: ARMS, USDA; (**b**) Median contract corn revenues for farmers located in counties with (treated group) and without (control group) an ethanol plant from 2006 to 2016. Source: ARMS, USDA.

The location of an ethanol plant is determined by factors such as access to feedstock, trends in corn production, government incentives, local amenities, and infrastructure [8–10]. Lambert et al. [9] find that infrastructure, markets, and subsidies can provide counties with a comparative advantage in charming ethanol plants to the county. The impact and duration of an ethanol plant location or ethanol production capacity on local corn markets can be diverse. Previous studies provide varying results on how ethanol plant location impact local corn supply [11–13], farmers' cropping decisions [14], and local corn market prices [4,5,15,16]. Upon closer examination of data from the Agricultural Resource Management Survey (ARMS) counties with an ethanol plant generally receive higher revenues than counties without, at the same corn prices (Figure 2a,b).

There is lack of consensus on how ethanol plant location and production capacity affect corn production. Using an acreage response model, Fatal and Thurman [11] find that an increase of 1 million gallons in ethanol production capacity increases planted corn acreage by 5.21 acres. Similarly, Motamed, McPhail, and Williams [12] find, in locations with previously low corn acreage, significant acreage response due to changes in the ethanol market. Du, Hennessy, and Edwards [14] show that increases in corn prices causes farmers to switch from non-cropped land to row crops (e.g., from forage crops to row crops). Distance to an ethanol plant also affects crop rotation decisions (e.g., from corn–soybean to corn–corn), allowing more corn to be produced [13]. The distance to the ethanol plant, though, may not always be the cause for changes in corn acreage. Ifft et al. [10] argue that land use change maybe mostly driven by general price changes rather than by the location of an ethanol plant.

Lack of consensus is also present with regard to the impact of ethanol plants on local corn basis. McNew and Griffith [5] find an inverse relationship between the distance of an ethanol plant to markets and corn prices: Facilities near terminal markets have a smaller impact on corn basis prices than those further away. Expanding on McNew and Griffith's [5] model, Fatal [17] finds a premium of $0.04/bushel on local corn, depending on the market, the number of ethanol plants, the total production capacity, and the distance from ethanol plants to corn suppliers. However, being in the vicinity of an ethanol plant is no guarantee of higher local corn prices [7]. Katchova [2] finds no effect from the ethanol plant location on local corn prices. O'Brien [6] finds a negative impact from ethanol plant proximity on corn prices using 2008 corn prices in Kansas. Furthermore, considering four ethanol producing states, Lewis [4] finds increases in corn basis in Michigan, Kansas, and Iowa, but decreases in corn basis in Indiana, following the opening of a new plant. In fact, Lewis [4] finds that the local corn basis decreases as the number of months the ethanol plant is in operation increases. In the case of this study, the data support the theory that the ethanol plant location does not impact corn contract prices but rather corn contract revenues.

Pre-existing market conditions and geographical location play an important role in determining changes in corn prices [16]. Corn is an important raw input in ethanol production. Haddad, Tayler, and Owusu [18] identify ethanol plants as a supply-oriented firm, given the high share that corn occupies in the production costs. Rask [19] finds a corn input price elasticity of (−3.03), meaning that small changes to corn prices provoke larger changes to the amount of ethanol produced. Hence, the demand for corn as an input is likely to be elastic. Furthermore, there is evidence that ethanol plants will halt production when corn prices go over a certain level and that corn farmers are price takers [20,21]. As such, the location of ethanol plants may occur in counties where there is excess corn supply such that an increased demand for corn has a limited effect on local corn prices. Ethanol plants may also be located in areas where farmers have limited bargaining power and are price takers. Katchova [22] finds that corn contracts in regions with larger spot markets have higher prices than in those with limited or no access to spot markets. When farmers have no influence on the corn contract price, they can increase their revenue by increasing the quantity supplied. This could explain why corn contract prices do not differ between counties with and without ethanol plants, but revenues do. Counties with ethanol plants may have a competitive advantage on corn production over those without one. This allows farmers in counties with a competitive advantage to have higher corn productivity at lower costs, thus receiving higher revenues at similar corn prices. As Porter [23] describes, cost leadership is a form of competitive advantage, which occurs when you can produce the same product at the lower cost than your competition [23].

The main objective of this study is to investigate the impact of ethanol plants on local corn contract revenues in more recent years. Understanding how ethanol plants may impact corn producers' welfare may provide guidance on possible impacts from the opening of other ethanol plants (e.g., ethanol plants that produce ethanol from sorghum) or agribusiness industries (e.g., the installation of dairy processing facilities) on the local county. For example, if the location of a plant is associated with surrounding corn producers receiving higher corn revenues, policy makers may decide to motivate the siting of ethanol plants in certain regions aiming to increase farmer's corn revenues. For instance, in Brazil, states in the Cerrado region attracted private ethanol facilities to their counties by investing in infrastructure and providing fiscal incentives [24]. For the farmer, the news of a new ethanol plant may help with their cropping decisions as they seek to maximize profits [15]. The ethanol industry may use the advantages from an ethanol plant siting to negotiate financial benefits from counties or states where they are deciding to build plants.

2. Materials and Methods

We assume farmers are price takers in this market (i.e., they cannot influence corn prices). In this study, since we observe marketing corn contracts, we assume that the farmer goes through a revenue maximization process to decide the quantity of corn to offer given the corn contract price they face. In order to check whether the higher revenues are due to competitive advantage or due to the proximity to the ethanol plant, total revenue is modelled as a function of farm business related factors (e.g., farm assets and operator characteristics), a dummy for whether there is an ethanol plant in the county and economic factors. The objective is to identify whether the higher revenues are due to economic growth, farm business-related factors, or the presence of an ethanol plant in the county.

We test this hypothesis by using difference-in-difference models where we can compare the outcome measures (i.e., corn revenues) of corn producers in counties with and without an ethanol plant. The farmer chooses the optimal quantity to offer in the contract taking into consideration the capital and labor available. We model corn contract revenues as a function of the capital available in the farm, an index of labor available in the county, characteristics of the farm operation (i.e., farm size and level of specialization), farm operator characteristics (i.e., age and education), as well as county economic characteristics.

The difference-in-differences (DID) estimation compares the difference in outcomes (e.g., revenues) between the treated observations (corn contracts located in counties with an ethanol plant) and control observations (corn contracts located in counties without an ethanol plant). Two differences are taken into account: 1) Differences in outcomes from time one to time two; 2) the difference between observations that received the treatment and those that did not. The difference-in-differences model is:

$$y_{it} = \alpha + \beta t_{it} + \gamma d_{it} + \delta t_{it} d_{it} + \phi x_{it} + e_{it}, \tag{1}$$

where y_{it} is the log of corn revenues for each contract i for the initial or second period, t_{it} is a dichotomous variable, which takes on the value of 1 if the observation is in the second period (i.e., 2016) and 0 otherwise (i.e., 2015), d_{it} is a dichotomous variable equal to 1 if the contract belongs to a county with an ethanol plant or more and 0 if it does not, $t_{it} d_{it}$ is the interaction term between the time and treatment dichotomous variables, x_{it} are characteristics that influence revenue (i.e., producer, county, and operation characteristics), and e_{it} is the error term. $t_{it} d_{it}$ is the difference-in-differences measure. It measures the effect of the treatment on the treated group. It controls for similar time differences between the treatment and control groups. It represents the percentage increase in corn revenues due to spatial closeness to an ethanol plant after general spatial (e.g., location in more rural areas) and time effects (e.g., changes in commodity prices) are accounted for.

By estimating the difference-in-differences model using the ordinary least squares estimator, we assume each observation to be independent of all other observations in the data. If observations are correlated inside of a cluster, then each observation will contain less exclusive information. In this study, data from farmers within the same county may be correlated. Therefore, the standard errors are corrected for the intraclass correlation within clusters using the robust cluster variance estimator:

$$V_{cluster} = (X'X)^{-1} \sum_{j=1}^{n_c} u'_j u_j (X'X)^{-1}, \tag{2}$$

where X consists of all variables and the constant, $u_j = \sum_{j \, cluster} e_i X_i$, and n_c contains all the clusters. In this study, the observations represent the farmers while the clusters the counties where their farm operations are located.

As an extra step, we apply difference-in-differences techniques for spatial data, which allow for spatially correlated treatments. The spatial DID model offers an advantage over the non-spatial DID model for it accounts for spatially correlated treatments and spatial interaction in treatment responses [25]. The method allows for the estimation of spillover effects (i.e., indirect treatment effects). The interaction term ($t_{it} d_{it}$) is re-written to incorporate a spatially weighted contiguity matrix (W_{ii}):

$$y_{it} = \alpha + \beta t_{it} + \gamma d_{it} + \delta t_{it} d_{it} + \tau W_{ii} d_{it} \cdot t_{it} + \phi x_{it} + e_{it}, \tag{3}$$

where $\tau = \rho \delta$ and $\cdot$ indicates element by element multiplication (Delgado and Florax 2015). In the absence of spatially correlated treatments, $\rho = 0$, and the model reverts to the non-spatial DID Equation (1) [25]. The average treatment effect (ATE), average direct treatment effect (ADTE), and average indirect treatment effect (AITE) can be estimated from the parameters obtained in (3) [25]:

$$ATE = \delta\left(1 + \rho \overline{WD}\right), \tag{4}$$

$$ADTE = \delta, \tag{5}$$

$$AITE = \tau \overline{WD}, \tag{6}$$

where $\overline{WD}$ is the average proportion of counties that border counties with ethanol plant. Spillover effects from the ethanol plant are given by AITE. The AITE coefficient measures the effect on local corn contract revenues in counties bordering those with an ethanol plant. The effect is on corn contract revenues in counties without an ethanol plant. ADTE captures the direct effect on corn contract revenues in a county with at least one ethanol plant.

Models are tested for misspecification and omitted variable bias. The omitted variable bias was tested using the Ramsey regression equation specification-error test [26]. The null hypothesis of the Ramsey RESET test is that there are no omitted variables. A rejection of the null hypothesis may point to possible endogeneity issues in the model. To check for misspecification, the link test was run. This test verifies whether the coefficient of the transformation of the dependent variable is statistically significant. As such, the null hypothesis is that the coefficient of the dependent variable squared is equal to zero [27,28]. The non-spatial difference in difference estimations and the statistical tests were conducted in Stata 14 through the NORC platform. The spatial estimations were conducted in R following the study and codes by Delgado and Florax [25].

Data and Descriptive Statistics

This study uses data from various sources. Information on the ethanol biorefineries was obtained from the Renewable Fuels Association, as well as the Nebraska Energy Office. We focus on ethanol plants that use corn as their primary input. The latitude and longitude of each county is obtained from the Census U.S. Gazetteer files. For county level weight matrix, an adaptation of the contiguity matrices for county levels from Merryman [29] was made. The choice of counties with ethanol plants is limited to the ones present in the ARMS data. ARMS are secondary data collected by the USDA's National Agricultural Statistics Service (NASS). In order to gain access to farmer level information, considered confidential information, the reader needs to contact NASS and fill out the appropriate forms. It surveys farmers and ranchers on various aspects of their farming practices at various stages of the year during different phases [30]. NASS samples farmers from two main sampling frames: List and area frames. ARMS is also stratified and probability weighted. Weights are assigned to each observation in order to account for the probability of the observation being selected. NASS uses a jackknife re-sampling process, consisting of 30 extra weights obtained from NASS to predict the variations for every data item [30]. As such, estimations conducted in this study use the probability weights provided by the ARMS dataset. All monetary values are indexed in 2016 dollars using the consumer price index.

The matching criteria used to join the two datasets were county for the location of the ethanol plants and farms. The analysis is conducted with data for the top ethanol producing states: Illinois, Indiana, Iowa, Kansas, Michigan, Minnesota, Missouri, Nebraska, North Dakota, Ohio, South Dakota, and Wisconsin. Together, these states produce close to 90% of the total U.S. corn production [31]. The DID model requires two periods to isolate general corn revenues changes over time. The initial period was 2015 and the ending period 2016, namely the most current data available. The DID model requires control and treatment to have constant trends over time. We believe that in one year, there will be less variation in the county and thus we can isolate the effect by the ethanol plant in comparison to when we use a longer timeframe. We believe that one year is enough to measure the impact of an ethanol plant on farmer's revenues for two reasons: (1) The decision on the amount of corn to plant is annual; (2) the impact of ethanol plant on corn contract prices has been found to change monthly [4].

The spatial weights matrix of N × N dimensions was constructed by adapting the contiguity matrices for county levels from Merryman [29]. The spatial weights matrix by Merryman [29] identifies all neighboring counties in the U.S. using contiguity matrices. As such, if county i and county j are neighbors, then the element w_{ij} of the matrix W has a value of 1 and a zero otherwise. In addition, a county is not considered to be its own neighbor (i.e., $w_{ii} = 0$). Given that not all counties are represented in ARMS and in our study area, we adapt the spatial weight matrix by eliminating the rows and columns identifying counties outside of our study area. By eliminating the rows and columns of counties outside of the sample, we avoid forcing non-neighbor counties to become neighboring counties. For example, if we start with the following matrix with counties A, B, C, D, and neighbors (A,B), (B,C), (C,D):

$$W_{ij} = \begin{bmatrix} 0 & 1 & 0 & 0 \\ 1 & 0 & 1 & 0 \\ 0 & 1 & 0 & 1 \\ 0 & 0 & 1 & 0 \end{bmatrix}, \tag{7}$$

and wish to transform the matrix to reflect the counties in our sample, say A and D, we would eliminate rows 2 and 3, as well as columns 2 and 3 from W_{ij}, ending up with the matrix (N_{ij}) where there are no neighbors:

$$N_{ij} = \begin{bmatrix} 0 & 0 \\ 0 & 0 \end{bmatrix}. \tag{8}$$

ARMS is not a panel data survey; as such, not all data on contracts are available in both years for all counties. We control for the unbalanced nature of the dataset by clustering standard errors at the contract level in a county. Information on the dependent variable, corn contract revenues, came from the ARMS. We consider as corn contract revenues the total dollar amount received from corn marketing contracts. Marketing contracts are agreements that determine price and quantities of a product to be delivered. Agricultural contracts make up a significant portion of U.S. agricultural production [32]. Corn costs represent 50%–70% of operational costs [33]; as such, ethanol plants may prefer to sign contracts with nearby corn producers as a protection against volatile corn markets and to ensure a steady processing flow [34]. Hence, we focus on corn contracts. In addition, farm characteristics such as a diversification entropy index, land tenure, farm size in terms of total assets, farm typology, operator age, and education also came from ARMS and were hypothesized to affect corn revenues [35]. The diversification entropy index ranges from 0 to 1 and provides a measure of farm diversification [36]. For example, a farm producing only corn would have a diversification entropy index of 0 and a fully diversified farm would have a diversification entropy index of 1. Specialization in corn production may lead to gains in economies of scale leading to higher revenues. County level characteristics included indicators for a farming dependent county, and a rural county index from the U.S. Census. These are included to control for economic factors affecting corn contract revenues (e.g., more rural counties may have lower corn contract revenues due to lower wages or lower corn production costs).

Table 1 displays information on the summary statistics of the data used in the model. The average corn contract revenues in 2015/2016 were $77,851. Among the counties in the sample, 28% have an ethanol plant and 18% are dependent on farming. Over half of the data (i.e., 56%) is from 2016, the rest of the observations are from 2015. The interaction term between the time and treatment dummy variables shows that 15% of the corn contracts in 2016 were in counties with an ethanol plant. The average farm size, measured as total assets, was about $3.5 million, with 37% of farms having assets over $3 million. The principal operator was, on average, 54 years old, operated on average a farm size between an intermediate and commercial farm (i.e., with a farm typology of 2.5), and had, on average, some college level education. Farm Typology is split into 1—Family farms, 2—Intermediate Farms, and 3—Commercial Farms, see Hoppe and MacDonald [37]. There were 629 counties included in the analysis.

Close to one-third of the observations are in the treated group (i.e., farmers with corn contracts located in counties with ethanol plants) and two-thirds are in the control group (i.e., farmers with corn contract revenues in counties without ethanol plants). In order to check for similarities between the treated and control groups, balancing tests are conducted (see Table 2). Results from the t-tests show that treatment and control groups had similar characteristics and that the control group was a valid counterfactual group. Except for operator education in 2015, for all variables, the hypothesis that the mean of the treatment group is different than that of the control group is rejected at a 5% statistical significance level. The statistically significant difference in education averages is not an issue since the averages, when rounded off belong to the same categorical level (i.e., 3).

Table 1. Descriptive statistics.

Variable	Definition	Mean	Standard Deviation
Ethanol plant Dummy	1 if the county of farmer has an ethanol plant	0.28	0.45
Time Dummy	1 if observation is in the second year	0.56	0.50
Ethanol plant dummy * time dummy	Interaction term measuring the difference-in-differences effect	0.15	0.35
Corn contract revenue	Total amount in received in dollars of 2016	$77,851	$126,626
Log corn contract revenue	Natural logarithm of the corn contract revenue	10.41	1.36
Entropy diversification index	Diversification index defined by ERS-USDA that ranges between 0 and 1. Higher values represents higher levels of diversification	0.24	0.11
Land tenure	Proportion of owned to total land	0.52	0.56
Land tenure (dummy)	1 if the ratio of the area owned to operated is 0.50 or more	0.45	0.50
Farm size	Total assets in 2016 dollars	$3,475,695	$5,455,123
Farm Size (dummy)	1 if the total assets are greater than 3 million	0.37	0.48
Operator age	Age of the operator in years	54.08	13.26
Operator education	Operator education in categories from 1 to 4	2.96	0.86
Farm Typology	Farm classification developed by the USDA ranging from 1-3	2.48	0.78
Farming dependent county	Based on U.S. Census Economic dependency index, where 1 if county is farming dependent	0.18	0.38
Low employment county	Index defined by the U.S. Census. It equals 1 when it is a low employment county	0.00	0.03
Urban-rural county index	Index defined by the U.S. Census ranging from 1–9. Higher numbers represent more rural locations	5.28	2.36
Number of observations	Number of corn contracts	3003	
Number of clusters	Number of distinct counties	629	

Notes: DID: Difference-in-difference. Farm size and tenure dummies were used in the estimation instead of the continuous variables because it produced a more robust model. Farm Typology is split into 1—Family farms, 2—Intermediate Farms, and 3—Commercial Farms. Operator education is split into 1—less than high school diploma, 2—High School, 3—Some college (includes associates degree) and, 4—4-year college graduate and beyond. The average corn contract price was $3.81.

Table 2. Balancing tests: *p*-values and averages.

Variable	2016 (Average)		*P*-Value	2015 (Average)		*P*-Value
	Control	**Treatment**		**Control**	**Treatment**	
Entropy	0.23	0.25	0.15	0.23	0.24	0.32
Land tenure	0.51	0.45	0.20	0.53	0.57	0.47
Farm size	$3356	$3714	0.27	$3277	$3768	0.14
Operator age	52	54	0.40	54	56	0.13
Operator education	2.93	2.96	0.73	3.06	2.88	0.03
Farm Typology	2.53	2.66	0.14	2.41	2.49	0.38
Farming dependent county	0.20	0.17	0.34	0.20	0.13	0.08
Urban-rural county index	5.28	5.25	0.88	5.27	5.54	0.21
Observations					Control	Treatment
Number of corn contracts					2118	885
Percentage of contracts in each group (%)					71	29

Note: Control group are counties without an ethanol plant while treated group are counties with an ethanol plant.

3. Results and Discussion

Non-spatial and spatial difference-in-differences models are estimated to analyze whether farmers located in the same county or adjacent to counties with ethanol plants incur additional corn revenues than those located in counties without an ethanol plant. Estimation results of the non-spatial and a spatial difference-in-differences (DID) models are presented in Table 3. Both models have an R^2 of 0.35, meaning that they explain 35% of the variation (i.e., the distance between the actual value and the mean). The presence of misspecification and omitted variable bias were tested for and rejected at a 5% statistical significance level. The standard errors are larger than standard errors produced by an ordinary least squares regression due to the clustering of the standard errors. The non-spatial and a spatial difference-in-differences (DID) models provided coefficients of similar magnitudes and of the same sign. In the spatial DID model, the coefficient associated with the spatial weight matrix and treatment dummy and the time dummy (Wdt) was statistically significant, meaning that the model did not revert to the non-spatial DID model [25]. Additionally, not controlling for the spatial treatment correlation would have rendered the treatment estimate of the non-spatial difference-in-difference model biased and inconsistent [25]. Since the dependent variable, corn revenue, was logged, the coefficients displayed in Table 3 are interpreted as the percentage change in corn revenue (e.g., if the farm is intermediate its corn revenue would be 33% higher than that of a small farm).

The difference-in-differences (DID) effect in both the non-spatial and spatial model were not statistically significant, indicating that we were unable to find a statistically significant impact from ethanol plants on corn contract revenues in their host counties (Table 3). The DID effect of the non-spatial difference-in-differences model is equivalent to the average treatment effect of the spatial difference-in-differences model. The average direct (ADTE) and average indirect (AITE) treatment effects of the spatial difference in differences model are displayed in the bottom section of Table 3. In the spatial model, ADTE was not statistically significant, but AITE was, meaning that ethanol plant siting has positive spillover effects onto adjacent counties (Table 3).

The results show that the higher revenues observed in counties with an ethanol plant are due to factors other than the presence of an ethanol plant. These may be counties with an excess supply of corn and hold a competitive advantage over corn production. These results eco the findings of Sarmiento et al. [8] and Lambert et al. [9] that find that the site selection of an ethanol plant is predominantly based on feedstock availability. Another possible explanation for our results is that farmers in these counties may also have lower bargaining power and be willing to accept a lower corn contract price, making up for their losses by increasing the quantity of corn supplied. Factors related to the farm business and the county have a statistically significant impact on local corn contract revenues. Farms with over three million dollars in assets received 65% more in corn revenues than those with smaller assets. Intermediate farms (those with sales up to $249,999) had 33% higher corn revenues than small family farms, while commercial farms (those with sales of $250,000 or more) earned 123% more in corn revenues than small family farms (for information on the farm typology, see Hoppe and MacDonald [37]). This difference in revenues may be a result of economies of scale in corn production. Given that ethanol plants are more likely to locate in regions with higher corn yields and with more areas of corn production [8], results seem to indicate that these regions are the ones attracting ethanol plants.

Table 3. Non-spatial and spatial difference-in-differences (DID) estimation results with treatment being counties with an ethanol plant from 2015 to 2016.

| | Log Corn Contract Revenue | | | |
| | DID OLS | | DID Spatial | |
	Coefficient	Standard Error	Coefficient	Standard Error
Time dummy	−0.019	0.110	−0.071	0.117
Ethanol plant dummy	−0.032	0.127	−0.034	0.127
Interaction of weight matrix with treatment and time dummy (Wdt)			0.010 **	0.004
Ethanol plant dummy * time dummy	0.162	0.185	0.155	0.183
Entropy	−0.585	0.536	−0.605	0.528
Tenure dummy	−0.455 ***	0.111	−0.460 ***	0.111
Farm size dummy	0.652 ***	0.086	0.650 ***	0.086
Intermediate farm	0.334 **	0.157	0.332 **	0.155
Commercial farm	1.226 ***	0.147	1.228 ***	0.147
Operator Age	0.002	0.005	0.002	0.005
Operator education	−0.048	0.054	−0.050	0.053
Farming dependent	0.060	0.127	0.056	0.127
Urban-rural county index	0.045 *	0.023	0.044 *	0.024
Low employment county	1.100 ***	0.155	1.096 ***	0.153
Constant	9.408 ***	0.456	9.444 ***	0.450
Average direct treatment effect (ADTE)				0.155
Average treatment effect (ATE)				0.187
Average indirect treatment effect (AITE)				0.032**
Number of observations		3003		3003
R^2		0.35		0.35
Counties		629		629

Notes: * 10%, ** 5%, and *** 1% levels of statistical significance. Standard errors are clustered at the county level. DID effect is equivalent to the average direct treatment effect (ADTE) in the spatial model. Wdt is the coefficient associated with the interaction of the weight matrix with the treatment and time dummy.

Farmers that owned over 50% of the land they operated received 46% less in corn revenues than those that operated mostly on rented land. This difference may be due to rental rates pressuring farmers that lease farmland to be more profitable [38]. Entropy, operator's age, and education did not have a statistically significant effect on local corn contract revenues. County characteristics also help explain the difference in revenues received by farmers. Corn revenues were higher in more rural locations. For instance, a change in the urban–rural index from 5 to 6 was associated with a 5% increase in corn revenues. This is likely a result of lower rental rates in less urban areas [39] and less off-farm jobs opportunities [40]. Lower rental rates may allow farmers to operate at a larger scale, and the lack of off-farm job opportunities may offer farmers more time to dedicate to farming. The positive effects from county characteristics (e.g., low employment county and urban-rural county index) aligned with farmers and farm characteristics may point to these counties with ethanol plants having a competitive advantage on corn production with respect to other counties.

The AITE shows that farmers in counties adjacent to counties with ethanol plants experienced an increase of 3% in their corn contract revenues (Table 3). A possible explanation for the positive effect on neighboring counties could be the expansion of corn production into these counties. This situation was, in fact, experienced in Brazil. During the 21st century sugarcane production expanded from the state of Sao Paulo, the leading state in ethanol and sugar production, to counties that bordered the state [41,42]. Farmers switched from grain and livestock production to sugarcane expanding sugarcane production to adjacent counties [41,42]. The size of the effect on corn contract revenues in adjacent counties resembles the changes in corn acreage found by Li et al. [43] due to increases in ethanol production capacity. The authors find that an increase in ethanol capacity can lead to a 3% increase in corn acreage. They also argue that corn price changes have little effect on corn acreage. As such, the benefits from being adjacent to counties with ethanol plants come from increases in the quantity of inputs sold and not from changes in prices, as Figure 2a,b indicate.

Costs associated with feedstock supply may influence the location of an ethanol plant, as such, observed ethanol plants may not be randomly assigned [24]. We repeated the analysis using instrumental variables for the endogenous variables: Ethanol plant dummy and ethanol plant dummy * time dummy. Similar to Miao [13] we use the ethanol production capacity aggregated at the state level. Although the state ethanol production capacity is correlated with the county-level ethanol production capacity, the state level is not dependent on the capacity of a single county [13]. We also use as instruments the interaction between the time dummy and the state ethanol production capacity and the U.S. Census index for low employment county in 2000. Information on the state ethanol production capacity comes from the Energy Information Administration [44]. Given the increase in the number of biofuel refineries since 2000 and the time it takes for a plant to be built and put into service, the general economy of the county in 2000 would influence an ethanol plant siting. Lower employment opportunities could mean lower labor costs for the ethanol plant. The index is not, however, expected to influence farmers' revenues in the years of 2015 and 2016 for the decision to grow corn can be made in a shorter time period (i.e., an annual decision). The results using the instrumental variables are similar to those in Table 3. The DID effect continues to be statistically insignificant and the statistically significant coefficients have the same signs and are similar in magnitude (Tables 3 and 4).

Table 4. Results from the two stage least square estimation for the years 2015 to 2016.

	Log Corn Contract Revenue		
	Coefficient		Standard Error
Time dummy	−0.204		0.504
Ethanol plant dummy	1.248		1.465
Ethanol plant dummy * time dummy	1.030		1.606
Entropy	−0.579		0.597
Tenure dummy	−0.393	***	0.134
Farm size dummy	0.488	***	0.150
Intermediate farm	0.349	*	0.201
Commercial farm	1.279	***	0.184
Operator age	−0.001		0.005
Operator education	−0.038		0.067
Farming dependent	0.102		0.226
Urban-rural county index	0.029		0.036
Constant	9.235	***	0.611
Kleibergen-Paap rk Wald F statistic			9.127
Hansen J statistic			1.966
			0.161
RMSE			1.36
Number of observations			3003
Counties			629

RMSE are the root mean squared errors. *P*-values are in the brackets. Standard errors are clustered at the county level. Instrumental variables are ethanol production capacity at the state level, the interaction between the time dummy and the state level ethanol production capacity, and the index for low employment county from the 2000 U.S. Census. Hansen J statistic test of the validity of over-identifying restrictions does not reject the null hypothesis that instruments are exogenous. The Kleibergen–Paap test is close to 10 and may indicate a good instrument [45].

The lack of an effect of ethanol plant location on local corn contract revenues supports ideas presented in other studies as to: (1) Other drivers rather than ethanol plant location affect changes in planted corn acreage [10]; (2) effects of the ethanol plant location may be temporary, dissipating in months [4] and as such would not be captured in this model; (3) counties with higher specialization in corn production attract a new ethanol plant [8]; or (4) there are no corn price premiums in counties with an ethanol plant [2]. It is important to highlight that our results are limited to revenues from corn contracts. We focused on this form of procurement since it has been shown to be an advantageous and main form of procurement of feedstock used by ethanol plants [46]. As such, our results may not be expanded to interpret the effects on other corn markets (i.e., spot markets). We leave this analysis to future studies.

4. Conclusions

The impact of ethanol plants on corn prices and corn acreage has been the topic of various studies. This study investigates a phenomenon observed between counties with ethanol plants and those without ethanol plants. That is, the average corn revenues in counties with an ethanol plant are higher than in counties without, at similar corn prices. Non-spatial and spatial difference-in-differences models are estimated using ARMS data. The models account for changes over two-years in corn contract revenues and estimate the impact of ethanol plants on local corn contract revenues. The effect from the presence of ethanol plants on the corn contract revenues in their counties was inconclusive. The higher revenues observed are probably due to factors related to the operator (i.e., capital and tenure), size of the operation, and characteristics of the county. Results suggest that ethanol plants may be located in areas with competitive advantage in corn production (i.e., areas specialized in corn production and with higher corn yields). While we do not see a statistically significant impact from ethanol plants on corn revenue in the county where the ethanol plant is located, we do see a positive impact in adjacent counties. It is likely that although the county where the ethanol plant is located may have a competitive advantage in corn production, its neighbors may not. As such, neighboring counties face gains in revenues with the location of the ethanol plant.

Author Contributions: Conceptualization, A.L.K. and A.C.S.A; methodology, A.L.K and A.C.S.A; data curation, A.L.K and A.C.S.A; formal analysis, A.L.K and A.C.S.A; writing—original draft preparation, A.L.K and A.C.S.A; writing—review and editing, A.L.K and A.C.S.A; visualization, A.C.S.A; supervision, A.L.K.

Funding: This research received no external funding.

Acknowledgments: We thank the anonymous reviewers and the editor. We thank the comments given by those present at the 2019 Australian Agricultural and Resource Economics Society Meeting. We thank Assistant Professor of Geography Dr. Gabriel Granco for his suggestions for the spatial weight matrix.

Conflicts of Interest: No conflicts of interest to declare.

References

1. RFA. Annual Industry Outlook. Available online: http://www.ethanolrfa.org/resources/publications/outlook/ (accessed on 29 May 2017).
2. Katchova, A.L. The spatial effect of ethanol biorefinery locations on local corn prices. In Proceedings of the Agricultural and Applied Economics Association's 2009 Annual Meeting, Milwaukee, WI, USA, 26–28 July 2009.
3. Miller, E.; Mallory, M.L.; Baylis, K.R.; Hart, C.E. Basis effects of ethanol plants in the U.S. Corn Belt. In Proceedings of the Agricultural & Applied Economics Association's 2012 AAEA Annual Meeting, Seattle, WA, USA, 12–14 August 2012; p. 23.
4. Lewis, K.E. *The Impact of Ethanol on Corn Market Relationships and Corn Price Basis Levels*; Michigan State University: East Lansing, MI, USA, 2010.
5. McNew, K.; Griffith, D. Measuring the Impact of Ethanol Plants on Local Grain Prices on JSTOR. *Rev. Agric. Econ.* **2005**, *27*, 164–180. [CrossRef]

6.   O'Brien, D.M. The Effects of the Micro-Market Structure for Kansas Grain Elevators on Spatial Grain Price Differentials. In Proceedings of the NCCC-134 Conference on Applied Commodity Price Analysis, Forecasting, and Market Risk Management, St. Louis, MO, USA, 20–21 April 2009.

7.   Behnke, K.; Fortenbery, T.R. The Impact of Ethanol Production on Local Corn Basis. In Proceedings of the NCCC-134 Applied Commodity Price Analysis, Forecasting, and Market Risk Management, St. Louis, MO, USA, 18–19 April 2011.

8.   Sarmiento, C.; Wilson, W.W.; Dahl, B. Spatial Competition and Ethanol Plant Location Decisions: Spatial Competition and Ethanol Plant Location. *Agribusiness* **2012**, *28*, 260–273. [CrossRef]

9.   Lambert, D.M.; Wilcox, M.; English, A.; Stewart, L. Ethanol plant location determinants and county comparative advantage. *J. Agric. Appl. Econ.* **2008**, *40*, 117–135. [CrossRef]

10.   Ifft, J.; Rajagopal, D.; Weldzuis, R. Ethanol Plant Location and Land Use: A Case Study of CRP and the Ethanol Mandate. *Appl. Econ. Perspect. Policy* **2019**, *41*, 37–55. [CrossRef]

11.   Fatal, Y.S.; Thurman, W.N. The Response of Corn Acreage to Ethanol Plant Siting. *J. Agric. Appl. Econ.* **2014**, *46*, 157–171. [CrossRef]

12.   Motamed, M.; McPhail, L.; Williams, R. Corn area response to local ethanol markets in the United States: A grid cell level analysis. *Am. J. Agric. Econ.* **2016**, *98*, 726–743. [CrossRef]

13.   Miao, R. Impact of Ethanol Plants on Local Land Use Change. *Agric. Resour. Econ. Rev.* **2013**, *42*, 291–309. [CrossRef]

14.   Du, X.; Hennessy, D.; Edwards, W.A. Does a rising biofuels tide raise all boats? A study of cash rent determinants for Iowa farmland under hay and pasture. *J. Agric. Food Ind. Organ.* **2008**, *6*. [CrossRef]

15.   Miller, E. How Does Changing Ethanol Capacity Affect Local Corn Basis? Available online: https://policymatters.illinois.edu/how-does-changing-ethanol-capacity-affect-local-corn-basis/ (accessed on 17 December 2019).

16.   Gallagher, P.; Wisner, R.; Brubacker, H. Price relationships in processors' input market areas: Testing theories for corn prices near ethanol plants. *Can. J. Agric. Econ.* **2005**, *53*, 117–139. [CrossRef]

17.   Fatal, Y.S. *Ethanol Plant Siting and the Corn Market*; North Carolina State University: Raleigh, NC, USA, 2011.

18.   Haddad, M.A.; Taylor, G.; Owusu, F. Locational Choices of the Ethanol Industry in the Midwest Corn Belt. *Econ. Dev. Q.* **2010**, *24*, 74–86. [CrossRef]

19.   Rask, K.N. Clean air and renewable fuels: The market for fuel ethanol in the US from 1984 to 1993. *Energy Econ.* **1998**, *20*, 325–345. [CrossRef]

20.   Blank, D. Will Farmers Benefit from Shelbyville's Ethanol Plant? Maybe. Available online: https://www.batesvilleheraldtribune.com/news/local_news/will-farmers-benefit-from-shelbyville-s-ethanol-plant-maybe/article_2d814894-e148-5304-9480-52150fe49d1f.html (accessed on 17 December 2019).

21.   Hargreaves, S. Calming ethanol-crazed corn prices. *CNNMoney*, 30 January 2007.

22.   Katchova, A.L. Agricultural Contracts and Alternative Marketing Options: A Matching Analysis. *J. Agric. Appl. Econ.* **2010**, *42*, 261–276. [CrossRef]

23.   Porter, M.E. *The Competitive Advantage: Creating and Sustaining Superior Performance*; Free Press: New York, NY, USA, 1985.

24.   Granco, G.; Sant'Anna, A.C.; Bergtold, J.S.; Caldas, M.M. Factors influencing ethanol mill location in a new sugarcane producing region in Brazil. *Biomass Bioenergy* **2018**, *111*, 125–133. [CrossRef]

25.   Delgado, M.S.; Florax, R.J.G.M. Difference-in-differences techniques for spatial data: Local autocorrelation and spatial interaction. *Econ. Lett.* **2015**, *137*, 123–126. [CrossRef]

26.   Ramsey, J.B. Tests for Specification Errors in Classical Linear Least-Squares Regression Analysis. *J. R. Stat. Soc.* **1969**, *31*, 350–371. [CrossRef]

27.   Pregibon, D. *Data Analytic Methods for Generalized Linear Models*; University of Toronto: Toronto, ON, Canada, 1979.

28.   StataCorp. *Stata 14 Base Reference Manual*; Stata Press: College Station, TX, USA, 2015.

29.   Merryman, S. *USSWM: Stata Module to Provide US State and County Spatial Weight (Contiguity) Matrices*; Statistical Software Components S448405; Boston College Department of Economics: Boston, MA, USA, 2008.

30.   ARMS Team. ARMS Farm Financial and Crop Production Practices—Documentation. Available online: https://www.ers.usda.gov/data-products/arms-farm-financial-and-crop-production-practices/documentation/#about (accessed on 8 November 2019).

31. Nebraska Energy Office Energy Statistics. Available online: http://www.neo.ne.gov/ (accessed on 1 October 2017).
32. MacDonald, J.M.; Perry, J.; Ahearn, M.C.; Banker, D.; Chambers, W.; Dimitri, C.; Key, N.; Nelson, K.E.; Southard, L.W. *Contracts, Markets, and Prices: Organizing the Production and Use of Agricultural Commodities*; United States Department of Agriculture: Washington, DC, USA, 2004.
33. Sesmero, J.P.; Perrin, R.K.; Fulginiti, L.E. A Variable Cost Function for Corn Ethanol Plants in the Midwest: A Variable Cost Function for Corn Ethanol. *Can. J. Agric. Econ.* **2016**, *64*, 565–587. [CrossRef]
34. MacDonald, J.M.; Korb, P. *Agricultural Contracting Update: Contracts in 2008*; Economic Information Bulletin; DIANE Publishing: Collingdale, PA, USA, 2011; p. 43.
35. Katchova, A.L.; Miranda, M.J. Two-Step Econometric Estimation of Farm Characteristics Affecting Marketing Contract Decisions. *Am. J. Agric. Econ.* **2004**, *86*, 88–102. [CrossRef]
36. Mishra, A.; Tegegne, F.; Sandretto, C.L. The Impact of Participation in Cooperatives on the Success of Small Farms. *J. Agribus.* **2004**, *22*, 59604.
37. Hoppe, R.A.; MacDonald, J.M. Updating the ERS farm typology. *USDA-ERS Econ. Inf. Bull.* **2013**. [CrossRef]
38. Garcia, P.; Sonka, S.T.; Yoo, M.S. Farm Size, Tenure, and Economic Efficiency in a Sample of Illinois Grain Farms. *Am. J. Agric. Econ.* **1982**, *64*, 119–123. [CrossRef]
39. Kuethe, T.; Ifft, J.; Morehart, M.J. The Influence of Urban Areas on Farmland Values. *Choices* **2011**, *26*, 1–7.
40. Mishra, A.K.; El-Osta, H.S.; Morehart, M.; Johnson, J.; Hopkins, J. *Income, Wealth, and the Economic Well-Being of Farm Households*; Resource Economics Division, Economic Research Service, U.S. Department of Agriculture: Washington, DC, USA, 2002; p. 77.
41. Granco, G.; Caldas, M.M.; Bergtold, J.S.; Sant'Anna, A.C. Exploring the policy and social factors fueling the expansion and shift of sugarcane production in the Brazilian Cerrado. *GeoJournal* **2017**, *82*, 63–80. [CrossRef]
42. Sant'Anna, A.C.; Shanoyan, A.; Bergtold, J.S.; Caldas, M.M.; Granco, G. Ethanol and sugarcane expansion in Brazil: What is fueling the ethanol industry? *Int. Food Agribus. Manag. Rev.* **2016**, *19*, 163–182. [CrossRef]
43. Li, Y.; Miao, R.; Khanna, M. Effects of Ethanol Plant Proximity and Crop Prices on Land-Use Change in the United States. *Am. J. Agric. Econ.* **2019**, *101*, 467–491. [CrossRef]
44. USEIA. U.S. Fuel Ethanol Plant Production Capacity. Available online: https://www.eia.gov/petroleum/ethanolcapacity/ (accessed on 26 July 2018).
45. Boberg-Fazlić, N.; Sharp, P. Does Welfare Spending Crowd Out Charitable Activity? Evidence from Historical England Under the Poor Laws. *Econ. J.* **2015**, *127*, 50–83. [CrossRef]
46. Schmidgall, T.J.; Tudor, K.W.; Spaulding, A.D.; Winter, J.R. Ethanol marketing and input procurement practices of US ethanol producers: 2008 survey results. *Int. Food Agribus. Manag. Rev.* **2010**, *13*, 137–156.

Article

Ammonia Volatilization Losses during Irrigation of Liquid Animal Manure

John P. Chastain

Department of Agricultural Sciences, Agricultural Mechanization and Business Program, Clemson University, McAdams Hall, Clemson, SC 29634, USA; jchstn@clemson.edu

Received: 16 September 2019; Accepted: 4 November 2019; Published: 5 November 2019

Abstract: Ammonia loss resulting from land application of liquid animal manure varies depending on the composition of the manure and the method used to apply manure to cropland. High levels of ammonia volatilization result in an economic loss to the farmer based on the value of the nitrogen and have also been shown to be a source of air pollution. Using irrigation as a method of applying liquid manure to cropland has generally been accepted as a method that increases the volatilization of ammonia. However, only three studies available in the literature measured the amount of ammonia lost during the irrigation process. Only one of the three studies concluded that ammonia loss during irrigation was significant. A pooled statistical and uncertainty analysis of the 55 available observations was performed to determine if ammonia loss occurred during irrigation of animal manure. Data on the total solids content of the manure were also included as an indicator of evaporation losses. Volatilization losses during irrigation were not found to be statistically significant, and evaporation losses were small, 2.4%, and agreed with previous studies on irrigation performance. Furthermore, the range of ammonia loss reported in previous studies was determined to be within the errors associated with the measurement of total ammoniacal nitrogen concentrations and the calculation of per cent differences.

Keywords: ammonia loss; land application; manure management; irrigation

---

1. Introduction

Ammonia loss to the atmosphere following manure application contributes to air pollution and is a loss of valuable fertilizer nitrogen. Broadcast application of liquid or slurry manure without incorporation can result in ammonia losses ranging from 11% to 70% of applied total ammoniacal nitrogen (TAN = NH_4^+-N + NH_3-N) [1,2]. A review of the literature by Meisinger, and Jokela [3] indicated that the main factors that determine the magnitude of ammonia loss were the total solids content of the manure (TS, %), the amount of time that elapsed following an application before incorporation into soil or rainfall, and whether the manure was applied to bare soil or a crop. In general, application of manure with a low TS (1% or less) to bare soil resulted in the lowest ammonia loss since a greater portion of the TAN in the manure infiltrated into the soil with water instead of remaining on the surface.

Several land application practices have been shown to reduce or nearly eliminate the ammonia losses associated with a broadcast application of slurry of manure (6% to 10% TS). The most common approaches were to use methods that provided incorporation into the soil using light tillage on the day of manure application, direct sub-surface injection of manure, use of implements that provided some means of immediate incorporation, or spreading of manure in bands [4–6]. The reduction in ammonia losses as compared to a surface application of manure varied from 30% for incorporation of manure on the same day to 98% for sub-surface manure injection. Many of the banding techniques (e.g., towed hose or shoe) provided ammonia loss reductions of 30% to 70% on grasslands [2,4,6].

Sprinkler irrigation of liquid animal manure onto crop, forage, or pasture land to recycle plant nutrients is a common practice in many regions of the United States. The practice is most common on dairy and swine farms that use large amounts of water to remove manure from the animal housing area on a daily (flush) or weekly basis (pit-recharge). Liquid manure from the buildings is treated and stored prior to reuse for manure removal from the animal housing area. Treatment systems can be configured in a variety of ways with the two most common being treatment in a single anaerobic lagoon and one or more stages of solid-liquid separation followed by a treatment lagoon. In some cases, two or more treatment lagoons are used in series to provide a higher level of biological treatment. Irrigation is the favoured method of liquid manure application due to lower labour cost, energy cost, reduced soil compaction, and higher speed of application as compared to application using a tractor and tank-type spreader [7] (p.99). The primary disadvantages of using irrigation equipment to land apply liquid manure are the high initial investment, the potential for increased odour generation, and the possibility of spraying manure outside the field onto a road or a neighbour's property. Proper design and operation of the irrigation equipment can minimize issues of over-spray to areas outside the field that is intended to be fertilized.

Unlike other methods of manure application, ammonia loss can potentially occur between the time the liquid exits the nozzle and lands on the soil or crop, that is during the irrigation process, and after the manure is applied to the ground. In one study, lagoon water with a TS of 0.57% or less resulted in ammonia losses of 0.4% to 3.6% of the TAN applied following application by irrigation [8]. These results suggest that most volatilization of ammonia occurred after manure was applied to the ground, and that irrigation of treated liquid manure (lagoon supernatant) facilitated the reduction in ammonia loss as compared to a broadcast of untreated liquid manure with a higher solids content. In some extension publications (e.g., [8], p. 99), and in a few research articles [9–11] it has been asserted that using traveling gun or centre pivot irrigation to apply manure to cropland increased the amount of ammonia volatilized to the air as compared to broadcast without incorporation. Westermann et al. [10] reported ammonia losses of 5.7% of the total ammoniacal nitrogen on the average with maximum losses of 24% when using a traveling gun to apply liquid swine manure. A previous study by Safley et al. [11] reported average ammonia losses of 2.9% to 4.9% of TAN applied when concentrations of the irrigation source was compared to concentrations in the manure obtained from containers on the ground used to catch the irrigated manure. When the data were analysed to make an estimate that included what was termed evaporation and drift they reported ammonia losses as high as 37% of the TAN contained in lagoon effluent [11]. Earlier work by Welsh [12] that also compared TAN concentrations in irrigated and ground collected samples concluded that volatilization losses during the irrigation of dairy slurry, liquid swine manure, and effluent from an oxidation ditch were insignificant. A more recent study by Montes [13] agreed with Welsh [12] and concluded that ammonia loss did not occur during sprinkler irrigation of swine lagoon water. However, the three studies that indicated that additional ammonia loss occurred during irrigation [9–11], along with endorsement of the idea in some cooperative extension literature [7] has led to a general acceptance of the idea that, regardless of the level of manure treatment implemented on a farm, using irrigation as a method to land apply liquid manure increased ammonia volatilization to the atmosphere and was to be avoided.

The level of physical (solid–liquid separation) and biological treatment (anaerobic lagoon or biological N removal) used prior to land application of dairy and swine manure has been shown to have a significant impact on the concentration of total solids, TAN, and the total Kjeldahl nitrogen (TKN = TAN + Organic-N) as shown by the data from several case studies provided in Table 1. Significant reductions in the TAN concentration reduced the mass of ammonia that could be lost during irrigation. The reduction in TS was also accompanied by a reduction in volatile solids which was often associated with a reduction in odour. Data from a dairy facility that used two stages of solid–liquid separation followed by a treatment lagoon resulted in 93% lower TS content and 54% lower TAN concentration than the manure flushed from the animal housing area [14]. Experiments were also performed to show that application of a polyacrylamide polymer (PAM) to screened dairy manure prior to settling could provide TS and TAN concentrations for the separated liquids that were similar to those achieved by lagoon treatment [14]. Treatment of swine manure in a single stage lagoon yielded a 75% reduction in TS concentration and a 69% reduction in TAN [15] as compared to untreated manure from the building. An advanced manure treatment on a swine farm that included chemically enhanced solid-liquid separation followed by biological treatment for nitrogen (nitrification and denitrification) and chemical treatment for phosphorous provided a 66% reduction in TS and a 96% reduction in TAN as compared to untreated manure [16]. In addition, comparison of surface water and agitated lagoon sludge and liquids from dairy and swine lagoons in South Carolina [17–20], California [21], Kansas [22], and Texas [23] indicated that lagoon treatment provided significant reductions in TS and TAN concentrations as compared to untreated manure. The highest TS and TAN concentrations shown in Table 1 were for lagoons located in regions of the USA with a dry climate [21–23] where evaporation tends to increase the concentrations due to loss of volume in the treatment system.

Table 1. Example concentrations of total solids (TS), total Kjeldahl nitrogen (TKN), total ammoniacal nitrogen (TAN), and TAN/TKN for liquid dairy and swine manure as removed from buildings and after various levels of treatment.

Description	TS (%)	TKN (mg/L)	TAN (mg/L)	TAN/TKN (%)
Multi-stage treatment on a dairy farm [14]				
As removed from animal housing	3.8	1433	661	46
After solid-liquid separation by screening	1.5	729	359	49
After gravity settling	1.1	703	399	57
After lagoon treatment—surface liquid	0.27	373	303	81
After screening and gravity settling with polymer (250 to 400 mg polyacrylamide polymer (PAM/L))	0.26–0.43	374–481	272–247	73–51
Single-stage lagoon treatment on a swine finishing farm [15]				
As removed from animal housing	2.0	2397	1666	70
After lagoon treatment—surface liquid	0.50	760	520	68
Advanced multi-step treatment on a swine finishing farm [16]				
As removed from animal housing	2.9	2007	1251	62
After polymer enhanced solid–liquid separation	1.4	1414	1190	84
After biological N treatment (nitrification and denitrification)	0.95	121	103	85
After phosphorus treatment	0.97	83	43	52
Effluent from treatment lagoons				
Dairy—surface water [17,21]	0.60–0.85	599–670	360–383	54–64
Dairy—agitated sludge and liquids [19–21,23]	4.0–7.5	918–2565	138–434	15–17
Swine—surface water [18,22]	0.37–1.3	576–1852	408–1506	71–87
Swine—agitated sludge and liquids [18,20]	2.2–3.7	960–2012	393–467	20–49

The TAN concentration is not the only factor that determines the amount of ammonia that could be lost during irrigation of liquid manure since only the fraction of total ammoniacal nitrogen that is in the ammonia form can be volatilized. The percentage of the TAN in the ammonia form has been shown to be a function of pH and temperature [24,25] as shown in Figure 1. Most liquid animal manure has a pH in the range of 7.0 to 8.0. Therefore, at a temperature of 25 °C the percentage of TAN in the ammonia form is in the range of 0.6% to 5.4%, Figure 1a. For liquid manure with a pH of 8.0, the percentage of TAN that could be lost as ammonia gas ranges from 5.4% at 25 °C to 13.4% at 40 °C, Figure 1b. The practical upper limit for ammonia loss as a percentage of TAN applied is 10% since most manure is not spread to fertilize cropland during hot weather. An ammonia loss of 10% during irrigation of liquid animal manure would require all ammonia in the liquid manure to be lost from the time it exits the sprinkler nozzle and before it strikes the ground. Therefore, ammonia losses during irrigation greater than 10% of the applied TAN in liquid manure were judged to be unlikely and nonhomogeneity in the liquid manure, or uncertainties in measurement or calculation may have confounded some of the observations.

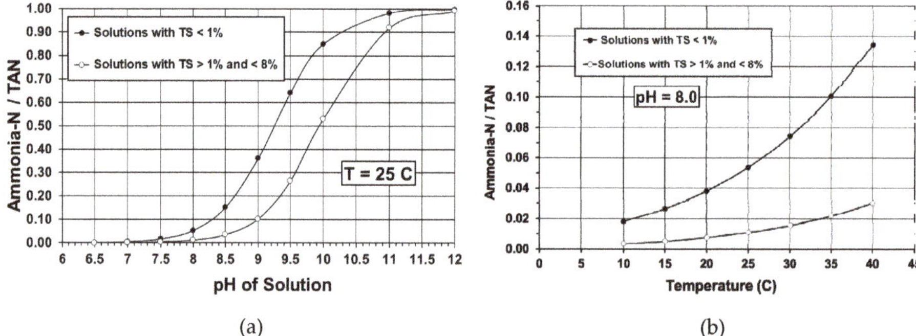

Figure 1. Impact of pH (**a**, temperature held at 25 °C) and temperature (**b**, pH held at 8.0) on the fraction of total ammoniacal nitrogen (TAN = NH_4^+-N + NH_3-N) in ammonia form for liquid animal manure (adapted from Denmead et al. [24] and Zhang [25]).

Only three studies [11–13] were found that had the quantification of ammonia loss during irrigation as a primary objective, and only one study concluded that ammonia loss was significant [12]. Only one of these studies [13] included statistical and error analyses. Therefore, the objective of this study was to perform a pooled statistical analysis of the available data related to ammonia volatilization and evaporation losses during sprinkler irrigation of liquid animal manure.

2. Materials and Methods

A summary of the available data on ammonia volatilization loss during irrigation of liquid animal manure is presented in Table 2. Ammonia volatilization losses were calculated from the data reported by the authors based on the difference between the irrigated ($[TAN_I]$) and ground collected, ($[TAN_G]$) concentrations of total ammoniacal nitrogen. Ammonia losses, as a percentage of TAN applied ($[TAN_I]$), ranged from −33% to 26%, and the mean ammonia loss ranged from −2.5% to 13% across all studies shown.

The values of pH reported in these studies ranged from 7.1 to 8.6 with an average of about 7.7. Data were not provided on air temperature or wind speed in these studies. Some of the pH values reflect an increase in pH during irrigation [11] while others simply reported a range. Comparison of the mean pH of 7.7 with the relationships provided in Figure 1 indicated that the fraction of the TAN in the ammonia form was in the range of 6% to 10%. Therefore, if all ammonia was lost as the liquid travelled through the air the ammonia loss would be 10% or less of the TAN applied.

Negative ammonia loss values implied that TAN concentrations increased during irrigation. This was only possible if evaporation of water was substantial. Overall, the data indicated that a significant amount of uncertainty in the quantification of ammonia losses existed. The factors that have been proposed to affect the magnitude of ammonia loss from the time manure exited the sprinkler nozzle until it was collected in containers on the ground include air temperature, relative humidity, irrigation operating pressure, drop diameter, spray velocity, TAN content of the irrigated material, and pH [9,24,26,27]. These factors have been suggested as the cause of the variability in measuring ammonia volatilization losses. However, most of the authors did not report data on these factors or perform an error analysis on their data collection procedures.

Table 2. Summary of available data on volatilization losses during sprinkler irrigation of liquid animal manure.

Description and Source	Irrigated TAN (ppm)	Irrigated TS (%)	pH	Ammonia Loss (% of TAN Applied)	n
Big Gun: Dairy, Beef (treated), Swine [12]	11 to 850	0.28 to 8.39	7.4 to 7.9	−2.5 (−12.4 to 9.8)	5
Centre Pivot: Swine [11]	299 to 327	0.14 to 0.17	7.4 to 7.5	4.9 (−2.1 to 18.4)	12
Big Gun: Swine [11]	214 to 510	0.11 to 0.37	7.1 to 7.7	2.9 (0.5 to 9.4)	6
Big Gun: Swine [10]	NR [1]	NR	NR	5.7 (−5.0 to 24)	3
Solid Set: Swine [9]	53	NR	NR	13	NR
Solid Set: Swine [13]	110 to 1183	0.04 to 0.57	7.6 to 8.6	0.3 (−33 to 26)	32

[1] NR = not reported.

In the investigation by Welsh [12], samples were taken from the storage or treatment structure before irrigation and from samples collected from several containers of unknown diameter on the ground following the irrigation event. The difference in average TAN concentration from the source and the containers was used to estimate NH_3-N loss that occurred between the time the manure exited the nozzle and when it struck the ground. This study, conducted in Minnesota, included four different liquid manure types with very different characteristics as was reflected by the large range in total solids and TAN concentrations, Table 2. The average ammonia loss was −2.5% and was reported as not significantly different from zero [12].

Safley et al. [11] studied ammonia losses during irrigation of swine lagoon supernatant using centre pivot and traveling gun irrigation equipment in North Carolina. Ammonia losses were estimated by calculating the difference in TAN concentration between samples taken from the top 0.6 m of depth in the lagoon, and samples taken from liquid caught on the ground during irrigation using rain gauges with a diameter of 95 mm. The TAN concentration difference between irrigated and ground collected samples in the data presented by Safley et al. [11] ranged from −2.1% to 18.4% with a mean of 2.9% for the large bore sprinkler (big gun) and 4.9% for the centre-pivot.

Montes [13] collected ammonia volatilization data for sprinkler irrigation from two swine lagoons in South Carolina. Montes collected irrigated lagoon water samples from a sampling port in the irrigation pipe on the discharge side of the irrigation pump. The ground collected samples were the composite of samples collected in 8 locations within the irrigated plots using short plastic containers with a diameter of about 152 mm.

The studies by Westermann et al. [10], and Sharpe and Harper [9] did not include all the data required to be included in the present study and were excluded. All data included in the analysis are tabulated in Appendix A.

The data from the studies by Welsh [12], Safley et al. [11], and Montes [13] were pooled into common linear regression analyses. The quantities that were included were concentrations of TS, TAN, and TKN. The change in TS between the irrigated and ground collected samples was included to provide a measure of evaporation loss. Both TAN and TKN were included since a significant reduction in TAN or loss of water by evaporation during irrigation would be expected to result in a change in TKN.

3. Results and Discussion

Pooled linear regression analyses were performed for the irrigated and ground collected concentrations of TAN, TKN, and TS. The least-squares best fit for each constituent was represented by the following equation form:

$$[C_G] = b\,[C_I] \tag{1}$$

where: $[C_I]$ = the concentration of TAN, TKN, or TS in the irrigated manure; $[C_G]$ = the concentration of TAN, TKN, or TS in the manure collected on the ground; and b = the slope of the line.

The y-intercept in Equation (1) was set to zero because it was impossible for the concentration of TAN, TKN, or TS in the manure collected from containers on the ground, [C_G], to have a value greater than zero if the corresponding concentrations in the irrigated manure, [C_I], were zero. Therefore, the analysis was performed to force Equation (1) through the origin and force all error into the value of the slope, *b*.

An analysis of variance (ANOVA) was performed for each regression [28]. The slope of the equation, *b*, was compared to 1.0 using a t-test at the 95% confidence level since a slope of 1.0 represented no change in concentration during irrigation. Correlations for irrigated versus ground collected TAN, TS, and TKN concentrations are provided in Figure 2. The results for the three analyses of variance are given in Table 3.

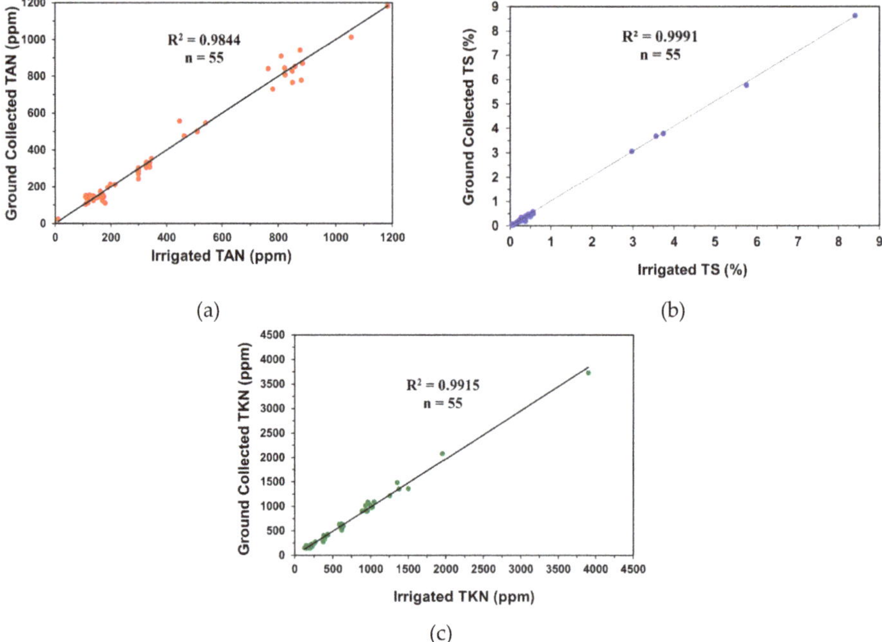

Figure 2. Comparison of irrigated and ground collected concentrations of TAN (**a**), TS (**b**), and TKN (**c**) for irrigated manure.

Table 3. Results of the analysis of variance of the regression using Equation (1) for comparison of irrigated and ground collected concentrations of TS, TAN, and TKN (n = 55, residual degrees of freedom = 54).

Irrigated Concentration	R^2	*b*	SE *b* [1]	C.I. (*b*) [2]	SE y [3]
[TS_I], 0.04% to 8.39%	0.9991	1.024*	0.004	± 0.008	0.046 %
[TAN_I], 11 to 1183 ppm	0.9844	0.9999	0.010	± 0.021	39.3 ppm
[TKN_I], 128 to 3900 ppm	0.9915	0.9846	0.009	± 0.018	56.0 ppm

[1] Standard error of *b*. [2] 95% confidence interval about *b*. [3] Standard error of the y-estimate. * Significantly different from 1.0 at the 95% level.

3.1. Influence of Irrigation on TAN—Ammonia Loss

The effect of the irrigation process on the TAN concentration of liquid animal manure is shown in Figure 2a, and the slope of the regression line was not significantly different from 1.0 at the 95% level (Table 3). As a result, the pooled analysis of 55 observations indicated that ammonia volatilization loss during irrigation was not statistically significant for manure with TS ranging from 0.04 to 8.39% TS, and TAN concentrations ranging from 11 to 1183 ppm.

The differences between TAN concentrations in irrigated and ground collected samples ([TAN$_I$] – [TAN$_G$]) were sometimes negative as indicated in Table 2 and Figure 2a. Since the statistical analysis indicated that the concentrations were not significantly different these negative values were due to the uncertainty, or lack of accuracy, in the measurements of TAN concentration.

The procedure to determine TAN concentration for irrigated and ground collected samples included the following potential sources of error: nonhomogeneity of the liquid manure, sampling in the field, sub-sampling in the laboratory to prepare aliquots for chemical analysis, and execution of the chemical analysis procedures. Each step had an associated error that contributed to the overall error in determining TAN concentration.

An estimate of the magnitude of overall error in determining TAN was made based on the variability in TAN concentration of samples taken from similar materials and conditions. The estimate of uncertainty in TAN measurements was based on the pooled variance of 965.3 (ppm)^2 based on 62 observations of TAN provided by Montes [13]. The estimate of uncertainty in TAN concentration was the pooled standard deviation of ± 31.1 ppm.

Calculation of the volatilization loss in per cent required taking the difference between the irrigated and ground collected concentrations. The uncertainty in the difference between two measured values was estimated as [29,30]:

$$u_{(a-b)} = \sqrt{(u_a)^2 + (u_b)^2} \qquad (2)$$

where: $u_{(a-b)}$ = uncertainty in knowing the difference between a and b; u_a = uncertainty in measuring *a*; and u_b = uncertainty in measuring *b*.

Using Equation (2) and the defined uncertainty for TAN (± 31.1 ppm), the uncertainty in per cent difference in concentrations between irrigated and ground collected samples (U $_{\Delta TAN}$) was estimated as:

$$U_{\Delta TAN} = (\pm\ 44\ ppm \div [TAN_I]) \times 100. \qquad (3)$$

The uncertainty interval for TAN loss defined by Equation (3) is plotted in Figure 3 with all the data included in the present study. These results indicated that volatilization losses were well distributed about the line of zero difference. Ten of the 55 data points were not contained within the uncertainty interval for TAN. These results support the statistical conclusion and indicate that volatilization losses were not significant within the errors induced by calculation of a per cent loss and the errors associated with measurement.

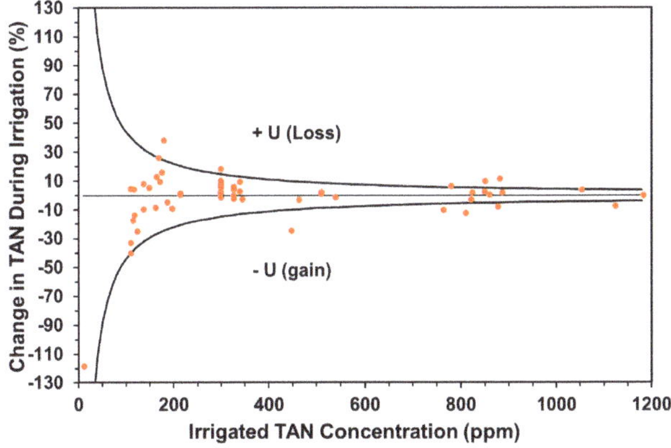

Figure 3. Comparison of the change in total ammoniacal nitrogen concentration during irrigation with the uncertainty associated with the calculation of per cent differences (± U $_{\Delta TAN}$, Equation (3)).

3.2. Influence of Irrigation on TS—Evaporation Loss

The correlation results between the ground collected and irrigated concentrations of total solids were given previously in Figure 2b and Table 3. A t-test on the slope for the TS relationship indicated that a slope of 1.024 was significantly different from 1.0. Therefore, evaporation during irrigation increased the TS of the ground collected sample by 2.4%. Both empirical and modelling studies have observed evaporation losses during irrigation in the range of 1.0% to 3.5% [31,32]. The observation from this study agreed with the literature.

3.3. Influence of Irrigation on TKN

Total Kjeldahl nitrogen is the sum of TAN and organic nitrogen. Therefore, the TKN concentration in the ground collected sample would be expected to be slightly higher, giving a slope greater than 1.0, even if ammonia volatilization did not occur due to small, but significant, evaporation losses. However, the correlation analysis summarized in Figure 2c and Table 3 indicated that the TKN concentrations were not significantly influenced by irrigation at the 95% level. It appears that the uncertainties associated with measuring TKN concentrations, similar to those discussed for TAN, overshadowed the impact of the small amount of evaporation that was observed.

3.4. Comparison of the Ammonia Loss Results with Efforts to Include Evaporation and Drift

Safley et al. [11] attempted to incorporate the influence of evaporation and drift into the estimation of ammonia losses during irrigation using a centre pivot equipped with impact sprinklers. They reported that the ammonia losses during irrigation of lagoon supernatant ranged from 13.9% to 37.3% of TAN applied if evaporation and drift were included. However, their concentration data indicated that volatilization losses averaged 4.9% for 12 observations (Table 2). The irrigate-catch technique to estimate volume loss during irrigation was used by Safley et al. [11]. The error in the irrigate-catch technique was described as a recovery error (R_E) defined as [31]:

$$R_E = [\,1 - (A_G/A_I)\,] \times 100 \tag{4}$$

where: A_G = measured application depth (cm); A_I = application depth (cm) based on flow measurements in the main irrigation pipe and the application area; and (A_G/A_I) = fraction of the actual irrigation depth (A_I) recovered in containers on the ground.

Safley et al. [11] incorporated Equation (4) in their calculation of ammonia loss (TAN $_{LOSS}$) during irrigation as:

$$\text{TAN}_{LOSS} = (1 - (A_G/A_I)\,([\text{TAN}_G]/[\text{TAN}_I])) \times 100. \tag{5}$$

If one notes that the irrigation depths in Equation (5) are for the same application area, the equation was an attempt to observe ammonia loss based on the mass of TAN collected on the ground versus the mass of TAN irrigated. Results obtained by this technique need to be interpreted with caution since all errors in (A_G/A_I) were counted as an irrigation recovery error (Equation (4)). The recovery error defined in Equation (4) included the following effects: (1) collection error, E_C; (2) error due to the lack of uniformity of the irrigation system, E_U; and (3) error caused by evaporation loss, E_E.

The collection error, E_C, was caused by liquid that drifted away from the collection containers, liquid that struck the collection containers but was not trapped, liquid lost by splashing out of the collection containers, and evaporation from the collection containers. A collection error related to the type of container used was explicitly measured by Kohl [33]. Kohl showed that the collection error for 76 mm diameter, funnel-type rain gauges (typical height of 304 mm) ranged from 85% at an application rate of 0.09 cm/h to 12% at a rate of 0.94 cm/h when compared to a precise collecting device ($E_C \approx 0$).

The error induced by lack of uniformity, E_U, was directly related to the design of irrigation equipment, and the number and distribution of collection containers used to capture the spray. Centre pivot irrigation equipment typically provides an application uniformity that varies from 70% to

90% [34]. For design purposes, 80% is typically used as the application uniformity which yielded an E_U of 20% [34].

Evaporation error from sprinkler spray, E_E, depended on system pressure and droplet size, and has been observed to be small when compared with the effects of irrigation uniformity [31]. Empirical and modelling studies have shown that evaporation losses from irrigation systems varied from 1.0% to 3.5% [31,32]. The value used for E_E in the present analysis was 2.0%.

The recovery error was estimated from the three common sources of irrigation calibration error to provide an independent estimate of the recovery error that was previously defined in Equation (4). This independent estimate of recovery error, R_E, was calculated as [29,30]:

$$RE = \sqrt{(E_C)^2 + (E_U)^2 + (E_E)^2}. \tag{6}$$

Safley et al. [11] used 95 mm rain gages to measure the application depth, A_G, from a centre pivot irrigation system with an average application rate of 1.1 cm/h. Assuming a collection error of 12%, a uniformity error of 20%, and an evaporation error of 2% in Equation (6) yielded a recovery error (R_E) of 23% for a centre pivot irrigation system. Evaporation from the sprinkler spray accounted for only 0.7% of the total recovery error while uniformity error contributed 73%.

Setting R_E equal to 23% in Equation (4) and solving for (A_G/A_I) indicated that one would expect to recover 0.77 A_I for a typical centre pivot irrigation system. The average fraction recovered observed by Safley et al. [11] was 0.77 indicating that their centre pivot performed as expected. Safley et al. [11] erroneously attributed the 23% recovery error, to evaporation and drift losses during irrigation.

As shown in Table 2, the average TAN loss for Safley's center pivot study was 4.9%, which sets ([TAN$_G$]/[TAN$_I$]) equal to 0.951, and the mean value of (A_G/A_I) was 0.77. As a result, the average TAN loss reported by Safley et al. [11] using Equation (5) was 26.8%. However, most of the average ammonia loss predicted using Equation (5) was due to volume collection error in the irrigate-catch technique and not evaporation and drift as assumed by Safley et al. [11].

4. Conclusions

Ammonia volatilization data from three independent studies [11–13] were pooled to determine if ammonia loss was significant during irrigation of liquid animal manure. The concentrations of TAN in the irrigated manure ranged from 11 to 1183 ppm. The corresponding range of total solids in the irrigated manure ranged from 0.04% to 8.39%. The following conclusions were developed based on the results.

1. Irrigation of liquid animal manure increased the TS concentration by 2.4%. Evaporation was small, but statistically significant and agreed with expectations for centre pivot irrigation using rain gauges to measure irrigation depth.
2. Irrigation of liquid animal manure did not significantly influence the concentration of TAN or TKN in the ground collected samples and the slopes of the regression equations were not significantly different from 1.0. Therefore, ammonia volatilization loss during irrigation was not statistically significant.
3. The per cent difference between irrigated and ground collected TAN concentrations was within the errors associated with measurement of TAN concentrations, and calculation of per cent differences.
4. Authors of a previous study attempted to calculate the impact of evaporation and drift on ammonia losses during center pivot irrigation of lagoon water. However, there efforts were confounded by irrigation volume measurement error (recovery error).

While the results of this study concluded that ammonia volatilization was not significant during irrigation, it does not imply that ammonia volatilization after the manure strikes the ground is to be ignored. The suitability of irrigation as a liquid manure application technique should be evaluated based on the level of treatment provided, the solids content of the manure, and the potential for odour impact on neighbours. The irrigation system should be designed and operated to prevent drift, or overspray onto roads, or adjacent property owned by neighbours. In addition, any method of manure application must be carried out to prevent manure from being applied so as to impair surface water. Irrigation may still be a suitable and cost-effective method to apply large quantities of liquid manure to utilize the plant nutrients for crop production in cases where physical and biological treatment is provided. Application methods that reduce ammonia loss following application, such as immediate incorporation, direct injection, band application, or similar methods that reduce ammonia loss are generally recommended if slurry manure or agitated lagoon sludge is to be used as a fertilizer substitute.

Author Contributions: The author is responsible for the content and data analysis provided in this paper.

Funding: This research received no external funding.

Conflicts of Interest: The author declares no conflict of interest.

Appendix A

The raw data used in the analysis are summarized in Tables A1 and A2.

Table A1. Concentrations of TS, TAN, and TKN in the irrigated manure and the ground-collected samples obtained from a studies conducted in Minnesota (Welsh [12]), and North Carolina (Safley et al [11]).

Source	$[TS_I]$ (%)	$[TS_G]$ (%)	$[TAN_I]$ (ppm)	$[TAN_G]$ (ppm)	$[TKN_I]$ (ppm)	$[TKN_G]$ (ppm)
Big Gun, Dairy, Beef, Swine [12]	8.39	8.65	850	767	3900	3733
	0.29	0.31	11	24	377	336
	3.73	3.80	345	354	1500	1358
	3.55	3.70	187	196	960	1090
	5.74	5.79	810	910	1953	2083
Big Gun, Swine Lagoon [11]	0.26	0.22	340	330	431	434
	0.26	0.36	340	308	431	414
	0.37	0.20	510	499	617	516
	0.37	0.25	510	504	617	538
	0.11	0.11	214	211	251	246
	0.11	0.10	214	213	251	243
Center Pivot, Swine Lagoon [11]	0.17	0.15	299	269	391	338
	0.17	0.17	299	274	391	359
	0.17	0.16	299	283	391	383
	0.17	0.17	299	279	391	388
	0.16	0.14	327	313	379	344
	0.16	0.17	327	307	379	351
	0.16	0.17	327	328	379	371
	0.16	0.17	327	334	379	406
	0.14	0.12	299	244	372	274
	0.14	0.14	299	291	372	329
	0.14	0.14	299	296	372	324
	0.14	0.14	299	303	372	335

Table A2. Concentrations of TS, TAN, and TKN in the irrigated manure and the ground-collected samples obtained from two swine lagoons in South Carolina (Montes [13]).

Source	$[TS_I]$ (%)	$[TS_G]$ (%)	$[TAN_I]$ (ppm)	$[TAN_G]$ (ppm)	$[TKN_I]$ (ppm)	$[TKN_G]$ (ppm)
Lagoon A, Solid-Set Impact Sprinkler	0.49	0.39	859	854	1026	985
	0.44	0.46	880	779	941	900
	0.44	0.42	780	731	929	915
	0.44	0.5	764	842	883	905
	0.47	0.45	849	828	951	907
	0.49	0.47	824	808	928	1014
	0.57	0.52	1054	1014	1252	1214
	0.5	0.48	1183	1183	1352	1485
	0.49	0.48	1124	1206	1378	1352
	0.54	0.53	876	943	1045	1093
	0.53	0.54	885	870	986	1037
	0.55	0.6	822	845	1009	977
	0.37	0.4	463	477	583	636
	0.37	0.42	447	557	614	640
	0.39	0.41	540	547	637	597
Lagoon B, Solid-Set, Impact Sprinkler	0.14	0.11	169	125	209	181
	0.22	0.2	164	143	232	193
	0.23	0.26	162	176	208	219
	0.21	0.23	197	215	269	274
	0.05	0.06	115	135	162	146
	0.05	0.05	137	126	174	153
	0.04	0.05	124	155	162	181
	0.04	0.08	175	147	216	224
	0.09	0.08	149	141	165	168
	0.06	0.08	118	134	128	155
	0.06	0.06	179	111	200	146
	0.05	0.08	110	154	161	176
	0.08	0.08	171	155	214	171
	0.07	0.06	137	150	152	201
	0.06	0.06	110	105	150	146
	0.06	0.06	117	112	137	162
	0.05	0.06	109	145	158	155

References

1. Sommer, S.G.; Olsen, J.E. Effects of dry matter content and temperature on ammonia loss from surface-applied cattle slurry. *J. Environ. Qual.* **1991**, *20*, 679–683. [CrossRef]
2. Sommer, G.S.; Friis, E.; Bach, A.; Schoriing, J. Ammonia volatilization from pig slurry applied with trail hoses or broad spread to winter wheat: Effects of crop developmental stage, microclimate, and leaf ammonia absorption. *J. Environ. Qual.* **1997**, *26*, 1153–1160. [CrossRef]
3. Meisinger, J.J.; Jokela, W.E. Ammonia Volatilization from Dairy and Poultry Manure. In *Managing Nutrients and Pathogens from Animal Agriculture*; NRAES-130; Natural Resource, Agriculture, and Engineering Service (NRAES), Cooperative Extension, Cornell University: Ithaca, NY, USA, 2000; pp. 334–354.
4. Pain, B.F.; Misselbrook, T.H. Sources of Variation in Ammonia Emission Factors for Manure Applications to Grassland. In *Gaseous Nitrogen Emissions from Grasslands*; Jarvis, S.C., Pain, B.F., Eds.; CAB Internat: Oxon, UK, 1997; pp. 293–301.
5. Frost, J. Effect of spreading method, application rate and dilution on ammonia volatilization from cattle slurry. *Grass Forage Sci.* **1994**, *49*, 391–400. [CrossRef]
6. Huijsmans, J.F.M.; Hol, J.M.G.; Bussink, D.W. Reduction of Ammonia Emission by New Slurry Application Techniques on Grassland. In *Gaseous Nitrogen Emissions from Grasslands*; Jarvis, S.C., Pain, B.F., Eds.; CAB Internat: Oxon, UK, 1997; pp. 281–285.

7. Dougherty, M.; Geohring, L.D.; Wright, P.; Lash, M. *Liquid Manure Application Systems Design Manual*; NRAES-89; NRAES, Cornell University Cooperative Extension: Ithaca, NY, USA, 1998.

8. Montes, F.; Chastain, J.P. Ammonia Volatilization Losses Following Irrigation of Liquid Swine Manure in Commercial Pine Plantations. In *Animal, Agricultural and Food Processing Wastes IX, Proceedings of the Ninth International Symposium, Research Triangle Park, NC, USA, 12–15 October 2003*; Burnes, R.T., Ed.; ASABE: St. Joseph, MI, USA, 2003; pp. 620–628.

9. Sharpe, R.R.; Harper, L.A. Ammonia and nitrous oxide emissions from sprinkler irrigation applications of swine effluent. *J. Environ. Qual.* **1997**, *26*, 1703–1706. [CrossRef]

10. Westerman, P.W.; Huffman, R.F.; Baker, J.C. Environmental and Agronomic Evaluation of Applying Swine Lagoon Effluent to Coastal Bermudagrass for Intensive Grazing and Hay. In Proceedings of the Seventh International Symposium, Chicago, IL, USA, 18–20 June 1995; pp. 150–161.

11. Safley, L.M.; Barker, J.C.; Westerman, P.W. Loss of nitrogen during sprinkler irrigation of swine lagoon liquid. *Bioresour. Technol.* **1992**, *40*, 7–15. [CrossRef]

12. Welsh, S.K. The Effect of Sprinkling on the Physical and Chemical Properties of Liquid Animal Waste. Master's Thesis, University of Minnesota, St. Paul, MN, USA, 1973.

13. Montes, F. Ammonia Volatilization Resulting from Application of Liquid Swine Manure and Turkey Litter in Commercial Pine Plantations. Master's Thesis, Clemson University, Clemson, SC, USA, 2002.

14. Chastain, J.P.; Vanotti, M.B.; Wingfield, M.M. Effectiveness of liquid-solid separation for treatment of flushed dairy manure: A case study. *Appl. Eng. Agric.* **2001**, *17*, 343–354. [CrossRef]

15. Chastain, J.P.; Lucas, W.D.; Albrecht, J.E.; Pardue, J.C.; Adams, J., III; Moore, K.P. Removal of solids and major plant nutrients from swine manure using a screw press separator. *Appl. Eng. Agric.* **2001**, *17*, 355–363. [CrossRef]

16. Vanotti, M.B.; Szogi, A.A.; Millner, P.D.; Loughrin, J.H. Development of a second generation environmentally superior technology for treatment of swine manure in the USA. *Bioresour. Technol.* **2009**, *100*, 5406–5416. [CrossRef] [PubMed]

17. Chastain, J.P.; Camberato, J.J. Dairy Manure Production and Nutrient Content. In *Confined Animal Manure Managers Certification Program Manual: Dairy Version*; Clemson University Extension: Clemson, SC, USA, 2004; pp. 3a-1–3a-16. Available online: https://www.clemson.edu/extension/camm/manuals/dairy_toc.html (accessed on 2 September 2019).

18. Chastain, J.P.; Camberato, J.J.; Albrecht, J.E.; Adams, J., III. Swine Manure Production and Nutrient Content. In *Confined Animal Manure Managers Certification Program Manual: Swine Version II*; Clemson University Extension: Clemson, SC, USA, 1999; pp. 3-1–3-17. Available online: https://www.clemson.edu/extension/camm/manuals/swine_toc.html (accessed on 2 September 2019).

19. Chastain, J.P.; Darby, J.A., Jr. A Thickening Process for Reducing the Cost of Utilizing Dairy Lagoon Sludge. In *Animal, Agricultural and Food Processing Wastes, Proceedings of the Eighth International Symposium, Des Moines, IA, USA, 9–11 October 2000*; Moore, A., Ed.; ASABE: St. Joseph, MI, USA, 2000; pp. 694–701.

20. Cantrell, K.B.; Chastain, J.P.; Moore, K.P. Geotextile filtration performance for lagoon sludges and liquid animal manures dewatering. *Trans. ASABE* **2008**, *51*, 1067–1076. [CrossRef]

21. Pettygrove, G.S.; Heinrich, A.L.; Eagle, A.J. Dairy Manure Nutrient Content and Forms. In *Manure Technical Bulletin Series*; University of California Cooperative Extension: Davis, CA, USA, 2010; Available online: http://manuremanagement.ucdavis.edu/files/134369.pdf. (accessed on 2 September 2019).

22. DeRouchey, J.M.; Goodband, R.D.; Nelssen, J.L.; Tokach, S.S.; Murphy, J.P. Nutrient composition of Kansas swine lagoon and hoop barn manure. *J. Anim. Sci.* **2002**, *80*, 2051–2061. [CrossRef] [PubMed]

23. Mukhtar, S.; Ullman, J.L.; Auvermann, B.W.; Feagley, S.E.; Carpenter, T.A. Impact of anaerobic lagoon management on sludge accumulation and nutrient content for dairies. *Trans. ASAE* **2004**, *47*, 251–257. [CrossRef]

24. Denmead, O.T.; Frenney, J.R.; Simpson, J.R. Dynamics of ammonia volatilization during furrow irrigation of maize. *Soil Sci. Soc. Am. J.* **1982**, *46*, 149–155. [CrossRef]

25. Zhang, R.H. Degradation of Swine Manure and a Computer Model for Predicting the Desorption Rate of Ammonia from an Under-floor Pit. Ph.D. Thesis, University of Illinois, Urbana-Champaign, IL, USA, 1992.

26. Pote, J.W.; Miner, J.R.; Koelliker, J.K. Ammonia losses during sprinkler application of animal wastes. *Trans. ASAE* **1980**, *23*, 1202–1206. [CrossRef]

27. Brunke, R.; Alvo, P.; Schuepp, P.; Gordon, R. Effect of meteorological parameters on ammonia loss from manure in the field. *J. Environ. Qual.* **1988**, *17*, 431–436. [CrossRef]

28. Steel, R.G.D.; Torrie, J.H.; Dickey, D. *Principles and Procedures of Statistics: A Biometrical Approach*, 3rd ed.; McGraw-Hill Book Company: New York, NY, USA, 1997.

29. Taylor, J.R. *An Introduction to Error Analysis: The Study of Uncertainties in Physical Measurements*, 2nd ed.; University Science Books: Sausalito, CA, USA, 1997.

30. Holman, J.P. *Experimental Methods for Engineers*, 6th ed.; McGraw-Hill Book Company: New York, NY, USA, 1993.

31. Heermann, D.F.; Kohl, A. Fluid Dynamics of Sprinkler Systems. In *Design and Operation of Farm Irrigation Systems*; ASAE Monograph No., 3; Jensen, M.E., Ed.; ASABE: St. Joseph, MI, USA, 1980.

32. Thompson, A.L.; Gilley, J.R.; Norman, J.M. A Sprinkler water droplet evaporation and plant canopy model. *Trans. ASAE* **1993**, *36*, 735–750. [CrossRef]

33. Kohl, R. Sprinkler precipitation gage errors. *Trans. ASAE* **1972**, *15*, 264–265. [CrossRef]

34. Kruse, E.G.; Bucks, D.A.; Von Bernuth, R.D. Comparison of Irrigation Systems. *Agronomy* **1990**, *30*, 475–508.

Review

Agricultural Sustainability: A Review of Concepts and Methods

Maria G. Lampridi [1,2,*], **Claus G. Sørensen** [3] **and Dionysis Bochtis** [1]

[1]   Center for Research & Technology Hellas—CERTH, Institute for Bio-economy and Agri-technology—iBO, 10th km Thessalonikis—Thermis, BALKAN Center, BLDG D, 57001 Thessaloniki, Greece; d.bochtis@certh.gr

[2]   Faculty of Agriculture, Forestry and Natural Environment, School of Agriculture, Aristotle University of Thessaloniki, 54124 Thessaloniki, Greece

[3]   Department of Engineering, Aarhus University, Finlandsgade 22, DK8200 Aarhus N, Denmark; claus.soerensen@eng.au.dk

*   Correspondence: m.lampridi@certh.gr; Tel.: +030-2311-257-650

Received: 21 August 2019; Accepted: 18 September 2019; Published: 19 September 2019

Abstract: This paper presents a methodological framework for the systematic literature review of agricultural sustainability studies. The framework synthesizes all the available literature review criteria and introduces a two-level analysis facilitating systematization, data mining, and methodology analysis. The framework was implemented for the systematic literature review of 38 crop agricultural sustainability assessment studies at farm-level for the last decade. The investigation of the methodologies used is of particular importance since there are no standards or norms for the sustainability assessment of farming practices. The chronological analysis revealed that the scientific community's interest in agricultural sustainability is increasing in the last three years. The most used methods include indicator-based tools, frameworks, and indexes, followed by multicriteria methods. In the reviewed studies, stakeholder participation is proved crucial in the determination of the level of sustainability. It should also be mentioned that combinational use of methodologies is often observed, thus a clear distinction of methodologies is not always possible.

Keywords: agricultural sustainability; sustainability assessment; review

---

1. Introduction

The world's population is rapidly increasing and, according to the most recent projections, it is expected to reach 9.8 million in 2050 and 11.2 million in 2100 [1]. To that end, the planet should be ready to cope with the expected rapid population growth. Producing and delivering adequate, high quality food will be one of the most important challenges for humanity in the next century [2]. The evolution of technology has led to intensification of agricultural production leading to increased productivity and (in most of the cases) quality of agriproducts as well. However, this intensification has significantly increased the environmental footprint of agriculture, leading to a number of environmental impacts associated with the extensive use of fertilizers, pesticides, water, changes in land use, etc. [3]. The environmental issues related to agriculture have drawn the attention of the scientific community, which is now turning towards exploring the definition of agricultural sustainability without having yet reached consensus [4,5].

Undoubtingly, defining agricultural sustainability, as with every other sustainability concept, is a challenging task. Nevertheless, it is a common agreement that agricultural sustainability should at least address the three basic pillars of sustainable development by appraising simultaneously environmental, economic, and social issues related to agricultural practices [6]. However, the sustainability assessment of agricultural practices, in general, can be a very challenging task since it involves many case-specific variables to be taken under consideration.

Figure 1 presents various processes, inputs, and outputs involved in agricultural production, demonstrating the difficulty and complexity in generalizing the sustainability assessment process. There are general cultivation guidelines and corresponding operations stages for almost all crops (e.g. seeding, irrigation, and harvesting). However, the agronomic practice, the machinery types, the technology level, as well as the quantities and type of materials used may vary, depending on the type of crop, the implementation practice, the country (even the region of the cultivation), and the prevailing climatic conditions. All of the aforementioned parameters affect the cultivation process and the respective inflows and outflows.

It is obvious that the standardization of the Agricultural Sustainability Assessment is a challenging task. Considering the growing interest in assessing the sustainability issues related to agriculture, several tools and methodologies have been developed [7,8]. Among those tools some have gained greater acceptance and are widely used by the majority of practitioners worldwide, such as life cycle assessment (LCA), which is standardized by ISO in ISO 14040:2006 and ISO 14044:2006 [9]. In addition, many indicator-based methods have been developed for the sustainability assessment of agricultural practices that use different approaches with regards to the overall objective, the intended users, and the definition of agricultural sustainability they employ [4].

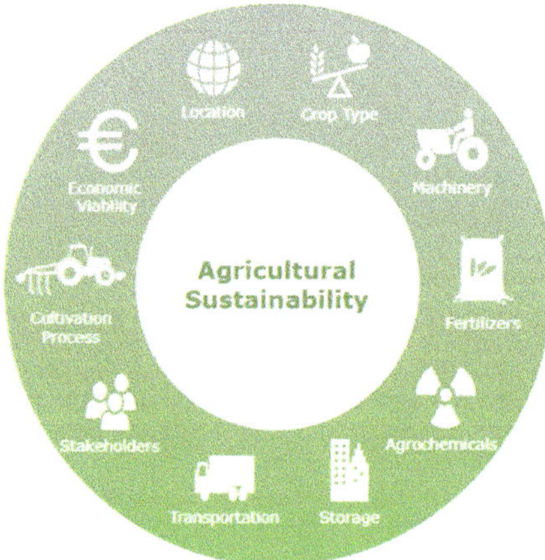

Figure 1. Variables involved in agricultural sustainability assessment.

Considering what was mentioned above and that there is not yet an established standardized methodology, it is very important for anyone attempting to assess agricultural sustainability to have an overview of the available and most usually used methodologies and tools to that scope. As a result, there is a need for a methodological framework that will help practitioners to evaluate the existing available tools and methods in order to select the appropriate one, for each specific task.

To that end, the present paper has a two-fold objective:

- To determine the evaluation criteria to systematically review agricultural sustainability assessment studies. To that end, several review papers were selected based on specific selection criteria and examined to determine the goal as well as the individual evaluation criteria adopted in each review. The ultimate goal is to critically synthesize a methodological framework for the systematic recording and evaluation of available agricultural sustainability assessment studies. Such systematic documentation can facilitate the comparison among the available studies as well as the development of a standard methodological framework for the sustainability assessment of agriculture.

- To implement the proposed methodology by investigating the available and mostly used methodologies to assess the sustainability of crop cultivations at the farm level. The methodological framework is applied to 38 Agricultural Sustainability studies published in peer-reviewed journals in the last decade (2009–2018).

2. Materials and Methods

2.1. Methodological Framework for the Systematic Review of Agricultural Sustainability Studies

2.1.1. Research Design

The evaluation process implemented to assess and select the criteria needed for the methodological framework of the systematic review on agricultural sustainability studies is presented in Figure 2. Initially, scientific literature published in Science Direct and Scopus was searched using the specific keywords and Boolean operators (AND/OR). The keywords were selected with respect to the integrated concept of "sustainability assessment", as well as the individual processes it consists of, namely, "environmental assessment", "economic assessment", and "societal assessment" (or "social assessment") combined with the keywords agriculture/farming using the Boolean Operator AND to exclude results that are not relevant to the field under examination. It should be added that the concept of "agricultural sustainability" was also included in the search.

The first sample of scientific papers that resulted from the initial search included 55 papers from peer-reviewed scientific journals. These papers were put through a screening process considering specific exclusion criteria presented in Figure 2. Specifically, studies that were not related to agriculture and especially focused on alternative agricultural processes were excluded. As a result, papers exclusively focused on aquaculture or organic farming studies, biofuels and biorefinery as well as review of studies comparing agronomic protocols were excluded from the present assessment. Additionally, review studies regarding soil quality, land management, food processing systems and discussions that did not specifically define the methods of the review conducted, were excluded. At this point, it should also be stated that in the context of agricultural sustainability studies, livestock farming was included in the search.

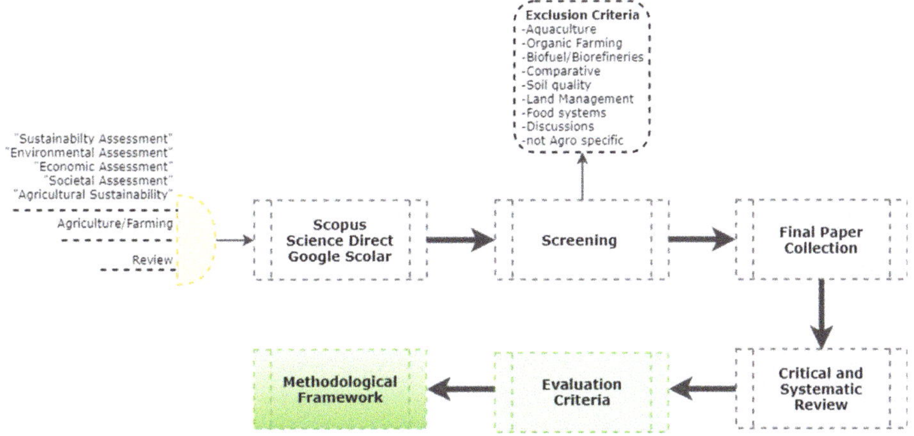

Figure 2. Evaluation process to introduce a methodological framework for the systematic review of agricultural sustainability studies.

The final paper collection comprises 16 review papers or studies that assess agricultural sustainability. It should be noted that the literature is relatively scarce regarding studies that consider all the three dimensions of sustainability with respect to other scientific fields, for example, the secondary production of goods. To that end, the sample includes studies considering the environmental aspect of agricultural sustainability which is the most often studied. The sample was then assessed in two ways, a systematic and critical [10]. The systematic way concerns the listing of the papers based on specifically defined criteria [11]. The initial listing criteria in the case of the presented framework, include the title and author of the paper, the year of publication as well as the spatial coverage of the study (Global or Regional) and the type of review (Critical or Systematic).

Critical reviews are thorough literature works that attempt to evaluate and assess the basic aspects or inputs and document the differences in methodology and implementation of scientific studies on a specific field [11]. In this case, the critical evaluation of the sample concerns the individual analysis of the selected studies with the purpose of extracting the individual evaluation criteria used in each study. The individual criteria with similar context were aggregated in a general table of criteria. Then, each paper was systematically reviewed as to whether each criterion was included in the review.

The resulting table is a comprehensive overview of the issues most frequently examined in a review study. The criteria that were used the most are the criteria that should be integrated in the methodological framework for the systematic review of agricultural sustainability studies. The rule followed in the present paper was to exclude criteria that were used in less than four papers. Following next is the sample presentation as well as the criteria frequency table along with a critical assessment of the sample used for the evaluation.

2.1.2. Systematic Approach

The 16 review papers that were extracted by the implementation of the first steps of the methodology, presented in the previous section are presented in Table 1 along with their classification with respect to their type and spatial coverage.

Table 1. Review studies examined.

a/a	Reference	Type of Review	Spatial Coverage
1	De Luca, A. I., et al., (2017) [10]	Critical and Systematic	Global
2	Binder, C.R., et al., (2010) [4]	Critical	Global
3	Peter, C., et al. (2017) [12]	Systematic	Global
4	Bockstaller, C., et al. (2009) [3]	Critical	Global
5	Roy, R., et al. (2012) [13]	Systematic	Country level - Bangladesh
6	Cerutti, A.K., et al. (2011) [7]	Systematic	Global
7	Acosta-Alba, I et al. (2011) [14]	Critical	Global
8	Baldini, C., et al. (2017) [15]	Critical and Systematic	Global
9	Morais, T.G., et al. (2016), [16]	Systematic	Country level - Portugal
10	McAuliffe, G.A., et al. (2016) [17]	Systematic	Global
11	Latruffe, L., et al. (2016) [18]	Critical	Global
12	Bockstaller, C., et al. (2008) [19]	Critical	Global
13	Payraudeau, S., et al. (2005) [20]	Systematic	Global
14	de Vries, M., et al. (2015) [21]	Systematic	Global
15	Yan, M.-J., et al. (2011) [9]	Systematic	Europe
16	Lebacq, T., et al. (2013) [22]	Critical	Global

As presented in Figure 3, during 2016–2017, the number of review papers has increased, indicating a boosted interest in the sustainability of agricultural practices. Payraudeau et al. (2005) first analyzed and systematically reviewed six (6) agricultural sustainability methods employed in eleven (11) case studies, indicating the variety of objectives, target groups, and methodologies used [20]. Bockstaller et al. (2008), followed by presenting a typology of indicators and the evolution of the methods used for their advancement [19], in 2009, critically evaluateing four (4) comparative studies to analyze the methods of the comparison, highlighting their main results [23]. Also focusing on indicators, Binder et al. (2010) presented an evaluation review framework that was used to review agricultural sustainability methods [4]. The framework assessed the normative, systematic and procedural aspects of the methods under evaluation.

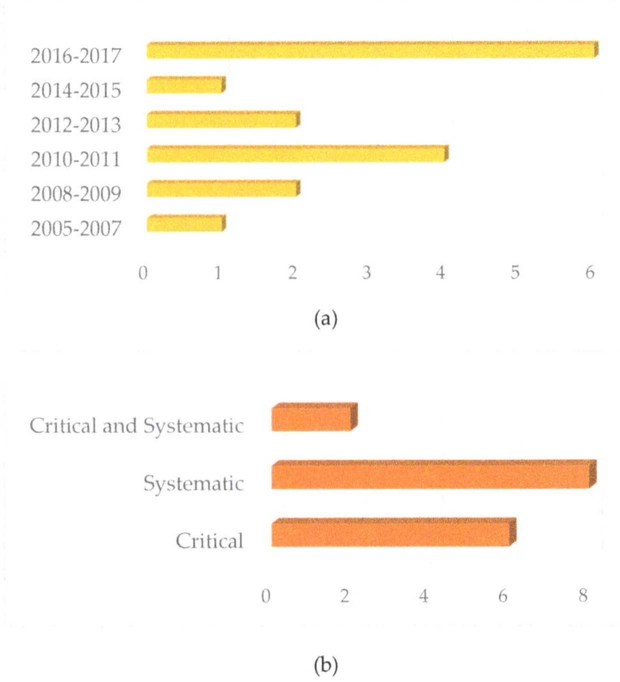

Figure 3. Sample presentation (review papers): (**a**) Number of papers per year and (**b**) type of paper.

Regarding the types of review papers and their classification to systematic or critical according to the definitions presented in the previous section [11], it is observed that, in principal, both categories are equally preferred by the researchers. However, in some cases, the distinction is not clear or a systematic and critical review is performed at the same time. Such example is the work of De Luca et al. (2017), where authors performed a critical and systematic review to determine, among other issues, which Multi Criteria Decision Analysis (MCDA) and participatory methods have been used along with LCA tools and the type of integration used in each case [10]. Also, Baldini et al. (2017) critically reviewed forty-four (44) LCA studies on milk production and systematically compared their methods and results to highlight issues requiring further discussion and investigation [15].

Considering the selected samples, it can be stated that in most cases systematic reviews are used in order to compare methodologies and results regarding a specific field of agricultural application. Towards this objective, Peter et al. (2017) performed a systematic evaluation of eighteen (18) carbon footprint calculators used in energy crop cultivations [12]. Cerutti et al. (2011) systematically reviewed twenty-two (22) fruit production sustainability assessment studies [7], whereas De Vries et al. (2015) systematically reviewed LCA studies on beef production [21]. Additionally, McAuliffe et al. (2015) conducted a chronological review of LCA studies in pig production, attempting to demonstrate how LCA has captured technological advancements in the field as well as the methodological issues observed [17].

On the contrary, the majority of the reviews that were characterized as critical are dealing with the evaluation of indicator-based methods or the classification of agricultural sustainability indicators, such as the work of Acosta-Alba et al. (2011), who reviewed eight (8) agricultural sustainability frameworks that use reference values for their indicators and analyzed the methods for the establishment of the reference values and investigating ways for their improvement [14]. Latruffe et al. (2016) provided a review of the available agricultural sustainability indicators, highlighting the relative high increase of environmental indicators as compared with the smaller interest in economic and social indicators [18]. Finally, Lebacq et al. (2013) reviewed the types of sustainability indicators and proposed indicative ground rules for the selection of agricultural sustainability indicators [22].

With respect to the spatial coverage of the reviews (Figure 4), the majority deals with studies from all around the world. Nevertheless, there are reviews assessing studies in specific countries or regions. For example, Roy et al. (2012), based on a systematic review and synthesis, presents a set of indicators that could be used to assess agricultural sustainability in Bangladesh, highlighting the need for integrated approaches and participatory processes during agricultural sustainability assessment [13]. Additionally, Morais et al. (2016) systematically reviewed twenty-two (22) agri-food-dedicated LCA studies in Portugal, revealing issues regarding the challenges faced and the lack of systematic regional approach in the country that could safeguard the accuracy and comparability of the results [16]. Lastly, Yan et al. (2011) reviewed thirteen (13) LCA studies on European milk production, indicating that direct comparison is challenging due to inconsistency regarding the used methodologies [9].

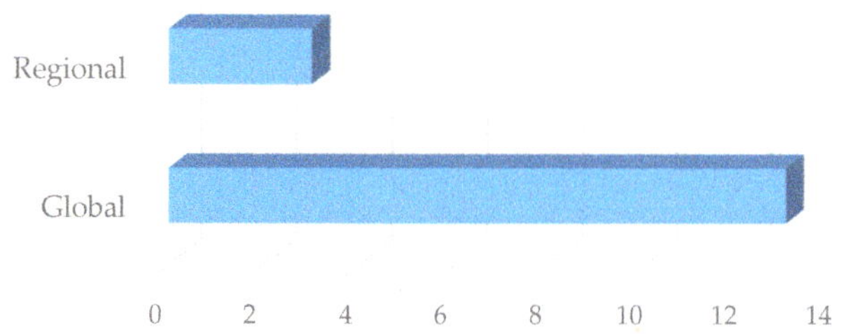

Figure 4. Spatial coverage of the papers reviewed.

2.1.3. Critical Approach

The selected sample, which was thoroughly described in the previous section, was screened, to extract the individual evaluation criteria used during each review. As some criteria had the same objective or were of the same context they were categorized accordingly. Also, some studies further analyzed the criteria including various subcriteria, but this is out of the scope of this paper since it is an issue related to the scrutiny of the review each author aims to achieve and the corresponding scope.

Figure 5 presents the criteria identified during the screening process and the frequency of their occurrence. A total of forty-four (44) different criteria were used in the sixteen (16) studies reviewed. The review criteria frequency table is presented in detail in Appendix A (Table A1). The first six criteria (beginning from the top of Figure 5) were common in most of the reviews examined and include the name and description of the assessment method or tool, the field of application, the country of application, and the year of issuing. The literature typology concerns the type of the document reviewed. For example, De Luca et al. (2017) classified the selected publications into three categories (Journal Article, Book Chapter, and Conference Proceedings paper) [10]. Baldini et al. (2017), on the other hand, refers to publication types classifying the sample according to whether the literature is an original article, a review, a research direction, or a scenario analysis [15].

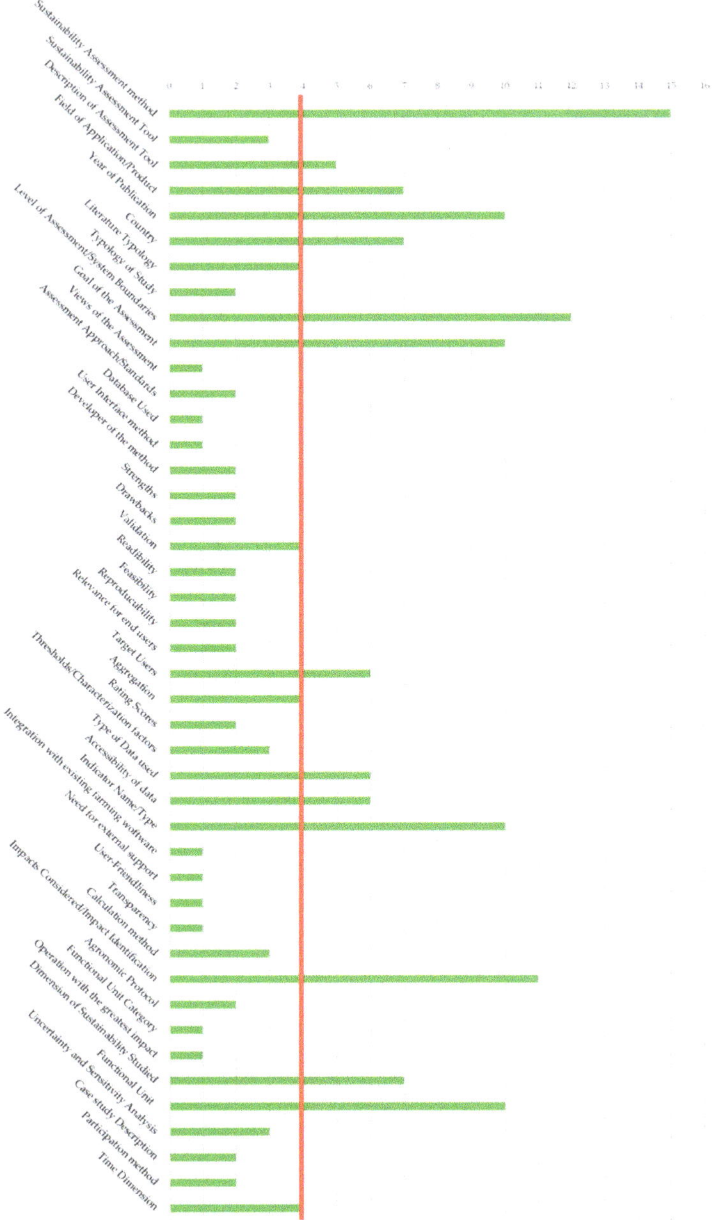

Figure 5. Criteria frequency use.

Regarding the level of assessment and the system boundaries, many different approaches were identified. Payraudeau et al. (2005), Roy et al. (2012), and de Luca et al. (2017) assessed the level of assessment of the studies reviewed regarding its spatial coverage (global, national, regional, local, and farm level) [10,13,20]. Lebacq et al. (2013) distinguished studies by whether they concern a farm or a region, and Bockstaller et al. (2009) examined the level of assessment giving two different scale options (field scale or farm scale) [22,23]. Cerutti et al. (2011) and Baldini et al. (2017) identified the system boundaries of the studies assessed [7,15], following a cradle-to-gate or cradle-to-market approach, whereas de Vries et al. (2015) reviewed studies at least from cradle-to-farm gate [21]. Lastly, Peter et al. (2017) examined both the level of assessment (global, regional, etc.) and the system boundaries (farm–gate or farm–gate–grave) of the studies they review [12].

The issue of the intended user of a method or tool is being considered in several of the studies reviewed. Binder et al. (2010) identified the target group of the examined methodologies [4], whereas de Luca et al. (2017) referred to the specific criterion as actors involved in the assessment process (i.e., local experts, scientists, workers, etc.) [10]. Bockstaller et al. (2008) classified the reviewed works according to the target user of the method reviewed, i.e. decision-maker, researcher, technician or farmer [19]. Considering the type and the accessibility of data criteria, Baldini et al. (2017) distinguish the data in experimental and model data [15]. The accessibility of data (or availability as expressed by Roy et al. 2012 [13]) is examined by Bockstaller et al. (2008) for three user groups, farmers, advisors, and administration [19].

With reference to the name and type of the indicators reviewed, many approaches were identified during the screening process. Lebacq et al. (2013) and Latruffe et al. (2016) categorized indicators based on their representation of the three pillars of sustainability (environmental, economic and social) [18,22]. Lebacq et al. (2013), however, extended the research scope by identifying means-based, system state, emission, and effect-based indicators [22]. Based on the calculation method Latruffe et al. (2016), Lebacq et al. (2013) and Bockstaller et al. (2008) distinguish indicators based on the method used for their calculation, i.e., single variables, emission factors, combination of variables, operational models, mechanistic models, etc. [18,19,22]. De Luca et al. (2017) categorize LCA indicators based on the impact categories they address [10].

Based on the rule set in the methodology section, the red line in Figure 5 presents the criteria exclusion threshold. Only criteria identified more than four times in the sample reviewed are included in the methodological approach for the systematic review of agricultural sustainability studies. A total of eighteen (18) criteria surpassed the exclusion threshold. These criteria are classified in groups with respect to their context and are presented in the subsequent section.

2.1.4. Methodological Framework Presentation

Following the criteria determination process described in the previous sections, Figure 6 presents the critical synthesis to systematically review agricultural sustainability related studies. The proposed methodological framework is based on a series of criteria and divided into five (5) underlying categories. The first two categories refer to the initial screening stage. During this preliminary stage, the studies are assessed to determine if the study will be included in the sample on the basis of the case-specific exclusion criteria determined with regards to the scope of the review.

The initial screening stage includes two categories (i.e., "method identification" and "general information") of criteria with respect to the basic description of each study. The general information of a study concerns the year of publication and the type of literature which can be journal article, conference proceedings paper, book chapter, technical report, etc., and the country that the study was conducted. The method identification category includes criteria that deal with the assessment method developed or employed. Therefore, the criterion description of the assessment tool describes the method or tool presented based on whether it is a presentation of a new methodology, the application of an existing method or tool or a combination namely a new methodology that is implemented with an application example. The last criterion is the level of the assessment performed, i.e., global, national, regional, or farm level, according to the approach introduced by Gomez-Limon et al. (2010) [24]. After the initial assessment and finalization, for the sample to be reviewed, phase is completed; the in-depth review stage follows. For this stage, three (3) categories of criteria have been defined.

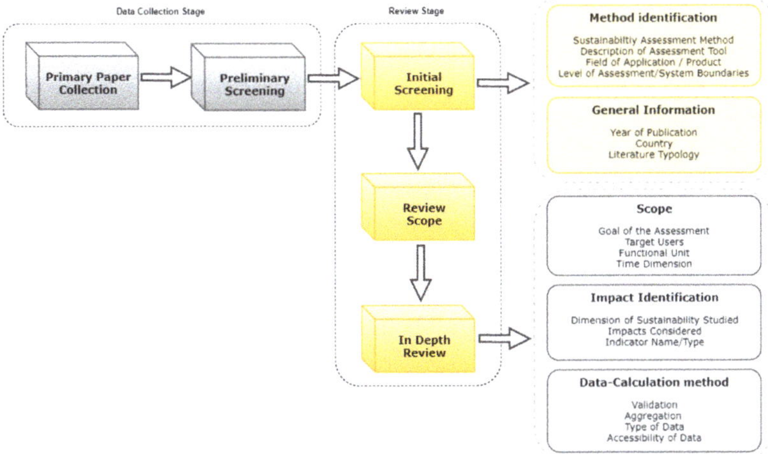

Figure 6. Methodological framework for the systematic review of agricultural sustainability studies.

The first category of criteria assesses the scope of the studies reviewed. The first criterion is the identification of the goal (or objective) of each assessment, so as it is feasible to perform comparative reviews among studies with the same objective. For that purpose, following the definition of Gaviglio et al. (2017), the papers are classified according to whether a method is "goal prescribing or "system describing" [25]. Other criteria proposed concern the determination of the target user, as well as the functional unit and the time dimension of the assessment.

The second category refers to the identification of impacts starting with the definition of the sustainability dimension examined in each study, continuing with the documentation of the impacts considered during the assessment expressed in indicators (name and type). The last category concerns the data and the calculation methods used for the assessment. The criterion type of data examines whether the data used are model or experimental. Furthermore, to examine the accessibility of data, the present study refers to the definition of Angevin et al. (2017) [26]. Therefore, depending on the data used, the assessment can be characterized as ex ante (indicating expectation and uncertainty) when focusing on assessing a new scenario or as ex post (indicating processing actual field data) when examining a current situation [26]. Additionally, for each study reviewed, the validation and aggregation methods should be examined too.

The proposed methodological framework aims at facilitating the comparison among studies in order to capture the research advancements and current practices in the field under examination. This is an issue of particular importance since the assessment of agricultural sustainability is not a standardized process and entails a plethora of different methods, tools, and frameworks that assess a large number of different indicators that represent an analogously large number of different impacts. Prior to designing any assessment model, an exhaustive review is mandatory to safeguard consistency and relevance with other works. Also, the systematic documentation of the advancements in the field is the only way to begin constructing a unified, commonly accepted methodology for agricultural sustainability assessment.

3. Crop Production Sustainability at Farm Level

3.1. Search Scheme

The methodology presented above was used to investigate the available and mostly used methodologies to assess the sustainability of crop cultivations at the farm level. The review begins with the collection of the initial sample of papers by searching within the most acknowledged databases and more specifically, Scopus and Science Direct. The search scheme is based on specific keywords and their combination as presented in Table 2, and the use of Boolean operators (OR and AND) to increase the efficiency of the search. The initial search resulted in 959 papers containing the keywords searched. The initial sample was then screened based on the inclusion/exclusion criteria of Table 2. This secondary assessment resulted in 387 papers which where, then reviewed against the initial screening criteria (Figure 6).

Table 2. Research keywords and inclusion/exclusion criteria.

Population	Intervention/Comparator	Inclusion	Exclusion
Agriculture	Sustainability Assessment	Primary Research	Review
Farming	Triple Bottom Line Assessment	In English	Not in English
Agricultural	Agricultural Sustainability	Published from 2009	Published before 2009
Farm	Environmental Assessment	Peer-reviewed	Book Chapters
Livestock	Economic Assessment	Agriculture /Livestock	Conference Proceedings
Husbandry	Societal/Social Assessment	Primary Production	Grey literature
Tillage	Life Cycle Assessment (or LCA)		
Agronomy	Multicriteria Decision Analysis (MCDA)		
Stockraising	Indicators (or KPI)		
	Environmental Impact Assessment (or EIA)		
	Cost–Benefit Analysis (CBA)		

As the purpose of this review is to examine studies assessing crop agricultural sustainability at the farm level, the 387-paper sample was filtered to select the peer-reviewed journal articles that fulfilled the following criteria. (a) Examine all three pillars of sustainability (environmental, economic, and social). (b) Examine production at the farm level. (c) Examine only crop cultivation. The final resulting collection of journal articles consists of 38 papers as presented in the Table 3 and Appendix A.

3.2. Results

3.2.1. Initial Screening

As presented in the previous section 387 papers were reviewed in the initial screening stage. The filtering of the reviewed sample according to the scope of the review under study, resulted in 38 peer-reviewed journal articles. This section presents the initial systematic review of the 38-paper sample with the use of descriptive statistics to gain further insight about the general information that derive from the reviewed sample. With respect to the general information, the majority of papers (21%) were issued in 2017, whereas only two papers (5%) fitting the review criteria ware published in 2012, 2011, and 2010 [27]. However, it is worth noting that 45% of the examined papers was issued during the last three years (2016–2018), indicating a boost in the scientific community's interest regarding integrated sustainability assessment (Figure 7).

Regarding the geographical origination, as presented in Figure 7, half of the assessments were performed in Europe (50%), whereas 16% were performed in Asia. Additionally, only three out of 38 assessments were performed in North America. With respect to the literature typology of the studies reviewed, as it was mentioned before only peer-reviewed journal articles were included in the reviewed sample.

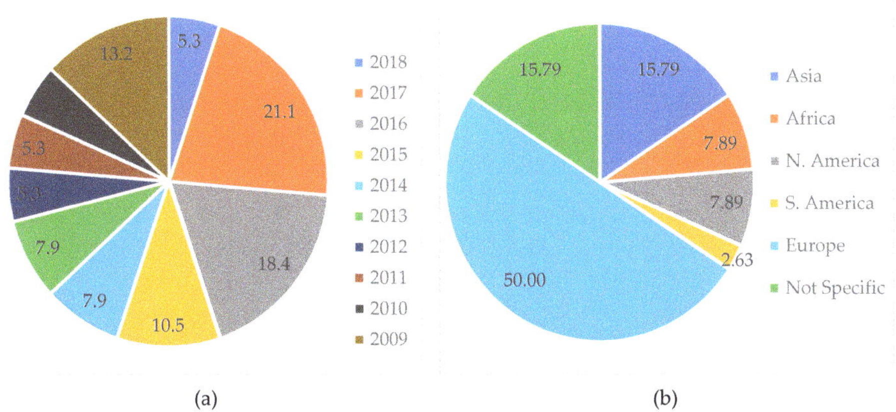

(a) (b)

Figure 7. (**a**) Year of publication (% percentage). (**b**) Geographical origination of publication (% percentage).

Regarding the method identification category, Table 3 presents all the methods and tools that were identified during the review process (the nomenclature is presented in Appendix A). All of the relevant methods will be presented in detail later. In the majority of the papers examined (66%), the methods or tools presented are also practically tested presenting the relevant examples (case studies). In 18% of the papers, an already existing methodology was applied and presented while 16% of papers presented a methodology without testing it in practice. Continuing with the level of assessment, in 79% of the works examined, the assessment was performed exclusively for the farm level, whereas for 21% of the works, the level of assessment was also broadened beyond the farm level by examining local, regional, or national sustainability. The most frequently examined crop is maize and wheat (examined in five cases each), followed by olive, spinach and rice (examined in two cases studies each). The other crops examined in the papers reviewed included legumes, lettuce, scallions, red radish, banana, soybean, grapes, cranberry, potato, and coffee. Additionally, different agronomic practices are examined as for example organic farms [28], greenhouse cultivations [29], and school gardens [30].

Table 3. Initial screening (38-paper collection).

a/a	Author	Country	Method/Tool	Methodology/ Application	Field of Application/Product	Level of Assessment/ System Boundaries
1	De Luca et al. (2018) [31]	Italy	LCSA AHP LCA LCC SLCA	M, A	Olive	F
2	Snapp et al. (2018) [32]	Malawi	INDICATORS	M, A	Maize, legume	F
3	Gaviglio et al. (2017) [25]	Italy	INDICATORS 4AGRO	M, A	Agricultural Park	F
4	Recanati et al. (2017) [33]	Palaistine	INDICATORS	M, A	Food Production	F, R
5	Bockstaller et al. (2017) [34]	-	CONTRA	M, A	-	F
6	Goswami et al. (2017) [35]	-	SFSI DPSIR SL	M	Smal Farms	F
7	Theurl et al. (2017) [36]	Austria	LCA CF INDICATORS	A	Lettuce, spinach, scallions, red radish	F, R
8	Chopin et al. (2017) [37]	FrenchWest Indies	MASC	A	Banana	F, N
9	Angevin et al. (2017) [26]	-	DEXiPM	M	-	F
10	Vasileiadis et al. (2017) [38]	Europe	DEXiPM SYNOPS-WEB CBA	A	Wheat, maize	F, R
11	Egea et al. (2016) [39]	Spain	MCDA ANP	M	Olive	F
12	Dong et al. (2016) [40]	USA	PCA DEA	M, A	Soybean	F
13	de Olde et al. (2016) [28]	Denmark	RISE	A	Organic Farms (vegetable, dairy, pig, poultry)	F
14	Yang et al. (2016) [29]	China	INDICATORS	M, A	Greenhouse vegetables	F
15	Sajjad et al. (2016) [41]	India	SLSI	A	-	F, R
16	Allahyari et al. (2016) [42]	Iran	INDICATORS	M	Paddy fields	F
17	Sottile et al. (2016) [30]	Kenya	SAEMETH-G PCA	M, A	School Gardens	F
18	El Chami et al. (2015) [43]	England	LCA CBA SCC	M, A	Winter Wheat	F
19	Santiago-Brown et al. (2015) [44]	-	INDICATORS	M	Grapes	F
20	Peano et al. (2015) [45]	-	SAEMETH	M, A	10 small agri-food systems	F
21	Dong et al. (2015) [46]	USA	PCA DEA	M, A	Cranberry	F
22	Yegbemey et al. (2014) [47]	Benin	INDICATORS	M, A	Maize	F

Table 3. *Cont.*

a/a	Author	Country	Method/Tool	Methodology/ Application	Field of Application/Product	Level of Assessment/ System Boundaries
23	Peano et al. (2014) [48]	Italy	INDICATORS	M, A	Agri-food system Slow food Presidia Project	F
24	Van Asselt et al. (2014) [49]	The Netherlands	INDICATORS	M, A	Potato	F
25	Colomb et al. (2013) [50]	France	MASC-OF DEXi	M, A	-	F, R
26	Vasileiadis et al. (2013) [51]	Europe	DEXiPM	A	Wheat, maize	F
27	Sami et al. (2013) [52]	Iran	INDICATORS	M, A	Wheat maize	F
28	Pelzer et al. (2012) [27]	France	DEXiPM	M, A	Arable crops and maize-based systems	F
29	Van Passel et al. (2012) [53]	Belgium	SVA MOTIFS	M, A	-	F
30	Reig-Martinez et al. (2011) [54]	Spain	DEA MCDA	M, A	-	F
31	Sharma et al. (2011) [55]	India	ASI	M, A	-	F
32	Rodriguez et al. (2010) [56]	-	APOIA-NovoRural	M, A	-	F
33	Gomez-Limon et al. (2010) [24]	Spain	SAFE AHP PCA	M, A	Rain-fed and irrigated	F
34	Gomes et al. (2009) [57]	Brazil	DEA	A	Rice, maize, coffee	F, R
35	Van Passel et al. (2009) [58]	Belgium	SVA	M	-	F
36	Sadok et al. ([59]	France	MASC DEXi	M, A	Cropping systems	F
37	Siciliano (2009) [60]	Italy	SMCE	M, A	Durum Wheat	F, L, R
38	Walter et al. (2009) [61,62]	Germany	INDICATORS	M, A	Spinach	F, R, G

M: Methodology; A: Application; F: Farm level; R: Regional level; N: National level; G: Global level.

3.2.2. In-Depth Review

This section presents the systematic review results against the in-depth review criteria initializing the presentation with the scope criteria category (Tables A2 and A3 of Appendix A). Regarding the goal of the assessment, 61% of the examined studies are system describing, whereas the other 40% attempts to identify and evaluate policies and techniques that could be used to improve agricultural sustainability performance. Regarding the target users of the methodologies proposed, the majority of the examined works is aimed at decision-makers, farmers, and researchers. More specifically, 40% of the studies identify decision-makers as their target users, whereas 26% aim at farmers and 21% aim at researchers. Continuing, only three (3) works define a functional unit as a basis for the assessment. In particular, De Luca et al. (2018), when examining the sustainability of olive growing systems, proposed "One hectare (1ha) of cultivated surface" as a functional unit [31]. On the other hand, Theurl et al. (2017) and El Chami et al. (2015) preferred functional units related to the weight of the final product ("kg of un-/packed fresh product at the point of sale-POS" and "1 tn fresh weight standardized to 86% dry matter, respectively") [36,43].

Concerning the criterion of the time dimension, in several studies the assessment was performed for a single year period [25,28,29,33,36,41,51,60]. However, there are also studies that perform the assessment for a range of years. Snapp et al. (2018) performed a 3-year trial, and Vasileiadis et al. (2017) extracted their data during a 4-year experiment [32,38]. Sharma et al. (2011) collected data from three separate decades from 1950 to present [55], and Gomes et al. (2009) collected data from 1986 to 2002 [57]. From another point of view, De Luca et al. (2018) expanded their assessment to the life cycle of an olive tree orchard (50 years) [31], and el Chami et al. (2015) projected the assessment to 2050 [43].

Regarding the Impact Identification category, as described above, the research scope contains only studies that attempt to examine all the three dimensions of sustainability, namely, the environmental, economic, as well as social pillar, contributing towards an integrated sustainability assessment evaluation. During the extensive review, all of the individual impacts—expressed as indicators—that were examined within the reviewed studies were extracted and documented. However, further thorough classification and commenting on the individual indicators used goes beyond the limits of this analysis and has already been investigated in several review studies in the past [4,13,18,19,22].

With respect to the data calculation method category of criteria, for 82% of the papers examined a validation process is not mentioned. Only 18% of the papers describe a validation process for the proposed methodologies. On the other hand, 74% of the studies mention the use of an aggregation technique or methodology aiming at the simplification and the generalization of the results. Regarding the type of data used for the assessments performed (Figure 8), the majority uses experimental data (68%), whereas a small percentage of works (18.4%) employ only model data for the sustainability assessment. Accordingly, 58% are ex post assessments attempting to evaluate current practices; whereas, in 31.6% of the papers, the evaluation of prediction scenarios is attempted.

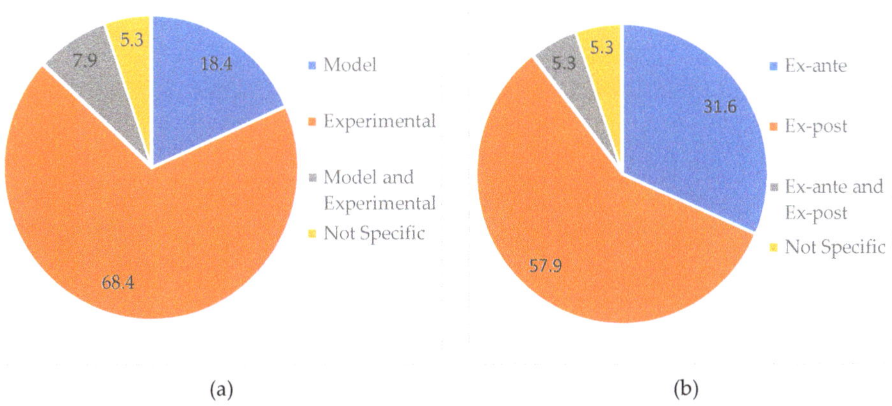

(a)   (b)

Figure 8. (**a**) Type of data (% percentage). (**b**) Accessibility of data (% percentage).

3.3. Agricultural Sustainability Methods and Tools

In the previous sections a descriptive qualitative analysis of the review criteria was presented. The aim was to examine the research trend of crop agricultural sustainability and specifically the trend of the criteria concerning the scope and the calculation methods used. In this section, the methodologies and tools, extracted as a result of the review conducted, are presented. Figure 9 demonstrates the methods and tools identified and the corresponding frequency of occurrence. These methods and tools were classified in five major categories based on the main scope of the assessment (as expressed by the authors), underlining the fact that the categories selected may overlap as part of the overall concept. A distinctive example is MCDA which is used to facilitate the assessment of multivariate problems that are expressed with indicators. Nevertheless, the scope of studies employing MCDA methods focus on the aggregation of the results while methods proposing indicator sets and indexes focus on determining the criteria of the assessment. Another example is the carbon footprint (CF) which is an indicator that is often met in Indicators sets and frameworks. Nevertheless, it is a very commonly used standalone methodology for environmental impact assessment.

To that end, LCA methods relate to the life cycle of the examined element. Environmental methods relate to the quantification of the environmental impact of the examined element, and economic methods refer to the use of financial methods in the impact assessment. Multicriteria methods are methods that employ multicriteria assessment for the evaluation of agricultural sustainability, and Indicator methods include indicator sets and frameworks for the assessment of agricultural sustainability. With respect to the individual methodologies that were identified, the term "indicators" refers to all those methodologies that were not given a specific name by their developers.

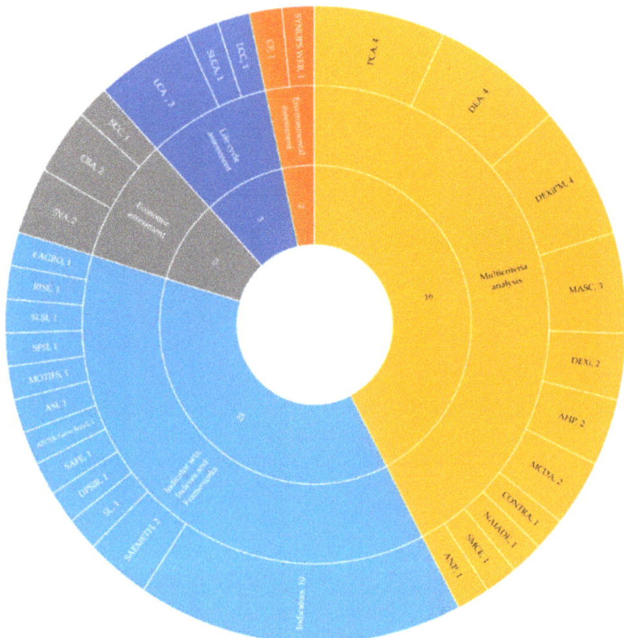

Figure 9. Crop agricultural sustainability at farm level (method categories and methods).

3.3.1. Life Cycle Assessment, Environmental, and Economic Methods and Tools

For 21% of the studies reviewed, methods belonging to LCA, environmental, or economic method categories are employed. El Chami et al. (2015) performed an integrated sustainability assessment comparing different irrigation scenarios of winter wheat production at the farm level by proposing a methodology that combined LCA, SCC, and CBA [43]. Following the concept of LCA, Theurl et al. (2017) assessed the environmental and socio-economic impacts of unheated soil-grown vegetables [36]. Theurl et al. performed a comparative assessment utilizing experimental field data and data collected from literature, calculating the GHG emissions with socio-economic indicators deriving from the Sustainability Assessment of Food and Agriculture Systems (SAFA) guidelines of the Food and Agriculture Organization (FAO) [36]. Theurl et al. combined methodologies from three of the five categories identified, namely the LCA, the environmental and the indicator methods.

From the most recent studies, De Luca et al. (2018) assess the sustainability of olive growing systems by focusing on scenarios differentiated in weeding [31]. For the assessment, authors combined a series of tools to evaluate the three pillars of sustainability, namely, LCA for the environmental pillar, LCC for the economic, and SLCA for the societal pillar. They integrated their results by employing the AHP method for multicriteria analysis [31]. From the economic methods category, Van Passel et al. (2009) proposed a methodological framework based on the sustainable value approach (SVA) to assess the sustainability on farm production level [58]. Van Passel et al. employed the SVA method attempting to correlate farm performance in respect to consumption of resources. The work represents a benchmarking approach since it does not focus on the evaluation of sustainability in absolute terms, but it assesses the performance compared to standards [58]. Van Passel et al. (2011) stated that to perform multilevel and multi-user assessments, a combination of methodologies can offer more advantages than integrated methodologies [53]. To that end, the SVA method was combined with the MOTIFS indicator tool. According to Van Passel et al. (2011), MOTIFS is a visual monitoring tool used for the aggregation of indicators of various themes, which creates benchmarks for the rescaling of the indicator values [53].

3.3.2. Multicriteria Assessment Methods and Tools

Within the multicriteria assessment methods that are used for assessing agricultural sustainability, the works examined can be classified into groups that employ and develop the same methodological framework. Such groups are the studies that use the MASC decision model developed by Sadok et al. (2009), which was built as part of the decision support system DEXi [59]. The MASC model is a hierarchical multiattribute decision support model designed for the ex ante assessment of cropping systems to address the need of in-field alternative scenario evaluation. Such models allow for the simplification of the decision problem by downscaling it to smaller and less complex problems expressed by designated variables [59]. The DEX methodology performs aggregation of qualitative attributes and utility functions using "IF-THEN" aggregation rules [59]. Colomb et al. (2013) building upon Sadok's et al. (2009) model, proposed the MASC-OF model to assess the strong and weak points of organic cropping systems in a regional context [50]. Pelzer et al. (2012) also following Sadok et al. (2009) presented the DEXiPM model which was particularly developed for integrated pest management systems (IPM). The DEXiPM model is built upon the MASC model and it is an ex ante methodology contributing towards the discussion around innovative systems. The model was implemented in winter crop and maize-based cropping systems and consists of seventy-five (75) basic and eighty-six (86) aggregated indicators [27].

Vasileiadis et al. (2013) used the DEXiPM model to compare the sustainability of innovative IPM-based systems [51]. Also, Vasileiadis et al. (2017) used Pelzer's et al. (2012) DEXiPM model, which was supplemented by Angevin et al. (2017), for the ex post assessment of the economic, environmental and social sustainability of conventional winter wheat and maize cropping systems [26]. The IPM-based systems were designed and tested in nine (9) locations in Europe [38]. They compared the sustainability of the examined systems, discussing the benefits or drawbacks of the IPM systems. Vasileiadis et al. (2017) also adopted methodologies from the environmental and economic categories. Economic data, with the use of a template, were collected from participants to perform cost–benefit analysis (CBA). Furthermore, an environmental risk assessment was performed by implementing the SYNOPS-WEB Tool [38]. Lastly, Chopin et al. (2017) adapted the MASC model in order to ex ante assess the sustainability in the area of local banana farming systems [37].

Multicriteria methods facilitate decision making while considering multiple variables, and such methods use weighting techniques in order to produce composite indices [24]. Among the studies examined, the most frequently used methods are the principal component analysis (PCA) and the data envelopment analysis (DEA). Specifically, Gomez-Limon et al. (2010) and Sottile et al. (2016) used the PCA method, whereas Gomes et al. (2009) and Reig-Martinez et al. (2011) used the DEA method to create a composite indicator [24,30,54,57]. Dong et al. (2015 and 2016) combined both methodologies attempting to construct a complex indicator composed of a large number of individual and interdepended variable [40,46].

Concluding with the multicriteria method category, Siciliano et al. (2009) used the social multicriteria evaluation (SMCE) framework, which was implemented through the NAIADE (novel approach to imprecise assessment and decision environments) software, to assess the sustainability of farming practices in a small rural area in Italy [60]. Egea et al. (2016) employed the analytic hierarchy process (AHP) in order to investigate the combination of protected destination of origin oil production system that leads to optimal sustainability [39]. Bockstaller et al. (2017) introduced the CONTRA tool, an innovative aggregation method that leads to the creation of decision trees using fuzzy sets [34]. Peano et al. (2014) proposed a multicriteria methodology to evaluate the effectiveness of the slow food presidia, which are organized structures aiming at the preservation of quality production at risk to extinction by following specific guidelines and protocols for each product category [48].

3.3.3. Indicator Sets, Indexed and Frameworks

This category of methods and tools contains indicator sets, indexes, and frameworks that were used in the reviewed works to assess agricultural sustainability at the farm level. Walter et al. (2009a; 2009b) proposes a new indicator-based method to assess the unsustainability of a system rather than its sustainability [62]. Their method borrows elements of the LCA methodology and was implemented in two stages. The first stage includes the creation of an issue inventory and its contextualization, while the second stage includes the standardization and sustainability valuation process [61,62]. Rodriguez et al. (2010) proposed the APOIA-NovoRural framework, which comprises a collection of basic and composite indicators covering five dimensions of sustainability: landscape ecology, environmental quality, sociocultural values, economic values, and management and administration [56]. Sharma et al. (2011) introduced a methodology based on questionnaires and surveys and composed an agricultural sustainability index (ASI) targeted to Bihar province (India) [55], and also calculated the sustainability parameters for a 60-year period.

Sami et al. (2013) selected six indicators that were considered appropriate to assess sustainability in a regional context. Additionally, in order to evaluate some of these indicators they used a selection of fuzzy submodels [52]. Van Asselt et al. (2014) propose a protocol for the collection and evaluation of indicators for the sustainability assessment of agri-food production systems [49]. Their proposed list covers a wide range of indicators related to the three pillars of sustainability, aiming at supporting policy makers in decision making by choosing the most relevant indicators. Yegbemey et al. (2014), proposed an innovative participatory approach that resulted in seventeen (17) indicators. All relevant data were collected through a household survey. The sustainability was evaluated with relative scores while the total sustainability level was based on the average scores of the individual indicators [47]. Peano et al. (2015), proposed the SAEMETH monitoring tool based on a set of qualitative indicators. The selection of the indicators was based on the criteria introduced by Meul et al. (2008) [63] and, for their evaluation, they set a minimum and maximum threshold based on reference values that was derived from best practices or through surveys [45]. Santiago-Brown et al. (2015) presented the process for selecting indicators to assess viticulture production sustainability. For the selection of the indicators, the adapted nominal group technique was used. The selected indicators were reduced according to their relevance [44] resulting in seventy-six (76) indicators hierarchized based on their importance.

Allahyari et al. (2016) selected five-hundred-and-eighty-eight (588) indicators through an extensive literature review. Following erasing duplicates and prioritizing the sample, it resulted in 62 indicators, which were used in an extensive survey among experts. The indicators were assessed based on their importance while the resulting data were assessed with the Minskowski fuzzy screening method [42]. Sajjad et al. (2016) examined the relevant agricultural sustainability at farm and regional scale using the sustainable livelihood security index (SLSI) [41]. Yang et al. (2016) assessed the sustainability of greenhouse vegetables using indicators. More specifically, to examine the greenhouse vegetable farming practices and the economic and social management conditions, they used rapid and participatory rural appraisal (RRA/PRA) tools combined with data derived from in-field measurements and parallel surveys [29]. In 2016, de Olde et al. proposed the sustainability assessment tool named response-inducing sustainability evaluation (RISE), which was implemented for the evaluation of organic farms in Denmark. The tool contains indicators for a total of 10 themes and 51 subthemes. The indicators were normalized and aggregated and each theme was evaluated based on the average score of the relevant subthemes [28].

Goswami et al. (2017) integrated the sustainable livelihood (SL) and the drivers–pressures–state–impact–response (DPSIR) framework, proposing a small farm sustainability index (SFSI) that could address the complexity of small-holder family farms under a participatory approach [35]. The proposed framework assesses sustainability in multiple levels assigning the relevant weights and resulting in the creation of an aggregated index for the entire system. They indicate that the introduction ICT technologies in agriculture (web-based platforms, wireless sensors, etc.) can facilitate data sharing among stakeholders and provide the basis for assessing the sustainability of farming systems.

Recanati et al. (2017) proposed an indicator-based framework for the assessment of sustainability of small-scale farming systems in water-limited regions. They implemented the framework by modeling an "average" farm based on a survey among 30 farmers [33]. Gaviglio et al. (2017), attempting to integrate various analytical techniques, introduced the 4AGRO tool, which is an online self-assessment tool based on indicators. It consists of 42 subindicators that are divided in 15 complex indicators, five for each pillar of sustainability [25]. The tool was demonstrated in an agricultural park in Italy. Finally, Snapp et al. (2018) proposed a methodology based on indicators that derived through a participatory approach involving a steering committee with multidisciplinary participants from eight (8) institutions [32]. The indicators were normalized based on max possible values.

4. Conclusions

To meet the ever-increasing interest towards agricultural sustainability, many methodologies and tools emerge, introducing integrated and holistic assessment approaches. However, there is still no consensus on the standardization of agricultural sustainability assessment as part of a unified concept of sustainable development. Newly introduced frameworks propose mostly case-specific tools that focus on resource use and their impact on the sustainability of farming practices. Combinational use of methodologies is observed in many cases; thus, a clear distinction of methodologies is not always possible. Contributing towards the indexing of the available methodologies, the present paper presented a methodological framework for the systematic literature review of agricultural sustainability studies. The framework synthesizes all the available literature review criteria and introduces a two-level analysis facilitating systematization, data mining, and methodology extraction.

The framework was implemented for the systematic literature review of crop agricultural sustainability assessment studies at farm-level for the last decade. The investigation of the methodologies used is of particular importance since there are no standards or norms for the sustainability assessment of farming practices. The chronological analysis revealed that the scientific community's interest in agricultural sustainability has been increasing during the last three (3) years, indicating a tendency to gradually progress from the theory of economic growth to the more comprehensive and inclusive concept of sustainable development. Nevertheless, the critical evaluation of effectiveness and the implications of the methods presented are outside the scope of the present work and are subjects of thorough future research.

The most used methods include indicator-based tools, frameworks and indexes followed by multicriteria methods. In the reviewed studies, stakeholder participation is proved crucial in the determination of the level of sustainability. However, a systematic assessment of the agricultural machinery's and operation management's contribution to the overall sustainability was not detected in the examined studies. The effect of resource use and input management is the most usually examined issue in the reviewed studies.

Author Contributions: Conceptualization, D.B. and M.G.L.; methodology, M.G.L.; validation, D.B. and C.G.S.; investigation, M.G.L.; data curation, C.G.S.; writing—original draft preparation, M.G.L.; writing—review and editing, D.B. and C.G.S.; visualization, M.G.L.; supervision, D.B.

Funding: The work was supported by the project "Research Synergy to address major challenges in the nexus: energy–environment–agricultural production (Food, Water, Materials)"—NEXUS, funded by the Greek Secretariat for Research and Technology (GSRT)—Pr. No. MIS 5002496.

Conflicts of Interest: The authors declare no conflicts of interest.

Nomenclature

LCA	Life Cycle Assessment
LCC	Life Cycle Costing
SLCA	Social Life Cycle Assessment
SYNOPS WEB	Environmental Risk Assessment Web tool
CF	Carbon Footprint

SCC	Social Cost of Carbon
CBA	Cost–benefit Analysis
SVA	Sustainable Value Approach
MCDA	Multi Criteria Decision Analysis
AHP	Analytic Hierarchy Process
ANP	Analytical Network Process
MASC	Multiattribute Assessment of Cropping Systems
DEXi	A Program for Multiattribute Decision Making
DEXiPM	DEXi Pest Management
SMCE	Social Multicriteria Evaluation
NAIADE	Novel approach to imprecise assessment and decision environments
PCA	Principle Components Analysis
CONTRA	the French acronym for 'design of transparent decision trees'
DEA	Data Envelopment Analysis
RISE	Response-Inducing Sustainability Evaluation
SLSI	Sustainable Livelihood Security Index
SFSI	Small Farm Sustainability Index
SAEMETH	Sustainable Agri-Food Evaluation Methodology
ANGT	Adapted Nominal Group Technique
MOTIFS	Monitoring Tool for Integrated Farm Sustainability
ASI	Agricultural Sustainability Index
APOIA-NovoRUral	A system for weighted environmental impact assessment of rural activities
SAFE	A hierarchical framework for assessing the sustainability of agricultural systems
DPSIR	Drivers, Pressures, State, Impact, Response

Appendix A

Table A1. Review criteria frequency table.

	Criteria	De Luca et al., 2017 [10]	Binder et al., 2010 [4]	Peter et al., 2017 [12]	Bockstaller et al., 2009 [3]	Roy et al., 2012 [13]	Cerutti et al., 2011 [7]	Acosta-Alba et al., 2011 [14]	Baldini et al., 2017 [15]	Morais et al., 2016 [16]	McAuliffe et al., 2016 [17]	Latruffe et al., 2016 [18]	Bockstaller et al., 2008 [19]	Payraudeau et al., 2005 [20]	de Vries et al., 2015 [21]	Yan et al., 2011 [9]	Lebacq et al., 2013 [22]
1	Sustainability Assessment method	+	+	+	+	+	+	+	+	+	+		+	+	+	+	+
2	Sustainability Assessment Tool	+		+		+											
3	Description of Assessment Tool		+	+	+		+							+			
4	Field of Application/Product	+			+	+	+	+		+					+		
5	Year of Publication				+	+			+	+	+			+	+	+	
6	Country	+		+		+	+		+						+	+	
7	Literature Typology	+		+					+	+							
8	Typology of Study	+														+	
9	Level of Assessment/System Boundaries	+	+	+	+	+	+	+	+				+	+		+	+
10	Goal of the Assessment		+	+				+	+	+			+	+		+	+
11	Views of the Assessment		+														
12	Assessment Approach/Standards		+							+							
13	Database Used			+													
14	User Interface method			+													
15	Developer of the method			+	+												
16	Strengths	+			+												
17	Drawbacks	+			+												

Table A1. *Cont.*

#	Criteria	De Luca et al., 2017 [10]	Binder et al., 2010 [4]	Peter et al., 2017 [12]	Bockstaller et al., 2009 [3]	Roy et al., 2012 [13]	Cerutti et al.,2011 [7]	Acosta-Alba et al.,2011 [14]	Baldini et al.,2017 [15]	Morais et al., 2016 [16]	McAuliffe et al., 2016 [17]	Latruffe et al.,2016 [18]	Bockstaller et al., 2008 [19]	Payraudeau et al.,2005 [20]	de Vries et al., 2015 [21]	Yan et al., 2011 [9]	Lebacq et al.,2013 [22]
18	Validation				+	+							+				+
19	Readability				+												+
20	Feasibility				+												+
21	Reproducibility				+												+
22	Relevance for end users				+												+
23	Target Users	+	+		+			+					+	+			
24	Aggregation	+			+								+				+
25	Rating Scores	+			+												
26	Thresholds/Characterization factors				+			+	+								
27	Type of Data used			+	+			+	+	+		+					
28	Accessibility of data				+	+		+		+			+				+
29	Indicator Name/Type	+	+	+	+	+	+					+		+	+		+
30	Integration with existing farming software				+												
31	Need for external support				+												
32	User-Friendliness				+												
33	Transparency				+												
34	Calculation method				+				+								+
35	Impacts Considered/Impact Identification	+		+	+	+	+		+	+	+				+	+	+
36	Agronomic Protocol							+						+			
37	Functional Unit Category							+									
38	Operation with the greatest impact							+									
39	Dimension of Sustainability Studied			+	+	+		+				+		+			+
40	Functional Unit	+			+			+	+	+	+	+		+	+	+	
41	Uncertainty and Sensitivity Analysis								+				+			+	
42	Case study Description												+	+			
43	Participation method	+	+														
44	Time Dimension				+	+								+		+	

Table A2. Scope (38-paper collection).

a/a	Author	Assessment Goal	Target Users	Functional Unit	Time Dimension
1	De Luca et al. (2018) [31]	SD	RE TE FA	1ha cultivated surface	50 years
2	Snapp et al. (2018) [32]	GP		-	3 years mother trial
3	Gaviglio et al. (2017) [25]	SD	DM FA RE	-	1 year
4	Recanati et al. (2017) [33]	GP		-	1 year
5	Bockstaller et al. (2017) [34]	GP	DM FA RE	-	-
6	Goswami et al. (2017) [35]	SD		-	
7	Theurl et al. (2017) [36]	GP		kg un-/pacjed fresh producty at the POS	10/2014-04/2015
8	Chopin et al. (2017) [37]	GP	DM RE FA	-	-
9	Angevin et al. (2017) [26]	GP	DM RE	-	-
10	Vasileiadis et al. (2017) [38]	GP	DM FA	-	4-year experiment

Table A2. *Cont.*

a/a	Author	Assessment Goal	Target Users	Functional Unit	Time Dimension
11	Egea et al. (2016) [39]	GP		-	-
12	Dong et al. (2016) [40]	SD		-	-
13	de Olde et al. (2016) [28]	SD	DM FA RE	-	2013-2014
14	Yang et al. (2016) [29]	SD		-	Fall 2011-spring 2012 (Sampling)
15	Sajjad et al. (2016) [41]	SD	-	-	2012-2013
16	Allahyari et al. (2016) [42]	SD	-	-	-
17	Sottile et al. (2016) [30]	SD	-	-	-
18	El Chami et al. (2015) [43]	GP	-	1tn fresh weight standardizes to 86% dry matter	Projection to 2050
19	Santiago-Brown et al. (2015) [44]	SD	DM FA RE	-	-
20	Peano et al. (2015) [45]	SD	FA TE	-	4-6 years in the ex post stage
21	Dong et al. (2015) [46]	SD	-	-	-
22	Yegbemey et al. (2014) [47]	SD	-	-	
23	Peano et al. (2014) [48]	GP	-	-	-
24	Van Asselt et al. (2014) [49]	GP		-	
25	Colomb et al. (2013) [50]	GP	DM	-	-
26	Vasileiadis et al. (2013) [51]	GP	-	-	2012
27	Sami et al. (2013) [52]	SD	-	-	-
28	Pelzer et al. (2012) [27]	GP	DM FA	-	-
29	Van Passel et al. (2012) [53]	SD		-	-
30	Reig-Martinez et al. (2011) [54]	SD	-	-	-
31	Sharma et al. (2011) [55]	SD		-	3 separate decades from 1950
32	Rodriguez et al. (2010) [56]	SD	DM RE FA TE	-	
33	Gomez-Limon et al. (2010) [24]	SD	DM	-	
34	Gomes et al. (2009) [57]	SD	-	-	1986-2002
35	Van Passel et al. (2009) [58]	SD	DM	-	-
36	Sadok et al. ([59]	SD	DM	-	-
37	Siciliano (2009) [60]	GP		-	2003
38	Walter et al. (2009) [61,62]	SD		-	

SD: System describing; GP: Goal prescribing; DM: Decision-maker; RE: Researcher; TE: Technician; FA: Farmer.

Table A3. Data calculation method (38-paper collection).

a/a	Author	Validation	Aggregation	Type of Data	Accessibility of Data
1	De Luca et al. (2018) [31]	-		E (Survey)	Ex ante
2	Snapp et al. (2018) [32]	-	-	E	Ex post
3	Gaviglio et al. (2017) [25]	-	Sum	E (Survey)	Ex ante
4	Recanati et al. (2017) [33]	-		M (Model Farm based on survey)	Ex ante
5	Bockstaller et al. (2017) [34]	-	Sum (rank of percentage of weight)	E, M	
6	Goswami et al. (2017) [35]	D, O, U	Linear, Geometric, Multicriteria function-based	-	Ex ante
7	Theurl et al. (2017) [36]	-		E (Data from field and survey)	Ex post

Table A3. *Cont.*

a/a	Author	Validation	Aggregation	Type of Data	Accessibility of Data
8	Chopin et al. (2017) [37]	-	Decision Rules and relative weightings (using direct scoring method)	M, E (When model not available)	Ex ante
9	Angevin et al. (2017) [26]	EU (Experts Evaluation)	"IF-THEN" Aggregation rules	E	ex post
10	Vasileiadis et al. (2017) [38]	-	"IF-THEN" Aggregation rules	E	ex post
11	Egea et al. (2016) [39]		YES (weighted sum)	E (Expert opinion)	Ex post
12	Dong et al. (2016) [40]	D	PCA and DEA (Weighting Factors)	E (Survey)	Ex post
13	de Olde et al. (2016) [28]	-	Arithmetic mean	E	Ex post
14	Yang et al. (2016) [29]	-	-	E (Survey)	Ex post
15	Sajjad et al. (2016) [41]	-	Not specific	E (Survey)	Ex post
16	Allahyari et al. (2016) [42]	-	Mean method	E (Survey to assess indicators)	Ex post
17	Sottile et al. (2016) [30]	-	PCA, Cluster Analysis	E (survey)	Ex post
18	El Chami et al. (2015) [43]	-	-	M	Ex ante
19	Santiago-Brown et al. (2015) [44]	-	-	-	-
20	Peano et al. (2015) [45]	O	Basic Indicators as equally important	E (Semi-structured interviews)	Ex ante and ex post
21	Dong et al. (2015) [46]	-	DEA model	E (Survey)	Ex post
22	Yegbemey et al. (2014) [47]		Simple and Linear aggregation technique	E(Survey)	Ex post
23	Peano et al. (2014) [48]	D	Equal weights or based on importance	E	Ex post
24	Van Asselt et al. (2014) [49]	-	Normalization between 0-100 then weights based on importance	E	Ex post
25	Colomb et al. (2013) [50]	NO	"IF-THEN" Aggregation rules	E	Ex ante
26	Vasileiadis et al. (2013) [51]	-	"IF-THEN" Aggregation rules	E and M	Ex ante
27	Sami et al. (2013) [52]	-	"IF-THEN" Aggregation rules	E	Ex post
28	Pelzer et al. (2012) [27]	U	"IF-THEN" Aggregation rules	M	Ex ante
29	Van Passel et al. (2012) [53]	YES	The indicators are integrated into MOTIFS graph	M (Farm Accountancy data)	Ex post
30	Reig-Martinez et al. (2011) [54]	-	DEA model	E (Survey)	Ex post
31	Sharma et al. (2011) [55]		Equal weights	E (Survey)	Ex post
32	Rodriguez et al. (2010) [56]	-	Normalization and weights then average for indices	E	Ex post
33	Gomez-Limon et al. (2010) [24]	-	Weighted sum, Product of weighted indicators	E (Survey)	Ex post
34	Gomes et al. (2009) [57]	-	-	E	Ex post
35	Van Passel et al. (2009) [58]	-	-	M (Empirical data)	Ex ante
36	Sadok et al. ([59]	-	"IF-THEN" Aggregation rules	M	Ex ante
37	Siciliano (2009) [60]	-	Not specific	E	Ex post and ex ante
38	Walter et al. (2009) [61,62]		Sum and Weighted sum	M	Ex ante

M: Model, E: Experimental.

References

1. United Nations Department of Economic and Social Affairs. *World Population Projection*; United Nations Department of Economic and Social Affairs: New York, NY, USA, 2017.
2. Viola, I.; Marinelli, A. Life Cycle Assessment and Environmental Sustainability in the Food System. *Agric. Agric. Sci. Procedia* **2016**, *8*, 317–323. [CrossRef]
3. Bockstaller, C.; Guichard, L.; Keichinger, O.; Girardin, P.; Galan, M.B.; Gaillard, G. Comparison of methods to assess the sustainability of agricultural systems. A review. *Agron. Sustain. Dev.* **2009**, *29*, 223–235. [CrossRef]
4. Binder, C.R.; Feola, G.; Steinberger, J.K. Considering the normative, systemic and procedural dimensions in indicator-based sustainability assessments in agriculture. *Environ. Impact Assess. Rev.* **2010**, *30*, 71–81. [CrossRef]
5. De Olde, E.M.; Moller, H.; Marchand, F.; McDowell, R.W.; MacLeod, C.J.; Sautier, M.; Halloy, S.; Barber, A.; Benge, J.; Bockstaller, C.; et al. When experts disagree: The need to rethink indicator selection for assessing sustainability of agriculture. *Environ. Dev. Sustain.* **2017**, *19*, 1327–1342. [CrossRef]
6. Van Pham, L.; Smith, C. Drivers of agricultural sustainability in developing countries: A review. *Environ. Syst. Decis.* **2014**, *34*, 326–341. [CrossRef]
7. Cerutti, A.K.; Bruun, S.; Beccaro, G.L.; Bounous, G. A review of studies applying environmental impact assessment methods on fruit production systems. *J. Environ. Manag.* **2011**, *92*, 2277–2286. [CrossRef] [PubMed]
8. De Olde, E.M.; Oudshoorn, F.W.; Sørensen, C.A.G.; Bokkers, E.A.M.; De Boer, I.J.M. Assessing sustainability at farm-level: Lessons learned from a comparison of tools in practice. *Ecol. Indic.* **2016**, *66*, 391–404. [CrossRef]
9. Yan, M.J.; Humphreys, J.; Holden, N.M. An evaluation of life cycle assessment of European milk production. *J. Environ. Manag.* **2011**, *92*, 372–379. [CrossRef]
10. De Luca, A.I.; Iofrida, N.; Leskinen, P.; Stillitano, T.; Falcone, G.; Strano, A.; Gulisano, G. Life cycle tools combined with multi-criteria and participatory methods for agricultural sustainability: Insights from a systematic and critical review. *Sci. Total Environ.* **2017**, *595*, 352–370. [CrossRef]
11. Grant, M.J.; Booth, A. A typology of reviews: An analysis of 14 review types and associated methodologies. *Health Inf. Libr. J.* **2009**, *26*, 91–108. [CrossRef]
12. Peter, C.; Helming, K.; Nendel, C. Do greenhouse gas emission calculations from energy crop cultivation reflect actual agricultural management practices?—A review of carbon footprint calculators. *Renew. Sustain. Energy Rev.* **2017**, *67*, 461–476. [CrossRef]
13. Roy, R.; Chan, N.W. An assessment of agricultural sustainability indicators in Bangladesh: Review and synthesis. *Environmentalist* **2012**, *32*, 99–110. [CrossRef]
14. Acosta-Alba, I.; van der Werf, H.M.G. The use of reference values in indicator-based methods for the environmental assessment of agricultural systems. *Sustainability* **2011**, *3*, 424–442. [CrossRef]
15. Baldini, C.; Gardoni, D.; Guarino, M. A critical review of the recent evolution of Life Cycle Assessment applied to milk production. *J. Clean. Prod.* **2017**, *140*, 421–435. [CrossRef]
16. Morais, T.G.; Teixeira, R.F.M.; Domingos, T. Regionalization of agri-food life cycle assessment: A review of studies in Portugal and recommendations for the future. *Int. J. Life Cycle Assess.* **2016**, *21*, 875–884. [CrossRef]
17. McAuliffe, G.A.; Chapman, D.V.; Sage, C.L. A thematic review of life cycle assessment (LCA) applied to pig production. *Environ. Impact Assess. Rev.* **2016**, *56*, 12–22. [CrossRef]
18. Latruffe, L.; Diazabakana, A.; Bockstaller, C.; Desjeux, Y.; Finn, J.; Kelly, E.; Ryan, M.; Uthes, S. Measurement of sustainability in agriculture: A review of indicators. *Stud. Agric. Econ.* **2016**, *118*, 123–130. [CrossRef]
19. Bockstaller, C.; Guichard, L.; Makowski, D.; Aveline, A.; Girardins, P.; Plantureux, S. Review article Agri-environmental indicators to assess cropping and farming systems. A review. *Agron. Sustain. Dev.* **2008**, *28*, 139–149. [CrossRef]
20. Payraudeau, S.; Van Der Werf, H.M.G. Environmental impact assessment for a farming region: A review of methods. *Agric. Ecosyst. Environ.* **2005**, *107*, 1–19. [CrossRef]
21. De Vries, M.; van Middelaar, C.E.; de Boer, I.J.M. Comparing environmental impacts of beef production systems: A review of life cycle assessments. *Livest. Sci.* **2015**, *178*, 279–288. [CrossRef]
22. Lebacq, T.; Baret, P.V.; Stilmant, D. Sustainability indicators for livestock farming. A review. *Agron. Sustain. Dev.* **2013**, *33*, 311–327. [CrossRef]

23. Bockstaller, C.; Guichard, L.; Keichinger, O.; Girardin, P.; Galan, M.B.; Gaillard, G. Comparison of methods to assess the sustainability of agricultural systems: A review. In *Sustainable Agriculture*; Springer: Berlin/Heidelberg, Germany, 2009; ISBN 9789048126651.

24. Gómez-Limón, J.A.; Sanchez-Fernandez, G. Empirical evaluation of agricultural sustainability using composite indicators. *Ecol. Econ.* **2010**, *69*, 1062–1075. [CrossRef]

25. Gaviglio, A.; Bertocchi, M.; Demartini, E. A Tool for the Sustainability Assessment of Farms: Selection, Adaptation and Use of Indicators for an Italian Case Study. *Resources* **2017**, *6*, 60. [CrossRef]

26. Angevin, F.; Fortino, G.; Bockstaller, C.; Pelzer, E.; Messéan, A. Assessing the sustainability of crop production systems: Toward a common framework? *Crop Prot.* **2017**, *97*, 18–27. [CrossRef]

27. Pelzer, E.; Fortino, G.; Bockstaller, C.; Angevin, F.; Lamine, C.; Moonen, C.; Vasileiadis, V.; Guérin, D.; Guichard, L.; Reau, R.; et al. Assessing innovative cropping systems with DEXiPM, a qualitative multi-criteria assessment tool derived from DEXi. *Ecol. Indic.* **2012**, *18*, 171–182. [CrossRef]

28. De Olde, E.; Oudshoorn, F.; Bokkers, E.; Stubsgaard, A.; Sørensen, C.; de Boer, I. Assessing the Sustainability Performance of Organic Farms in Denmark. *Sustainability* **2016**, *8*, 957. [CrossRef]

29. Yang, L.; Huang, B.; Mao, M.; Yao, L.; Niedermann, S.; Hu, W.; Chen, Y. Sustainability assessment of greenhouse vegetable farming practices from environmental, economic, and socio-institutional perspectives in China. *Environ. Sci. Pollut. Res.* **2016**, *23*, 17287–17297. [CrossRef] [PubMed]

30. Sottile, F.; Fiorito, D.; Tecco, N.; Girgenti, V.; Peano, C. An interpretive framework for assessing and monitoring the sustainability of school gardens. *Sustainability* **2016**, *8*, 801. [CrossRef]

31. De Luca, A.I.; Falcone, G.; Stillitano, T.; Iofrida, N.; Strano, A.; Gulisano, G. Evaluation of sustainable innovations in olive growing systems: A Life Cycle Sustainability Assessment case study in southern Italy. *J. Clean. Prod.* **2018**, *171*, 1187–1202. [CrossRef]

32. Snapp, S.S.; Grabowski, P.; Chikowo, R.; Smith, A.; Anders, E.; Sirrine, D.; Chimonyo, V.; Bekunda, M. Maize yield and profitability tradeoffs with social, human and environmental performance: Is sustainable intensification feasible? *Agric. Syst.* **2018**, *162*, 77–88. [CrossRef]

33. Recanati, F.; Castelletti, A.; Dotelli, G.; Melià, P. Trading off natural resources and rural livelihoods. A framework for sustainability assessment of small-scale food production in water-limited regions. *Adv. Water Resour.* **2017**, *110*, 484–493. [CrossRef]

34. Bockstaller, C.; Beauchet, S.; Manneville, V.; Amiaud, B.; Botreau, R. A tool to design fuzzy decision trees for sustainability assessment. *Environ. Model. Softw.* **2017**, *97*, 130–144. [CrossRef]

35. Goswami, R.; Saha, S.; Dasgupta, P. Sustainability assessment of smallholder farms in developing countries. *Agroecol. Sustain. Food Syst.* **2017**, *41*, 546–569. [CrossRef]

36. Theurl, M.C.; Hörtenhuber, S.J.; Lindenthal, T.; Palme, W. Unheated soil-grown winter vegetables in Austria: Greenhouse gas emissions and socio-economic factors of diffusion potential. *J. Clean. Prod.* **2017**, *151*, 134–144. [CrossRef]

37. Chopin, P.; Tirolien, J.; Blazy, J.M. Ex-ante sustainability assessment of cleaner banana production systems. *J. Clean. Prod.* **2016**, *139*, 15–24. [CrossRef]

38. Vasileiadis, V.P.; Dachbrodt-Saaydeh, S.; Kudsk, P.; Colnenne-David, C.; Leprince, F.; Holb, I.J.; Kierzek, R.; Furlan, L.; Loddo, D.; Melander, B.; et al. Sustainability of European winter wheat- and maize-based cropping systems: Economic, environmental and social ex-post assessment of conventional and IPM-based systems. *Crop Prot.* **2017**, *97*, 60–69. [CrossRef]

39. Egea, P.; Pérez y Pérez, L. Sustainability and multifunctionality of protected designations of origin of olive oil in Spain. *Land Use Policy* **2016**, *58*, 264–275. [CrossRef]

40. Dong, F.; Mitchell, P.D.; Knuteson, D.; Wyman, J.; Bussan, A.J.; Conley, S. Assessing sustainability and improvements in US Midwestern soybean production systems using a PCA-DEA approach. *Renew. Agric. Food Syst.* **2016**, *31*, 524–539. [CrossRef]

41. Sajjad, H.; Nasreen, I. Assessing farm-level agricultural sustainability using site-specific indicators and sustainable livelihood security index: Evidence from Vaishali district, India. *Community Dev.* **2016**, *47*, 602–619. [CrossRef]

42. Allahyari, M.S.; Daghighi Masouleh, Z.; Koundinya, V. Implementing Minkowski fuzzy screening, entropy, and aggregation methods for selecting agricultural sustainability indicators. *Agroecol. Sustain. Food Syst.* **2016**, *40*, 277–294. [CrossRef]

43. El Chami, D.; Daccache, A. Assessing sustainability of winter wheat production under climate change scenarios in a humid climate—An integrated modelling framework. *Agric. Syst.* **2015**, *140*, 19–25. [CrossRef]

44. Santiago-Brown, I.; Metcalfe, A.; Jerram, C.; Collins, C. Sustainability assessment in wine-grape growing in the New World: Economic, environmental, and social indicators for agricultural businesses. *Sustainability* **2015**, *7*, 8178–8204. [CrossRef]

45. Peano, C.; Tecco, N.; Dansero, E.; Girgenti, V.; Sottile, F. Evaluating the sustainability in complex agri-food systems: The SAEMETH framework. *Sustainability* **2015**, *7*, 6721–6741. [CrossRef]

46. Dong, F.; Mitchell, P.D.; Colquhoun, J. Measuring farm sustainability using data envelope analysis with principal components: The case of Wisconsin cranberry. *J. Environ. Manag.* **2015**, *147*, 175–183. [CrossRef] [PubMed]

47. Yegbemey, R.N.; Yabi, J.A.; Dossa, C.S.G.; Bauer, S. Novel participatory indicators of sustainability reveal weaknesses of maize cropping in Benin. *Agron. Sustain. Dev.* **2014**, *34*, 909–920. [CrossRef]

48. Peano, C.; Migliorini, P.; Sottile, F. A methodology for the sustainability assessment of agri-food systems: An application to the slow food presidia project. *Ecol. Soc.* **2014**, *19*, 1–11. [CrossRef]

49. Van Asselt, E.D.; Van Bussel, L.G.J.; Van Der Voet, H.; Van Der Heijden, G.W.A.M.; Tromp, S.O.; Rijgersberg, H.; Van Evert, F.; Van Wagenberg, C.P.A.; Van Der Fels-Klerx, H.J. A protocol for evaluating the sustainability of agri-food production systems—A case study on potato production in peri-urban agriculture in the Netherlands. *Ecol. Indic.* **2014**, *43*, 315–321. [CrossRef]

50. Colomb, B.; Carof, M.; Aveline, A.; Bergez, J.E. Stockless organic farming: Strengths and weaknesses evidenced by a multicriteria sustainability assessment model. *Agron. Sustain. Dev.* **2013**, *33*, 593–608. [CrossRef]

51. Vasileiadis, V.P.; Moonen, A.C.; Sattin, M.; Otto, S.; Pons, X.; Kudsk, P.; Veres, A.; Dorner, Z.; van der Weide, R.; Marraccini, E.; et al. Sustainability of European maize-based cropping systems: Economic, environmental and social assessment of current and proposed innovative IPM-based systems. *Eur. J. Agron.* **2013**, *48*, 1–11. [CrossRef]

52. Sami, M.; Shiekhdavoodi, M.J.; Almassi, M.; Marzban, A. Assessing the sustainability of agricultural production systems using fuzzy logic. *J. Cent. Eur. Agric.* **2013**, *14*, 318–330. [CrossRef]

53. Van Passel, S.; Meul, M. Multilevel and multi-user sustainability assessment of farming systems. *Environ. Impact Assess. Rev.* **2012**, *32*, 170–180. [CrossRef]

54. Reig-Martínez, E.; Gómez-Limón, J.A.; Picazo-Tadeo, A.J. Ranking farms with a composite indicator of sustainability. *Agric. Econ.* **2011**, *42*, 561–575. [CrossRef]

55. Sharma, D.; Shardendu, S. Assessing farm-level agricultural sustainability over a 60-year period in rural eastern India. *Environmentalist* **2011**, *31*, 325–337. [CrossRef]

56. Rodrigues, G.S.; Rodrigues, I.A.; de Almeida Buschinelli, C.C.; de Barros, I. Integrated farm sustainability assessment for the environmental management of rural activities. *Environ. Impact Assess. Rev.* **2010**, *30*, 229–239. [CrossRef]

57. Gomes, E.G.; Soares De Mello, J.C.C.B.; Souza, G.D.S.E.; Angulo Meza, L.; Mangabeira, J.A.D.C. Efficiency and sustainability assessment for a group of farmers in the Brazilian Amazon. *Ann. Oper. Res.* **2009**, *169*, 167–181. [CrossRef]

58. Van Passel, S.; Van Huylenbroeck, G.; Lauwers, L.; Mathijs, E. Sustainable value assessment of farms using frontier efficiency benchmarks. *J. Environ. Manag.* **2009**, *90*, 3057–3069. [CrossRef] [PubMed]

59. Sadok, W.; Angevin, F.; Bergez, J.-E.; Bockstaller, C.; Colomb, B.; Guichard, L.; Reau, R.; Messéan, A.; Doré, T. MASC, a qualitative multi-attribute decision model for ex ante assessment of the sustainability of cropping systems. *Agron. Sustain. Dev.* **2009**, *29*, 447–461. [CrossRef]

60. Siciliano, G. Social multicriteria evaluation of farming practices in the presence of soil degradation. A case study in Southern Tuscany, Italy. *Environ. Dev. Sustain.* **2009**, *11*, 1107–1133. [CrossRef]

61. Walter, C.; Stützel, H. A new method for assessing the sustainability of land-use systems (II): Evaluating impact indicators. *Ecol. Econ.* **2009**, *68*, 1288–1300. [CrossRef]

62. Walter, C.; Stützel, H. A new method for assessing the sustainability of land-use systems (I): Identifying the relevant issues. *Ecol. Econ.* **2009**, *68*, 1275–1287. [CrossRef]
63. Meul, M.; Passel, S.; Nevens, F.; Dessein, J.; Rogge, E.; Mulier, A.; Hauwermeiren, A. MOTIFS: A monitoring tool for integrated farm sustainability. *Agron. Sustain. Dev.* **2008**, *28*, 321–332. [CrossRef]

 *sustainability*

Review

From Laboratory to Proximal Sensing Spectroscopy for Soil Organic Carbon Estimation—A Review

Theodora Angelopoulou [1,2,*], Athanasios Balafoutis [1], George Zalidis [2,3] and Dionysis Bochtis [1]

[1] Centre of Research and Technology-Hellas (CERTH), Institute for Bio-Economy and Agri-Technology (iBO), Thessaloniki, 57001 Thermi, Greece; a.balafoutis@certh.gr (A.B.); d.bochtis@certh.gr (D.B.)

[2] Laboratory of Remote Sensing, Spectroscopy, and GIS, Department of Agriculture, Aristotle University of Thessaloniki, 54124 Thessaloniki, Greece; zalidis@agro.auth.gr

[3] Interbalkan Environment Center (i-BEC), 18 Loutron Str., 57200 Lagadas, Greece

* Correspondence: d.angelopoulou@certh.gr

Received: 21 November 2019; Accepted: 2 January 2020; Published: 7 January 2020

Abstract: Rapid and cost-effective soil properties estimations are considered imperative for the monitoring and recording of agricultural soil condition for the implementation of site-specific management practices. Conventional laboratory measurements are costly and time-consuming, and, therefore, cannot be considered appropriate for large datasets. This article reviews laboratory and proximal sensing spectroscopy in the visible and near infrared (VNIR)–short wave infrared (SWIR) wavelength region for soil organic carbon and soil organic matter estimation as an alternative to analytical chemistry measurements. The aim of this work is to report the progress made in the last decade on data preprocessing, calibration approaches, and system configurations used for VNIR-SWIR spectroscopy of soil organic carbon and soil organic matter estimation. We present and compare the results of over fifty selective studies and discuss the factors that affect the accuracy of spectroscopic measurements for both laboratory and in situ applications.

Keywords: reflectance spectroscopy; soil spectral libraries; VNIR-SWIR; soil organic matter; carbon sequestration

1. Introduction

Food production requires fertile soils that can be deteriorated by intensive agricultural practices [1]. In order to have sustainable agricultural systems, economic viability for both farmers and society should be considered, but environmental costs also need to be taken into account, avoiding the implementation of practices that could lead to irreversible soil degradation [2]. Sustainable land use management can be achieved by maintaining and enhancing ecosystem services that include environmental, physical, and socioeconomic aspects [3,4]. The consequences of soil degradation have been recognized by the scientific community, but it was not until 1987 that the United Nations Environment Programme (UNEP) made an agreement with the International Soil Reference and Information Centre (ISRIC) for the execution of the Global Assessment of Soil Degradation (GLASOD) project that produced a world map of human-induced soil degradation [5]. Although a potentially nonrenewable natural resource, the EU Thematic Strategy for Soil Protection was issued in 2006, identifying eight main threats to soils i.e., organic matter decline, erosion, compaction, decline of biodiversity, salinization, contamination, sealing, and landslides [6]. In addition to that, the Sustainable Development Goals, adopted by the United Nations on September 2015, identified the importance of preserving soil resources from degradation in order to achieve such goals [7,8]. A key indicator of soil fertility, and thus soil quality, is soil organic matter (SOM) content [9]. Soil organic carbon (SOC) as a component of SOM affects soil fertility, but also has an impact on climate change, as soil represents one of the largest terrestrial carbon pools [10]. SOC losses could be induced from global warming that stimulates the rate of SOM decomposition, land use changes, and plant cover reduction [11,12]. To that effect, estimating SOC stocks is considered crucial for greenhouse gas (GHG) emission reports, which is not a very feasible task on large scales. Precision agriculture technologies could positively contribute to GHG emissions mitigation [13]. The various methods for SOC content determination can be categorized as analytical methods (i.e., dry and wet combustion), remote sensing based methods (i.e., space-borne and airborne), spectroscopy methods (i.e., visible near-infrared and short wave infrared, VNIR-SWIR, and mid-infrared, MIR spectroscopy), and laser-induced breakdown spectroscopy and inelastic neutron scattering methods [14]. Yet, there are still challenges that need to be addressed concerning data harmonization between different sources due to diversions in spatial and temporal data resolution, as well as a lack of the necessary information about soil sampling design and, in numerous cases, an absence of important auxiliary variable measurements for SOC stock estimation (i.e., bulk density) [15].

All analytical SOC measuring protocols have been thoroughly described in the literature [16], while remote sensing techniques for SOC estimation have been recently reviewed in detail [17]. However, there is a need to review the research work in laboratory and proximal sensing techniques in the VNIR-SWIR region, due to their significance on the development of nondestructive SOC estimation.

Therefore, this work aims to introduce the basic principles and approaches used for the analysis of spectroscopic measurements for SOC and/or SOM estimation, and not the humus fraction. Specifically, we present advances in soil reflectance spectroscopy from laboratory to proximal sensing applications, including a broad range of different approaches to inform the reader about progress made in the past decade. To that end, we cite selected papers that, to our knowledge, cover most of the methodologies used for modelling the acquired spectroscopic data, regarding multivariate statistical methods and preprocessing techniques, as well as the efforts made to address the effects of external factors on the accuracy of SOC/SOM predictions.

2. Laboratory VNIR-SWIR Spectroscopy

Over recent decades, the use of soil reflectance spectroscopy in laboratory conditions has gained much attention. Due to the principles of energy–matter interactions, a material can reflect, absorb, scatter, and emit electromagnetic radiation in a characteristic manner that depends on its molecular composition and shape, resulting in a unique spectral signature [18]. A sensor could measure the reflectance of an object at a wide area of wavelengths, providing information about its constituents.

Soil spectroscopy in the VNIR-SWIR region (400–2500 nm) has been evaluated as a possible alternative method for monitoring soil parameters, to address the need for continuous information about soil's condition, while reducing the cost of soil analysis [19–21]. The nondestructive nature of this technique enables simultaneous and repeatable measurements to be made, giving it a significant advantage over conventional laboratory measurements [22]. Exploiting the direct spectral responses of SOM—mainly due to overtones, as well as the bending and stretching of NH, OH, and CH groups [23]—has become one of the most studied and accurately measured soil properties regarding soil reflectance spectroscopy applications [23,24]. In addition to this, the presence of SOM affects the color of the soil, and hence, could be directly related to the visible region of the spectrum [25]. Research was initially focused on evaluating this technique in controlled laboratory conditions, which would make it less time consuming and expensive compared to conventional soil analysis [26]. In addition to that, the use of chemical reagents is not required, which is safe for the environment and provides safer working conditions [27]. Nevertheless, soil samples that are measured in laboratory conditions need the conventional preparation of drying and sieving prior to spectroscopic measurements [28].

One of the first studies that identified the potential to measure SOM content was made by Krishnan et al. [29], who observed that the slope of the spectral curve at around 800 nm increased with increasing SOM content. Dalal and Henry, on the other hand, observed that variations in soil moisture content and soil texture were factors that affected the entire spectral signature of soil samples with the same OC content [30].

2.1. Preprocessing Techniques

Soil matrix consists of numerous parameters that are characterized by low concentrations and have overlapping absorptions that interfere with the spectroscopic measurements. For that reason, soil spectra in the VNIR-SWIR region are considered nonspecific, broad, and with low signal. To that end, spectral analysis requires the use of multivariate approaches to extract hidden information [31]. Commonly, prior to calibration, raw soil spectral data are subjected to preprocessing techniques to remove or minimize noise, enhance their signal, and select characteristic spectral features to improve the subsequent calibration models [32]. Preprocessing techniques are mathematical procedures that transform reflectance measurements and are able to remove variability from light scattering effects and enhance spectral features. These techniques can be categorized in scatter correction methods and spectral derivatives [33]; however, selecting the proper preprocessing technique is not a common procedure. Dotto et al. applied seven of the most common preprocessing techniques (i.e., Savitzky-Golay first derivative, normalization by range, standard normal variates (SNV), multiplicative scatter-correction, continuum removed reflectance (CRR), and the transformation to absorbance and application of a Savitzky-Golay first derivative with a first-order polynomial and a window size of 5 nm) and compared their performance within a wide range of multivariate calibration algorithms [34]. CRR was indicated as the most proper preprocessing technique for SOC prediction, irrespective of the multivariate method applied.

Another aspect of consideration is outlier detection and removal [35]. Though most studies remove extreme values from the modelling procedure, there is a notion that they should not be excluded from the dataset for a robust model to be able to detect such values [36]. An example is the work of De Santana et al. that aimed at providing adequate information about the methodology used for the evaluation of outliers [37], and proposed a new method for their detection using the random forest (RF) technique. It was found that RF generated a lower number of outliers from partial least square regression (PLSR).

2.2. Multivariate Calibrations

Spectral data analysis is performed with the use of multivariate statistical methods, the success of which is highly dependent on the selected calibration method [38]. PLSR is the most commonly-used linear method to describe the relationship between spectral data and soil properties due to its interpretability and low computation time [39,40]. Thus, it has been observed that the aforementioned relationship is not always linear, and that therefore PLSR could be considered insufficient for modelling soil properties [41]. Several studies have attempted to address this problem using machine learning techniques, in the hope that this would produce more accurate results than linear regression [27,28,42,43]. As Davis et al. [16] stated, the diverse and various methodologies for SOC estimation limit the researchers' ability of the to compare the results of their studies in order to generate conclusive results. Therefore, laboratory-based studies focus on finding the most accurate and effective approach for such a calibration. Hence, differences in these studies are mainly in the protocol used and the selection of the preprocessing technique and multivariate statistical method. The accuracy of a model can be assessed using a variety of different methods and metrics. The most common are (i) the coefficient of determination (R^2), which measures the percentage of variance of the dependent variable, as explained by the independent variable, (ii) the root mean square error (RMSE), (iii) the residual prediction deviation (RPD), which is the ratio of the standard deviation of the measured reference values to RMSE, and more recently, (iv) the ratio of performance to interquartile range (RPIQ).

For robust VNIR-SWIR soil spectral measurements, it is important to construct a database that is representative of soil variability for each studied area [44]. For the generation of the required data variability, Sithole et al. [45] acquired samples from various depths and different tillage systems that produced a robust PLSR predictive model for SOC with R^2 = 0.993, RMSEP = 0.157, and RPD = 2.55, i.e., with very high accuracy in terms of its ability to predict SOC. The modified PLSR was utilized by Heinze et al. [46], as it was considered more accurate than PLSR. Soil samples were acquired from three long-term experimental sites that resulted in high diversity in SOC content (3.4–18 mg g $^{-1}$) and soil texture range. The results for SOC predictions were found to be very accurate, with R^2 above 0.96 for the two of the three sites, while poorer predictions were found for the site with a high sand content and smaller SOC range. Similarly, results were found by Stenberg et al. claiming that a high sand content reduces the accuracy of SOC estimation [39]. In contrast to previous studies, Bikindou et al. reported the ability of soil reflectance spectroscopy to accurately determine SOC in sandy soils with low SOM concentrations [47].

The accuracy and variability of results, depending on the multivariate statistical technique used for modelling SOC predictions, is evident in large number of studies that have evaluated and compared different algorithms. The available spectroscopic data have increased, and along with the use of machine learning techniques, created new opportunities for data analysis in the agricultural sector [48]. Spectroscopic models based on the M5 algorithm [49] were evaluated by Rossel et al. [50] using the Australian soil library. The library comprises over 10,000 samples with SOC measurements, providing an adequate sample size, which is an important parameter for the application of machine learning techniques. SOC models performed well, with an RPD of 2.17, and provided information about the important wavelengths used to partition the data and to predict the SOC content. Comparing PLSR, support vector machines regression (SVMR) and multivariate adaptive regression splines (MARS), in salt affected soils, Nawar et al. [32] reported different prediction accuracies that were dependent both on the multivariate technique and the preprocessing method. Overall, MARS, in combination with continuum removal for spectral preprocessing, showed better results ($R^2 = 0.81$), while PLSR performed better than SVMR. The study also showed a high level of correlation of SOM and raw spectra between the 550–680 nm regions, probably related to soil color. Due to the complex relationship between spectral data and soil parameters that is not always considered linear, de Santana et al. [37] compared RF to PLSR; the results were marginally better for RF, as they were able to identify outliers using a proximity matrix. In the same line, SVMR was found to perform better from linear multivariate methods (principal component analysis-PCA, PLSR) and back-propagation neural networks [51]. The selection of the most appropriate data pretreatment and calibration can be a laborious procedure due to the various combinations that can be applied. Gholizadeh et al. [52] compared some of the most common calibration techniques i.e., PLSR, RF, BRT, SVMR, and MBL with a new data-mining engine PARACUDA-II$^®$ [53]. PARACUDA-II$^®$ is based on the all possibilities approach (APA) and a conditional Latin hypercube sampling (cLHs) algorithm. It also has the capability of parallel programming, to assess all the possible combinations of eight different spectral preprocessing techniques against the original reflectance and the chemical data prior to model development. The performance was better compared to all other five multivariate techniques, with a $R^2 = 0.80$. The development of large soil spectral libraries has also given rise to the opportunity to explore the potential of deep learning in the analysis of such data. Padarian et al. [54] presented one of the first studies to evaluate convolutional neural networks (CNN), while introducing the concept of spectrograms to visualize soil spectra; it was found that CNN outperformed both the PLSR and Cubist models.

2.2.1. Feature Selection

Variable selection algorithms are applied to reduce the complexity of spectral data, and preselecting the useful features of the spectrum for calibration could increase the predictive ability of a model. Among the various techniques that could be used when selecting an optimum set of features, genetic algorithms have been shown to perform well for SOC prediction models [41,55,56]. Considering the complexity of genetic algorithms and the risk of overfitting when there are more than 200 variables, Peng et al. [42] evaluated a successive projection algorithm (SPA) combined with SVMR and PLSR to select the important wavelengths. This resulted in selecting 28 wavelength regions mainly found in the near infrared region, which is highly correlated with SOC content. The best predictive model was the combination of SPA-SVMR ($R^2 = 0.73$) compared to SPA-PLSR ($R^2 = 0.62$). To make a step further, Chen et al. [57] proposed a combination of RF and a back propagation network to refine the selective informative wavelengths. It has been also suggested that selecting the inputs of a least squares SVMR model based on the regression coefficients obtained by PLS analysis could provide better prediction models [58]. Recently, Raj et al. [59] utilized a series of variable indicators (Pearson's correlation coefficient (r), biweight midcorrelation (bicor), mutual information based adjacency (AMI), variable importance in the projection (VIP)) combined with an ordered predictor selection (OPS) method to choose the optimum number of spectral variables. The best results were obtained with the use of the AMI indicator combined with SVMR ($R^2 = 0.68$) compared to the use of the full spectrum PLSR model. Bayer et al. [60] compared a physically-based model, where spectral feature analysis is coupled with multiple linear regression techniques (MLR), with a PLSR model. The results showed that model calibrations using significant spectral wavelengths could provide similar results with PLSR for SOC predictions.

2.2.2. Calibration Dataset

Apart from the proper preprocessing and calibration techniques, the stratification of local spectral libraries by homogeneity criteria of the spectral data regarding soil classes and land uses could lead to more accurate SOC predictions, as Moura-Bueno et al. [61] noted in their study. In the same context, Gupta et al. [36] applied a locally-weighted PLSR algorithm and selected an appropriate distance metric for calculating the similarity between a test sample and the calibration samples, and compared it to global calibrations with Lasso and ridge regression. Clingensmith et al. [62] evaluated four different calibration subsetting techniques coupled with PLSR, sparse partial least squares regression (SPLSR), and the heteroscedastic effects model (HEM). It was found that although the SPLSR and HEM algorithms can reduce the number of predictors, they yielded minor improvements in SOC predictions, probably due to the detritus nature of SOM.

Calibration sample size is also a factor that has been proven to affect model calibration; considering the use of soil reflectance spectroscopy as a cost-efficient method, it is important to determine the optimum number of samples for soil modelling. In general, a larger calibration sample size contributes to more robust models. Debaene et al. [63], after evaluating different sampling schemes on a field scale level, concluded that the use of K-means clustering provided better results over random sampling selection, analyte concentration, and PCA scores. However, it was observed that increasing the number of samples after a certain number did not provide proportional results in the decrease of RMSE, and thus, was not significant. In the same scope, Lucà et al. [64] found different levels of model performance variation with a selected calibration size that ranged from 14 to 144 samples and different modelling approaches, i.e., PCR, PLSR, and SVMR. Overall the SVMR model provided the best performance with 72 samples and an average RMSE of prediction of 1.10. It was also reported that creating a robust model depends on the representativeness of the data that can be selected based on the spectral characteristics and/or the physicochemical properties. Hence, the quality of the calibration dataset is affected by more factors than its size, but also by the spatial variability of the studied area.

2.3. Soil Moisture Effects

Since laboratory measurements are mainly conducted on air-dried and sieved soil samples, transferring such models to field conditions entails issues to be addressed, notably soil moisture that is known to affect the soils spectral signature [20,23,65,66]. Therefore, several studies have attempted to create prediction models for different levels of soil moisture [67–69]. The main absorption peaks of water content can be found near 1400, 1900, and 2200 nm due to the overtones and vibrations of the O–H group [70]. To quantify the effect of soil water content on the prediction models, Marakkala Manage et al. [71] used six different matrix potentials that could also provide information about the relationship between soil moisture, SOC, and soil texture. A typical decrease in reflectance was observed with the increase of water content [72], though the most significant change in the spectral signature was found in the transition from dry to wet soils, where water retention forces change from adsorptive to capillary forces. The results indicated that for in situ measurements of SOC content, it is best for soils to be close to dry. Nocita et al. [73] proposed the normalized soil moisture index (NMSI) [74] to classify artificially-moistened soil samples to estimate SOC content from samples with unknown soil moisture contents. It was observed that there was an overlap in the mean spectra in the regions from 400 to 1000 nm at soil moisture levels above 0.15 gW gS^{-1}, while in the region of 1100–2500 nm, there were minor but detectable differences among samples. It was also found that calibration models developed from moist samples had better accuracy for SOC predictions than models created from dry samples. Hence, in the development and exploitation of national and global spectral libraries of dry soil samples, there is a need to remove the effect of soil moisture when moist samples are to be measured.

Several methods have been proposed to remove the effect of soil moisture from soil spectra [75]. The most common are direct standardization (DS) [76] and external parameter orthogonalization (EPO) [77]. Minasny et al. [78] reported that with increasing soil moisture, the accuracy of the PLSR model for SOC was decreasing, but with the use of the EPO algorithm, the accuracy of the wettest soil sample was comparable to that of a dry soil sample. Specifically, PLSR results ranged from R^2 = 0.52–0.78, while EPO-PLSR ranged from R^2 = 0.75–0.80. They concluded that it is preferable for the calibration sample size to be greater than 100 samples, and suggested further investigation to validate this method in open-field conditions. Similar results were reported in the study of de Santana [79]. Comparing DS and EPO, Roudier et al. [80] concluded that both approaches could mitigate the effect of soil moisture when using soil spectral libraries to develop calibration models, and that both performed similarly. It was also observed that the number of calibration samples affected the performance of DS and EPO, with the latter performing better with a limited number of samples. In addition to that, different soil types affected the prediction models (i.e., soils with uniform textures generated better results compared to soils formed from complex parent materials). Soil roughness is another factor that affects spectroscopic measurements, as it causes light scattering, and thus decreases reflectance. Rodionov et al. [81] aimed to classify the effects of soil moisture and roughness and use these finding for in situ applications. To do so, they created different datasets with disturbed and undisturbed soil samples. The disturbed samples were artificially wetted at six different soil moisture contents and sieved at seven different classes (the size of their aggregations did not exceed 30 mm). The undisturbed samples were air dried for two weeks; during that time, VNIR-SWIR spectra were recorded. They concluded that for sieved samples, it was possible to predict SOC at different soil moisture levels. Regarding soil roughness, an overestimation of SOC content was observed with an increase of aggregate size.

2.4. Soil Spectral Libraries for Local Calibrations

One of the first studies that tried to unlock the potential of soil spectral libraries (SSLs) and develop accurate calibration models was conducted by Shepherd and Walsh [82]. Consequently, several attempts have been made towards the creation of regional, continental, and global SSLs [20,83–87]. Considering the local character of predictions in most studies, Stevens et al. [83] evaluated the potential of using the LUCAS soil database [88] to possibly cover soil heterogeneity. It was observed that the accuracy of the predictions depended on soil classes (i.e., cropland, grassland, woodland mineral, and organic soils) and the use of auxiliary predictors (i.e., sand and clay). Better prediction models were developed for croplands, grasslands, and mineral soils, while sand content most affected the error of predictions when the SOC content was low. In a similar manner, Liu et al. [89] evaluated whether the classification of the Chinese SSL, according to the spectrally- and conventionally-derived soil type, could improve SOC estimations. Overall SOC prediction was improved by soil type stratification. The model's accuracy was slightly lower when soil types were spectrally derived, probably due to the misclassification of some soil types. From another perspective, the use of large-scale SSLs for accurate soil properties predictions on local scales could be challenging due to their diversity and heterogeneity. Lobsey et al. [90] developed a new method, named ReSampling-Local (RS-LOCAL), that selects a subset from global SSL spiked with a small representative sample from a local site. They concluded that the proposed method could not only improve results, but also reduce the number of calibration samples, thus reducing the cost. In the same line, Gogé et al. [91] aimed to identify the best strategy to utilize a national SSL. They compared the models generated from a national soil database and the same database spiked with subsets from a local library using a fast Fourier transform local weighted method, and reported moderate improvements in accuracy.

2.5. Current Trends

Though VNIR-SWIR soil measurements are considered cost effective compared to analytical chemical analyses, purchasing a spectrometer is still expensive. Currently, there have been efforts to create smaller and cheaper spectrometers. The reduced cost is due to the smaller spectral range these sensors can cover. Therefore, it is imperative to specify the important wavelengths that are correlated with the estimation of a specific soil's properties. Evaluations of the accuracy of such devices is still at a very early stage, with Barthès et al. [92] comparing a JDSU MicroNIR 2200 spectrophotometer (Milpitas, CA, USA) equipped with an InGaAs array detector and a spectral range of 1151–2186 nm with a Foss NIRSystems 5000 (Laurel, MD, USA). The results were promising, though more research needs to be done.

2.6. Summary of VNIR-SWIR Spectroscopy Research Results on a Laboratory Scale

The work analyzed above on VNIR-SWIR spectroscopy under laboratory conditions for SOC/SOM estimations has been summarized in Table 1, in order to provide easy access to the methodology used and the results achieved in each case.

Table 1. Literature review comparing different multivariate methods for SOC/SOM estimations under laboratory conditions.

	Reference	Spectral Range (nm)	Multivariate Method	R^2	RMSE	RPD
1	Vohland et al. (2011) [41]	400–2500	PLSR GA-PLSR SVMR	0.10–0.72 0.13–0.68 0.73–0.81	0.37–0.65% 0.41–0.64% 0.38–0.86%	1.01–1.75 - -
2	Minasny et al. (2011) [78]	350–2500	PLSR EPO-PLSR	0.56–0.83 0.82–0.89	0.50–1.72 (log[g 100 g^{-1}]) 0.26–0.46 (log[g 100 g^{-1}])	- -
3	Rossel et al. (2012) [50]	350–2500	M5 algorithm	-	0.26 (log$_{10}$C)	2.16
4	Bikindou et al. (2012) [47]	1100–2500	PLSR	0.91	0.045%	-
5	Bayer et al. (2012) [60]	350–2500	PLSR MLR	0.69 0.74	0.45% 0.36%	1.53 1.93
6	Nocita et al. (2013) [73]	350–2500	PLSR	0.25–0.63	12.17–30.21 (g kg^{-1})	1.63–1.89
7	Stevens et al. (2013) [83]	400–2500	SVM Cubist	0.67–0.86 0.76–0.89	4.0–15 (g kg^{-1}) 6.4–50.6 (g kg^{-1})	1.74–2.62 1.99–2.88
8	Xuemei (2013) [58]	325–1075	PLSR SVM	0.77–0.80 0.83–0.86	3.32–3.91 (g kg^{-1}) 3.21–3.70 (g kg^{-1})	- -
9	Heinze et al. 2013 [46]	400–2500	MPLS	0.41–0.99	-	1.25–8.02
10	Debaene et al. 2014 [63]	350–2220	PLSR	0.42–0.72	0.12–0.27%	1.0–2.0
11	Rodionov et al. (2014) [81]	350–2500	PLSR	0.84–0.88	0.64–0.75 (g kg^{-1})	2.49–2.92
12	Rienzi et al. (2014) [67]	340–2220	PLSR	0.63–0.88	4.02–7.13 (g kg^{-1})	-

Table 1. *Cont.*

	Reference	Spectral Range (nm)	Multivariate Method	R^2	RMSE	RPD
13	Gogé et al. (2014) [91]	400–2500	PLSR, FFT-LW	0.10–0.58	3.39–7.63 (g kg^{-1})	0.57–1.37
14	Peng et al. (2014) [42]	350–2500	SVMR	0.72	2.83 (g kg^{-1})	1.86
			SPA-PLSR	0.62	3.23 (g kg^{-1})	1.63
			SPA-SVMR	0.73	2.78 (g kg^{-1})	1.89
15	Wijewardane et al. (2016) [68]	350–2500	PLSR	0.40–0.71	0.73–1.01%	-
16	Morellos et al. (2016) [43]	305–2200	PCR	0.72	0.08%	1.89
			PLSR	0.71	0.08%	1.86
			LS-SVM	0.84	0.06%	2.25
			Cubist	0.79	0.07%	2.15
17	Nawar et al. (2016) [32]	350–2500	PLSR	0.50–0.79	0.28–0.42%	1.41–2.16
			SVR	0.51–0.75	0.26–0.37%	1.43–2.0
			MARS	0.66–0.81	0.22–0.33%	1.74–2.27
18	Roudier et al. (2017) [80]	350–2500	PLSR	-	0.93–1.60%	-
19	Lobsey et al. (2017) [90]	350–2500	PLSR,cubist	0.78–0.84	0.48–1.16%	-
20	Luca et al. (2017) [64]	350–2500	PCR	0.69	0.88%	-
			PLSR	0.79	0.71%	-
			SVMR	0.82	0.68%	-
21	Jiang et al. (2017) [93]	350–2500	PLSR	0.79–0.90	0.54–0.88%	-
22	Hong et al. (2018) [69]	350–2500	PLS-SVM	0.70–0.76	-	1.87–2.06
23	Xu et al. (2018) [51]	350–2500	PCR	0.81	6.01 (g g^{-1})	2.31
			PLSR	0.85	5.48 (g g^{-1})	2.54
			BPNN	0.86	5.16 (g g^{-1})	2.69
			SVMR	0.88	4.85 (g g^{-1})	2.84
24	de Santana et al. (2018) [37]	350–2500	RF	0.8	5.46 g/dm^3	-
			PLSR	0.75	6.19 g/dm^3	-
25	Raj et al. (2018) [59]	350–2500	SVM	0.68	0.62%	1.79
			PLSR	0.65	0.64%	1.72
26	Gholizadeh et al. (2018) [52]	350–2500	PLSR	0.63	0.29%	-
			RF	0.65	0.23%	-
			BRT	0.68	0.25%	-
			SVMR	0.71	0.20%	-
			MBL	0.78	0.20%	-
			PARACUDA	0.80	0.12%	-
27	Gupta et al. (2018) [36]	350–2500	PLSR	0.60–0.70	0.18–0.20%	-
28	Liu et al. (2018) [89]	350–2500	PLSR	0.51–0.82	1.63–3.18 (g kg^{-1})	1.44–2.37
29	Sithole et al. (2018) [45]	450–2500	PLSR	0.99	0.16%	2.55
30	Ludwig et al. (2018) [55]	400–2500	GA-PLSR	-	0.04%	2.58
			SVMR	-	0.04%	2.67
			improved GA-PLSR	-	0.04%	2.89
31	Marakkala Manage et al. (2018) [71]	350–2500	PLSR	0.48–0.82	0.001–0.003 (kg kg^{-1})	-
32	Vibhute et al. (2018) [40]	350–2500	PLSR	0.72–0.89	3.51–5.64 (g kg^{-1})	-
33	de Santana et al. (2019) [79]	1150–2500	PLSR	0.86	1.85 (g/dm^3)	2.59
			EPO-PLSR	0.81–0.85	2.15–2.16 (g/dm^3)	2.02

Table 1. Cont.

	Reference	Spectral Range (nm)	Multivariate Method	R^2	RMSE	RPD
34	Padarian, et al. (2019) [54]	350–2500	PlS	0.35	130.5 (g kg^{-1})	-
			Cubist	0.79	43.75 (g kg^{-1})	-
			CNN	0.88	32.14 (g kg^{-1})	-
			CNN multi	0.69	16.81 (g kg^{-1})	-
35	Moura-Bueno et al. (2019) [61]	350–2500	PLSR	0.74	0.52%	-
			MLR	0.72	0.57%	-
			SVM	0.72	0.55%	-
			RF	0.72	0.56%	-
36	Clingensmith et al. (2019) [62]	350–2500	PLSR	0.42–0.53	0.32–0.48%	1.30–1.47
			SPLSR	0.45–0.65	0.31–0.42%	1.34–1.69
			HEM	0.44–0.63	0.31–0.43%	1.34–1.64
37	Barthes et al. (2019) [92]	1151–2186	PLSR	0.64–0.82	-	1.4–2.3

3. Proximal Soil Sensing

Regarding the need for continuous monitoring in the agricultural domain and for site specific applications, in situ soil sensors have been developed for rapid, cost-effective, and quasi real-time data acquisition that can provide high-resolution maps [94]. Proximal soil sensing is defined as the use of field-based sensors that are in close proximity with the ground, i.e., within a maximum distance of two meters [94]. To that end, proximal sensing directly in the field became a challenge, and gained interest due to its potential advantages [95]. These technologies concern either on-the-go sensors, mounted on agricultural vehicles, or hand-held instruments that can be used for site-specific management (e.g., variable rate applications) [96]. Due to the high sampling density these sensors provide, they are considered more effective in capturing field variability, hence addressing the problem of selecting the correct soil sampling strategy that will ensure representative soil samples. The accuracy of a sensor can be measured by means of measurement repeatability at the same time and place, and the correlation between reference measurements of soil properties [97]. Although promising, in situ soil reflectance spectroscopy applications require proper environmental conditions and various pretreatment methods to mitigate the effect of moisture content, soil roughness, and vegetation cover [98].

The use of proximal soil sensing techniques could increase the number of measured soil samples that are needed for an adequate characterization of soil heterogeneity on a field scale. To that end, researchers have focused on developing or updating already-developed sensors [96,99]. Achieving successful measurements would be of great benefit for agriculture, as they can assist on delineating management zones and control the inputs applied in the environment using precision agriculture techniques [100]. In this section, we present several prototypes that have been developed in research studies, as well as evaluations of commercially-used VNIR-SWIR sensors, and discuss the limitations and hindrances of their use.

3.1. Commercial Available In situ Soil Sensors

Bricklemyer and Brown [101] tested a commercially-available, on-the-go soil sensor (Veris Technologies Inc., Salina, KS, USA) in eight fields. Two approaches for model calibration were evaluated: (i) using measurements of the seven out of eight fields to independently predict SOC from the remaining field and continuing this procedure until each of the eight fields was predicted, (ii) adding nine randomly-selected samples from the remaining field to the calibration dataset. The accuracy was poor for both approaches with an RPD = 0.1, probably due to low SOC variability. Due to spectral correlation between soil properties, combining different sensors could potentially improve the accuracy of in situ measurements. To that end, Kweon et al. [102] utilized a dual-wavelength (660 and 940 nm), on-the-go soil optical sensor coupled with electrical conductivity measurements. The relationship between these two wavelengths and SOM at levels below 5% was considered linear, and therefore, MLR was used for model calibration while slope, curvature, and elevation were also used as independent variables. The various combinations of the predictor values provided different results, with R^2 ranging from 0.55 to 0.94. These variations were attributed to the soil moisture content and temperature. In the same line, Knadel et al. [103] compared the predictive ability between a single sensor and a fusion of sensors (VNIR-SWIR, EC and temperature), and found that SOC estimation could be improved by the use of sensor fusion. Wetterind et al. [104] tested the P4000 drill rig-mounted spectrometer (Veris® Technologies, Salina, KS, USA) and also reported that predictions for SOM were marginally improved with a sensor combination. Similar results were reported by Pei et al. [105]. Utilizing the same sensor, Veum et al. [106] aimed to address the effect of soil moisture by applying the EPO transformation in conjunction with the Bayesian Lasso, along with additional covariate information. The accuracy for models trained on dry spectra and tested on field moist spectra was improved. Kuang et al. [107] compared the accuracy between PLSR and artificial neural networks (ANN) for online spectral measurements, and concluded that ANN provided better results in cross validation (R^2 = 0.83) compared to PLSR (R^2 = 0.71). Though using an independent dataset for calibration, the model's performance dropped significantly for both ANN and PLSR calibrations, with R^2 = 0.49 and R^2 = 0.46, respectively. Sorenson et al. [108] performed in situ spectral measurements with the same sensor as Wetterind et al. [104]. For model calibration, they evaluated different multivariate techniques, i.e., MARS, ANN, SVM, PLSR, RF, and Cubist. The lowest RMSE values were found with the use of Cubist and the highest with ANN. They also reported that a model's accuracy increases when it is developed from a homogeneous dataset, though this comes at the expense of model transferability.

3.2. Experimental Prototypes of In situ Soil Sensors

Apart from commercially-available equipment, researchers aimed to develop prototypes for in situ applications. To this end, Kodaira and Shibusawa [109] upgraded a prototype real-time soil sensor (RTSS) (SAS 1000, SHIBUYA MACHINERY Co., Ltd., Japan). The system was mounted on a tractor and comprised a sensor unit, a touch panel, a soil penetrator, and a set of probes. The sensor unit had two spectrophotometers, with spectral range from 350–1100 nm and 950–1700 nm respectively, and a differential global positioning system. The sensor was designed to acquire spectra at adjusted depths every 4 s which, based on the speed of the tractor (0.56 m s^{-1}), resulted in 144 measurements from 31.48 ha; this sampling density was considered adequate for high resolution mapping. For model calibration, the PLSR method with full cross calibration resulted in an accuracy of R^2 = 0.90 for SOM. Field conditions are highly variable and should be taken into consideration during measurements therefore, Kuang and Mouazen [110] utilized an online sensor developed by Mouazen and Ramon [111] that penetrated soil to a depth of 15 cm. Comparing the PLSR models generated for processed soils, fresh soils, and online measurements, it was observed that MC and increased clay content decreased the model's accuracy, while models created from dry samples with high clay content increased the accuracy. To address disturbing factors of in-field spectral measurements (i.e., illumination conditions), Rodionov et al. [112] developed a closed chamber with a commercial VNIR-SWIR spectrometer mounted on a tractor. The system was tested in a small-scale, bare soil field and at different spectral acquisition modes, i.e., stop-and-go and continuous. The modelling approach also accounted for soil moisture and roughness according to Rodionov et al. [81]. Regarding the continuous mode, spectral discontinuities were observed, and thus, the stop-and-go mode was considered more appropriate, providing an R^2 = 0.65. Although the results were significantly lower from laboratory measurements (R^2 = 0.94), they provided the potential for SOC pattern recognition. On-the-go spectral measurements are also affected by sensor movement and heterogeneous soil; hence, Franceschini et al. [113] aimed to mitigate these effects by comparing the results of utilizing the EPO, DS, and orthogonal signal correction (OSC) [114]. The in-situ measurements were performed in relatively dry soil conditions to minimize the effect of soil moisture. It was observed that the albedo of field measurements was generally lower than that of laboratory measurements, probably due to differences in the sensor configuration and illumination conditions during acquisition. Although an improvement in the model's accuracy was observed with the use of OSC, the results were substantially inferior compared to those of laboratory-generated models, indicating that factors like soil-to-sensor distance and angle, gravels or straws, and changes in the illumination conditions should also be considered.

While proximal sensors have been largely developed and used to overcome the constraints of conventional laboratory chemical analyses, in some cases, the use of a single sensor is inadequate for the simultaneous estimation of several soil properties [115]. To address this, Rossel et al. [116] developed an integrated Soil Condition Analysis System (SCANS). SCANS is a fusion of proximal sensing technologies, analytic statistics, and smart engineering that can analyze a 1.20 m soil core after extracting it from the field. A VNIR-SWIR spectrometer, an active γ-ray attenuation sensor for bulk density estimation, a visible camera, and a Lepton long wave infrared camera mounted on the platform made it possible to estimate approximately thirteen physical, biochemical, and mineralogical soil properties. A total of 150 soil cores were derived and measured from a 600-ha cattle grazing farm. Model calibration was performed with the Cubist algorithm combined with the RS-LOCAL algorithm, providing a more efficient exploitation of the local spectral libraries. The EPO algorithm was also used to eliminate the effect of water in the soil spectra in order to achieve the transferability of the calibrated models with models from spectral libraries of dry soil. The results were promising for SOC estimation with a good level of accuracy ($R^2 = 0.83$) for a repeated 10-fold cross validation and $R^2 = 0.81$ for independent validation. The advantage of SCANS was that it provided accurate, fast, and cost-efficient predictions at high spatial resolution and from various depths.

3.3. Handheld Proximal Sensors

Apart from spectroradiometers mounted on vehicles, portable instruments equipped with a handheld contact probe are also used for direct in situ measurements. Cozzolino et al. [117] used the ASD FieldSpec III spectroradiometer with spectral analysis from 350–2500 nm (Analytical Spectral Devices, ASD, Boulder, CO, USA) and utilized the range 350–1850 nm. The spectral measurements were performed in 68 soil samples whose surface was flattened. The PLSR model calibration using only the NIR (950–1800 nm) wavelength region reported an RPD = 1.8 for SOC that suggests moderate accuracy. Gras et al. [118] evaluated seven different practices for acquiring spectra in the field using an ASDLabSpec 5000 spectrometer (Analytical Spectral Devices, Boulder, CO, USA). These procedures were applied either directly to the soil surface (in cores extracted with an auger) or to clods crumbled from the cores. No external validation was performed, as the aim was to select the most efficient approach for spectral acquisition. Different procedures gave different RPD values, and specifically for SOM estimations, RPD ranged from 2.1 for clods crumbled from cores after half a day, to 2.8 in raw core measurements. The same spectrophotometer was also used by Cambou et al. [119]; the spectral measurements were made on the outer core of samples collected with a manual auger. PLSR and seven different preprocessing techniques were used for model calibration and resulted in $R^2 = 0.75$ for SOC estimations and SOC. The poor results obtained could possibly be related to the fact that spectral measurements were not performed on the same samples that were conventionally analyzed.

3.4. Photosynthetic and Nonphotosynthetic Vegetation Affecting In situ Measurements

Although bare soil conditions are the most favorable for in situ measurements, in real field conditions, the presence of either green vegetation or straw cover is very common, and may lead to overestimations of SOC [120]. Rodionov et al. [121] aimed, in their study using the same spectra, to distinguish SOC from photosynthetic and nonphotosynthetic vegetation. They conducted an experiment under laboratory conditions to estimate the effect of vegetation fractional cover. Soil samples were placed in petri dishes with an increasing degree of plants and straws (24 soil samples with 13 coverage degrees). To characterize straw cover, they used the Cellulose Absorbance Index (CAI), and for green vegetation various, known indices were estimated, such as the Normalized Difference Vegetation Index (NDVI). The aforementioned indices were subtracted from field measurements to prevent SOC overestimation. Increasing vegetation coverage resulted in higher reflectance in the whole spectral range and distinctive absorption peeks at 600–700 nm. Because the spectral response of straw was not very distinctive, both the use of CAI and NDVI was recommended. Specifically, CAI values $\geq$ 0.70 and NDVI values $\geq$ 0.80 were indicative of a SOC overestimation. Although this study confirmed the overestimation of SOC content with the presence of green vegetation and straw coverage, the model gave average predictions with R^2 = 0.58–0.66. The authors suggested the importance of estimating the vegetation coverage degree prior to SOC estimations.

3.5. Spiking Techniques

To address the lack of adequate local data for SOC and to better exploit existing SSls, the spiking technique is frequently evaluated. It is mostly used to augment an existing SSL with local spectra to improve the models accuracy [122–124]. Nawar et al. [125] showed that by using spiking for model calibration, the sample selection method and the number of selected samples also affects the model's accuracy. Specifically, a similarity analysis gave better predictive models with a minimum number of spiked samples compared to Kennard stone and random sampling. Guerrero et al. [126] suggested that an extra-weighted spiking subset could increase the prediction accuracy over spiked calibrations. Collecting field data and creating new SSLs for unprocessed field samples is a laborious procedure. To that end, Kühnel and Bogner [127] proposed the use of the synthetic minority oversampling technique (SMOTE) [128] to address the problem of limited field data. The technique resides in the principal of giving extra weight to existing soil samples to generate new synthetic spectra, hence augmenting existing datasets. This method aims to compensate for the loss of significant spectral data when using an EPO or DS algorithms to remove the effect of MC or for the time-consuming measurement of samples at different MC levels. It was observed that synthetic spectra gave more similar spectral responses with in situ spectra rather than spectra from air-dried, sieved samples. The augmentation of the spectral libraries with the synthetic spectra provided lower RMSE and higher RPD compared to spiking. However, the importance of selecting a representative site samples is significant.

3.6. Summary of VNIR-SWIR Soil Proximal Sensing Research Results in Field Scale

The work analyzed above on VNIR-SWIR soil proximal sensing under laboratory conditions for SOC/SOM estimations has been summarized in Table 2, in order to provide easy access to the methodology used and the results achieved in each case.

Table 2. Literature review comparing different multivariate methods for SOC/SOM estimations under field conditions.

	Reference	Spectral Range (nm)	Multivariate Method	R^2	RMSE	RPD
1	Bricklemyer and Brown (2010) [101]	350–2500	PLSR	0.00–0.42	-	1.0–1.3
2	Cozzolino et al. (2013) [117]	350–1850	PLSR	0.81	-	1.8
3	Kodaira and Shibusawa (2013) [109]	400–1700	PLSR	0.9	0.35%	2.9
4	Kweon et al. (2013) [102]	660 and 940	MLR	0.55–0.94	0.11–0.77%	1.50–4.27
5	Kuang and Mouazen, (2013) [110]	305–2200	PLSR	-	1.29–1.90 (g kg^{-1})	2.01–2.24
6	Gras et al. (2014) [118]	350–2500	MPLS	0.77–0.86	-	2.1–2.8

Table 2. *Cont.*

	Reference	Spectral Range (nm)	Multivariate Method	R^2	RMSE	RPD
7	Knadel et al. (2015) [103]	350–2200	PLSR	0.57–0.94	-	1.4–3.9
8	Rodionov et al. (2015) [112]	350–2500	PLSR	0.65	-	-
9	Wetterlind et al. (2015) [104]					
10	Ji et al. (2015) [75]	350–2500	PLSR	0.63–0.70	0.21–0.27 (g kg^{-1})	1.44–1.79
11	Kuang et al. (2015) [107]	305–2200	PLSR ANN	0.37–0.81 0.39–0.90	1.46–3.88% 1.22–3.66%	1.15–2.29 1.22–3.01
12	Rodionov et al. (2016) [121]	350–2500	PLSR	0.84	0.73%	2.53
13	Cambou et al. (2016) [119]	350–2500	PLSR	0.75	-	2
14	Viscarra Rossel et al. (2017) [116]	350–2500	CUBIST	0.81	0.0041%	-
15	Kühnel and Bogner (2017) [127]	350–2500	smote/PLSR	0.40–0.86	1.90–16.63 (mg g^{-1})	-
16	Sorenson et al. (2017) [108]	350–2200	MARS ANN SVMR PLSR RF Cubist	0.76 0.01 0.75 0.54 0.78 0.8	0.66% 1.56% 0.67% 0.90% 0.62% 0.60%	2 0.9 2 1.5 2.1 2.2
17	Veum et al. (2018) [106]	350–2200	PLSR	0.23–0.82	0.19–0.96 g 100 g^{-1}	-
18	Nawar et al. (2018) [125]	305–2200	PLSR	0.74–0.78	0.16–0.18%	1.97–2.14
19	Nawar et al. (2019) [124]	305–2200	RF	0.12–0.75	0.17–0.33%	1.08–2.04
20	Pei et al. (2019) [105]	343–2222	PLSR NN RT RF	0.8 0.86 0.69 0.58	- - - -	- - - -

4. Discussion

Soil properties' monitoring and estimation are essential processes for site-specific management practices, though there is a lack of up-to-date and consistent information on their spatial distribution [129]. Considering the laborious and costly procedure of collecting a sufficient number of samples to develop high-resolution maps, more efficient methods have been evaluated as alternatives to support or, to some extent, replace time-consuming chemical soil analysis. The interaction of soil properties with electromagnetic radiation in the VNIR-SWIR spectral region (400–2500 nm) has provided the grounds by which to assess this method for soil properties estimations. The first attempts began in controlled laboratory conditions, and subsequently, in-field applications appeared. SOC, as one of the most important soil quality parameters, has gained much attention due to its characteristic interactions with electromagnetic radiation.

The procedure for spectroscopic measurement, modelling, and prediction of SOC requires well-defined steps, though the procedures in each step vary among studies. Firstly, soil samples from the studied area are analyzed with conventional laboratory measurements. Even these standardized procedures could differ in each study [32,119], as it was observed that laboratory measurements could also entail errors [130] that affect the predictive ability of the models, since they are dependent upon these specific laboratory reference measurements [131]. Laboratory soil spectroscopy has been evaluated for more than twenty years, and despite the development of soil spectral libraries on a global scale, there is still not an agreed upon protocol for performing such measurements. That raises issues when it comes to sharing and comparing soil spectral data [132]. Although laboratory measurements are held in controlled conditions, each laboratory has a different measurement protocol (e.g., sample preparation, instrument configuration, and different instrument). Soil spectra are affected by systematic (i.e., different instruments) and nonsystematic effects (i.e., random noise and instabilities). To overcome such hindrances, Ben Dor et al. [132] proposed the internal soil standard and a simple protocol to measure soil properties under laboratory conditions; thanks to these efforts, soil spectral libraries from different sources could be merged. Prior to the multivariate analysis, preprocessing techniques are utilized to remove physical phenomena from spectra, enhance spectral features, and improve the model's performance [33]. The most efficient and commonly-used preprocessing techniques seem to be the first derivative [133], standard normal variate [134], and continuum removal [34].

Several studies have demonstrated that important wavelength selection could improve the accuracy of a model by removing unnecessary information and improving a model's interpretability [41,42,55,56,135]. Another factor that affects the accuracy of a model is the calibration size. At present, it is gaining much attention, because it could also prevent soil sampling from being a waste of resources [83]. However, finding the optimum calibration size is not an easy task [64]. It has been reported that with an increasing calibration size, the model's accuracy is increased up to a certain point were no significant improvements can be further observed. Moreover, deciding the calibration sampling strategy could define the representativeness of the calibration dataset. To that end, Kennard–Stone sampling, Fuzzy c-means sampling, and cLHS could provide better results than random sampling [63].

The use of local models for estimating soil properties of a different region could be proven insufficient due to underrepresented ranges of soil property values and differences in soil types in a given area [136]. To cover global soil diversity and variation, many countries have started initiatives to create national, regional, and global SSLs (Australia, European Union, Brazil, and China) [20,83–87], while there are also projects providing publicly-available data, containing metadata regarding soils' key properties and their spectral signature [88,137]. Building these large databases is more cost effective than using conventional analytical methods for the determination of soil properties. The increased size of the data generated by the creation of SSLs, along with improvements in computational resources, enabled the use of machine learning techniques for model calibration [138]. Machine learning and data mining techniques could generate information by exploring the relationships between existing data, and has the ability to learn without being strictly programmed. Using global or national models for local calibrations could also prove to be problematic because it usually captures general trends with a greater level of generalization. To address this shortcoming, several methods are now being successfully evaluated to increase the accuracy of such predictions, like spiking [93,126] and RS-LOCAL [90]. These techniques can augment an SSL with site-specific samples minimizing the required reference measurements while capturing the respective variability.

Although acquiring the spectroscopic data is a feasible and rapid procedure, information is still hidden in the data. The potential of VNIR-SWIR spectroscopy to predict SOC is highly dependent on the selection of the multivariate calibration technique. Apart from the most commonly-used PLSR method, large SSLs provide the opportunity for machine learning techniques (ANN, RF, MARS, SVM, Cubist) to be increasingly evaluated, and in most cases they outperformed PLSR. Specifically, PLSR and other linear approaches may fail to successfully represent the complex relationship between the measured reflectance spectra and soil attributes, while they are also prone to model noise. Compared to PLSR, SVM has demonstrated better performance, probably thanks to its tolerance in the presence of outliers and spectral noise [22]. SVM use kernel functions to project the data onto a new hyperspace where complex nonlinear patterns can be simply represented; therefore, it is capable of learning in a high-dimensional feature space with fewer training data [58]. MARS is a nonparametric method that is also used to estimate complex nonlinear relationships, and in most cases, has outperformed both PLSR and SVM. MARS is a generalization of recursive partitioning regression approaches which generates piece-wise functions that are aggregated in terms of an additive model; therefore, the accuracy of the predictions is higher when the underlying function is continuous [27]. Random forests is also a nonparametric algorithm that is composed of several regression or classification trees that are a combination of bootstrap aggregation and random feature selection. By that means, it is able to rank controlling factors of SOM and provide the relative importance of spectral variables [139]. The Cubist is a rule-based model that is able to generate more comprehensive predictive accuracy than conventional statistical methods with minimized risk of over fitting. It can include various predictor variables (categorical or continuous), and the importance of a variable can be automatically obtained to interpret the variable contribution mechanism in the final predictor model [140]. Deep learning algorithms such as ANN are based on the ability to learn during the training procedure in which they are presented with inputs and a set of expected outputs. Although they have shown good accuracy, their use is limited due to factors such as the selection of the hidden layer size, the requirement of a large number of training data, and the tendency to overfit the data [141]. In a similar manner, CNN is a neural network that includes one or more convolutional layers in its architecture, and has the ability to exploit local correlations to make them more attractive than the fully-connected neural networks [54]. Despite the superiority shown in these techniques, there is a lack of consistency in the results; hence, PLSR is still utilized, as it provides sufficient accuracy [45]. Variations in the study areas and in the selection of calibration data and their size, numerous combinations of preprocessing, and multivariate methods are some of the reasons that no calibration method has achieved universal acceptance.

When it comes to the validation procedure, cross validation [117,123] is the most commonly-used method due to the small sample size of the dataset, though one should keep in mind that there is the risk of overfitting the model; therefore, testing its performance on independent datasets usually provides less accuracy [46].

Considering in situ soil measurements, they also share the same issues with laboratory soil spectroscopy when it comes to model calibration, but with the additional hindrance of external factors affecting the measurements. The most frequently-addressed parameter is soil moisture. Soil moisture has a distinct impact on soil reflectance by decreasing the albedo [113]. To that end, methods to eliminate this effect such as EPO and DS have been proposed with relatively good results [51,142]. Attempts to create samples with variable soil moisture content have also been made, though they are somewhat time consuming [73,80,143]. Additional factors that could interfere with in situ measurements are ambient conditions such as temperature, wind, and precipitation [144,145], soil roughness that could be altered by field operations, stones, plant residues and movement during spectral acquisition, i.e., vibrations [112,113,121]. Moreover, it should be noted that in-field measurements are usually performed in the top layer of the soil, with an exception being studies where the sensor penetrates the soil, which may impact the model's quality because laboratory reference measurements concern samples taken from a layer. Different systems with various configurations regarding the mode of spectral acquisition, the sensor distance from the ground, and illumination conditions vary among studies, increasing the obstacle of comparing the results from one study to another. Regarding the experimental stage of most studies, it is difficult to create a common protocol for in situ, on-the-go measurements.

Differences in evaluations of results are also observed among studies. The most frequent statistical indicators are RMSE, RPD, and R^2. It should be noted that the error should be referred to using the same units as those of the analyzed soil property [146]; most studies do not use the same evaluation methods, as shown in Tables 1 and 2. In addition to that, the normalization of the data due to skewness makes it even harder to compare studies.

5. Conclusions

The objective of this review was to summarize the progress made during the last decade using laboratory and proximal soil sensing in the VNIR-SWIR region for SOC estimations. Laboratory measurements are considered to be a well-established method for soil properties estimations, but there is not yet a commonly accepted universal model with broad application. The attention that soil spectroscopy has gained is evident on the efforts of creating SSLs worldwide, albeit, with a nonagreed protocol. Research has also been made towards the use of proximal soil sensors for in situ applications that could estimate SOC in real time and on larger spatial scales. Overall, the results are promising for SOC estimations, and more research needs to be done in terms of selecting the proper spectral range of the sensor, the preprocessing methods, and the calibration techniques. Detrimental effects such as soil moisture, soil roughness, vegetation cover, and others that affect SOC spectral response need to be addressed for model transferability from laboratory to in field applications. Due to the various inconsistencies among studies, it is suggested that articles should include more information about the experimental design, the criteria used for the selection of the chemometric approach, and the pre- and post- processing procedures to facilitate comparisons of results among studies. The advent of much smaller and more affordable spectrometers could potentially provide even more rapid soil properties estimations to assist farmers in selecting the most appropriate management practices with respect to soil preservation.

Author Contributions: Conceptualization, T.A., D.B., and G.Z.; methodology, T.A., A.B.; formal analysis, T.A. and D.B.; investigation, T.A. and D.B.; resources, T.A. and A.B.; writing—original draft preparation, T.A., A.B.; writing—review and editing, D.B. and G.Z.; supervision, D.B. All authors have read and agreed to the published version of the manuscript.

Funding: This research was funded by the project "Research Synergy to address major challenges in the nexus: energy-environment-agricultural production (Food, Water, Materials)"—NEXUS, funded by the Greek Secretariat for Research and Technology (GSRT)—Pr. No. MIS 5002496.

Conflicts of Interest: The authors declare no conflict of interest.

Abbreviations

AMI	mutual information-based adjacency
ANN	artificial neural network
APA	all possibilities approach
BRT	boosted regression trees
CAI	cellulose absorbance index
cLHs	conditional Latin hypercube sampling
CNN	convolutional neural networks
CRR	continuum removed reflectance
DS	direct standardization
EC	electrical conductivity
EPO	external parameter orthogonalization
HEM	heteroscedastic effects model
MARS	multivariate adaptive regression splines
MBL	memory bases learning
MC	moisture content
MIR	mid infrared
MLR	multivariate linear regression
NDVI	normalized difference vegetation index
NMSI	normalized soil moisture index
OPS	ordered predictor selection
OSC	orthogonal signal correction
PCA	principal component analysis
PLSR	partial least square regression
R^2	coefficient of determination
RF	random forest
RMSE	root mean square error
RPD	residual prediction deviation
RS-LOCAL	re-sampling-local
SCANS	soil condition analysis system
SMOTE	synthetic minority oversampling technique

SNV	standard normal variate
SOC	soil organic carbon
SOM	soil organic matter
SPA	successive projection algorithm
SPLSR	sparse partial least squares regression
SSLs	soil spectral libraries
SVMR	support vector machines regression
SWIR	short wave infrared
VIP	variable importance in the projection
VNIR	visible near infrared

References

1. Johnston, A.E.; Poulton, P.R.; Coleman, K. Chapter 1 Soil Organic Matter: Its Importance in Sustainable Agriculture and Carbon Dioxide Fluxes. *Adv. Agron.* **2009**, *101*, 1–57.

2. Cécillon, L.; Barthès, B.G.; Gomez, C.; Ertlen, D.; Genot, V.; Hedde, M.; Stevens, A.; Brun, J.J. Assessment and monitoring of soil quality using near-infrared reflectance spectroscopy (NIRS). *Eur. J. Soil Sci.* **2009**, *60*, 770–784. [CrossRef]

3. Bogunovic, I.; Muñoz-Rojas, M.; Brevik, E.C. Soil ecosystem services, sustainability, valuation and management. *Curr. Opin. Environ. Sci. Heal.* **2018**, *5*, 7–13.

4. Lampridi, M.G.; Sørensen, C.G.; Bochtis, D. Agricultural Sustainability: A Review of Concepts and Methods. *Sustainability* **2019**, *11*, 5120. [CrossRef]

5. Oldeman, L.R.; Hakkeling, R.T.A.; Sombroek, W.G. *World Map of the Status of Human-Induced Soil Degradation: An Explanatory Note*, 2nd. rev. ed.; ISRIC: Wageningen, The Netherlands, 1991.

6. Panagos, P.; Montanarella, L. Soil Thematic Strategy: An important contribution to policy support, research, data development and raising the awareness. *Curr. Opin. Environ. Sci. Heal.* **2018**, *5*, 38–41. [CrossRef]

7. Keesstra, S.D.; Bouma, J.; Wallinga, J.; Tittonell, P.; Smith, P.; Cerdà, A.; Montanarella, L.; Quinton, J.N.; Pachepsky, Y.; Van Der Putten, W.H.; et al. The significance of soils and soil science towards realization of the United Nations sustainable development goals. *SOIL* **2016**, *2*, 111–128. [CrossRef]

8. Visser, S.; Keesstra, S.; Maas, G.; de Cleen, M.; Molenaar, C. Soil as a Basis to Create Enabling Conditions for Transitions Towards Sustainable Land Management as a Key to Achieve the SDGs by 2030. *Sustainability* **2019**, *11*, 6792. [CrossRef]

9. Carter, M.R. Soil quality for sustainable land management: Organic matter and aggregation interactions that maintain soil functions. *Agron. J.* **2002**, *94*, 38–47. [CrossRef]

10. Kirschbaum, M.U.F. Will changes in soil organic carbon act as a positive or negative feedback on global warming? *Biogeochemistry* **2000**, *48*, 21–51. [CrossRef]

11. Qiu, L.; Wei, X.; Zhang, X.; Cheng, J.; Gale, W.; Guo, C.; Long, T. Soil organic carbon losses due to land use change in a semiarid grassland. *Plant Soil* **2012**, *355*, 299–309. [CrossRef]

12. Martínez-Mena, M.; Alvarez Rogel, J.; Castillo, V.; Albaladejo, J. Organic carbon and nitrogen losses influenced by vegetation removal in a semiarid mediterranean soil. *Biogeochemistry* **2002**, *61*, 309–321. [CrossRef]

13. Balafoutis, A.; Beck, B.; Fountas, S.; Vangeyte, J.; Wal, T.; Soto, I.; Gómez-Barbero, M.; Barnes, A.; Eory, V. Precision Agriculture Technologies Positively Contributing to GHG Emissions Mitigation, Farm Productivity and Economics. *Sustainability* **2017**, *9*, 1339. [CrossRef]

14. FAO. Soil Organic Carbon the Hidden Potential. 2017. Available online: https://www.amazon.com/Soil-Organic-Carbon-Hidden-Potential/dp/9251096813 (accessed on 2 September 2019).

15. Jandl, R.; Rodeghiero, M.; Martinez, C.; Cotrufo, M.F.; Bampa, F.; van Wesemael, B.; Harrison, R.B.; Guerrini, I.A.; deB Richter, D., Jr.; Rustad, L.; et al. Current status, uncertainty and future needs in soil organic carbon monitoring. *Sci. Total Environ.* **2014**, *468–469*, 376–383. [CrossRef] [PubMed]

16. Davis, M.R.; Alves, B.J.R.; Karlen, D.L.; Kline, K.L.; Galdos, M.; Abulebdeh, D. Review of soil organic carbon measurement protocols: A US and Brazil comparison and recommendation. *Sustainability* **2017**, *10*, 53. [CrossRef]

17. Angelopoulou, T.; Tziolas, N.; Balafoutis, A.; Zalidis, G.; Bochtis, D. Remote sensing techniques for soil organic carbon estimation: A review. *Remote Sens.* **2019**, *11*, 676. [CrossRef]

18. Shaw, G.A.; Burke, H.K. Spectral Imaging for Remote Sensing. *LINCOLN Lab. J.* **2003**, *14*.

19. Nocita, M.; Stevens, A.; van Wesemael, B.; Aitkenhead, M.; Bachmann, M.; Barthès, B.; Ben Dor, E.; Brown, D.J.; Clairotte, M.; Csorba, A.; et al. *Chapter Four—Soil Spectroscopy: An Alternative to Wet Chemistry for Soil Monitoring*; Sparks, D.L.B.T.-A., Ed.; Academic Press: San Diego, CA, USA, 2015; Volume 132, pp. 139–159. ISBN 0065-2113.

20. Brown, D.J.; Shepherd, K.D.; Walsh, M.G.; Mays, M.D.; Reinsch, T.G. Global soil characterization with VNIR diffuse reflectance spectroscopy. *Geoderma* **2006**, *132*, 273–290. [CrossRef]

21. Soriano-Disla, J.M.; Janik, L.J.; Viscarra Rossel, R.A.; Macdonald, L.M.; McLaughlin, M.J. The Performance of Visible, Near-, and Mid-Infrared Reflectance Spectroscopy for Prediction of Soil Physical, Chemical, and Biological Properties. *Appl. Spectrosc. Rev.* **2014**, *49*, 139–186. [CrossRef]

22. Pasquini, C. Near infrared spectroscopy: A mature analytical technique with new perspectives—A review. *Anal. Chim. Acta* **2018**, *1026*, 8–36. [CrossRef]

23. Viscarra Rossel, R.A.; Walvoort, D.J.J.; McBratney, A.B.; Janik, L.J.; Skjemstad, J.O. Visible, near infrared, mid infrared or combined diffuse reflectance spectroscopy for simultaneous assessment of various soil properties. *Geoderma* **2006**, *131*, 59–75. [CrossRef]

24. Vasques, G.M.; Grunwald, S.; Sickman, J.O. Modeling of Soil Organic Carbon Fractions Using Visible—Near-Infrared Spectroscopy. *Soil Sci. Soc. Am. J.* **2009**, *73*, 176. [CrossRef]

25. Baumgardner, M.F.; Kristof, S.; Johannsen, C.J.; Zachary, A. Effects of organic matter on the multispectral properties of soils. *Proc. Indian Acad. Sci.* **1970**, *79*, 413–422.

26. Cécillon, L.; Cassagne, N.; Czarnes, S.; Gros, R.; Vennetier, M.; Brun, J.J. Predicting soil quality indices with near infrared analysis in a wildfire chronosequence. *Sci. Total Environ.* **2009**, *407*, 1200–1205. [CrossRef] [PubMed]

27. Rossel, R.A.V.; Behrens, T. Using data mining to model and interpret soil diffuse reflectance spectra. *Geoderma* **2010**, *158*, 46–54. [CrossRef]

28. Wetterlind, J.; Stenberg, B.; Rossel, R.A.V. Soil analysis using visible and near infrared spectroscopy. *Methods Mol. Biol.* **2013**, *953*, 95–107.

29. Krishnan, P.; Alexander, J.D.; Butler, B.J.; Hummel, J.W. Reflectance Technique for Predicting Soil Organic Matter1. *Soil Sci. Soc. Am. J.* **1980**, *44*, 1282. [CrossRef]

30. Dalal, R.C.; Henry, R.J. Simultaneous Determination of Moisture, Organic Carbon, and Total Nitrogen by Near Infrared Reflectance Spectrophotometry1. *Soil Sci. Soc. Am. J.* **1986**, *50*, 120. [CrossRef]

31. Shi, T.; Guo, L.; Chen, Y.; Wang, W.; Shi, Z.; Li, Q.; Wu, G. Proximal and remote sensing techniques for mapping of soil contamination with heavy metals. *Appl. Spectrosc. Rev.* **2018**, *53*, 783–805. [CrossRef]

32. Nawar, S.; Buddenbaum, H.; Hill, J.; Kozak, J.; Mouazen, A.M. Estimating the soil clay content and organic matter by means of different calibration methods of vis-NIR diffuse reflectance spectroscopy. *Soil Tillage Res.* **2016**, *155*, 510–522. [CrossRef]

33. Rinnan, Å.; van den Berg, F.; Engelsen, S.B. Review of the most common pre-processing techniques for near-infrared spectra. *TrAC Trends Anal. Chem.* **2009**, *28*, 1201–1222. [CrossRef]

34. Dotto, A.C.; Dalmolin, R.S.D.; ten Caten, A.; Grunwald, S. A systematic study on the application of scatter-corrective and spectral-derivative preprocessing for multivariate prediction of soil organic carbon by Vis-NIR spectra. *Geoderma* **2018**, *314*, 262–274. [CrossRef]

35. Bellon-Maurel, V.; Fernandez-Ahumada, E.; Palagos, B.; Roger, J.-M.; McBratney, A. Critical review of chemometric indicators commonly used for assessing the quality of the prediction of soil attributes by NIR spectroscopy. *TrAC Trends Anal. Chem.* **2010**, *29*, 1073–1081. [CrossRef]

36. Gupta, A.; Vasava, H.B.; Das, B.S.; Choubey, A.K. Local modeling approaches for estimating soil properties in selected Indian soils using diffuse reflectance data over visible to near-infrared region. *Geoderma* **2018**, *325*, 59–71. [CrossRef]

37. de Santana, F.B.; de Souza, A.M.; Poppi, R.J. Visible and near infrared spectroscopy coupled to random forest to quantify some soil quality parameters. *Spectrochim. Acta Part A Mol. Biomol. Spectrosc.* **2018**, *191*, 454–462. [CrossRef] [PubMed]

38. Geladi, P. Chemometrics in spectroscopy. Part 1. Classical chemometrics. *Spectrochim. Acta Part B At. Spectrosc.* **2003**, *58*, 767–782. [CrossRef]

39. Stenberg, B.; Viscarra Rossel, R.A.; Mouazen, A.M.; Wetterlind, J. Visible and near infrared spectroscopy in soil science. *Adv. Agron.* **2010**, *107*, 163–215.

40. Vibhute, A.D.; Kale, K.V.; Mehrotra, S.C.; Dhumal, R.K.; Nagne, A.D. Determination of soil physicochemical attributes in farming sites through visible, near-infrared diffuse reflectance spectroscopy and PLSR modeling. *Ecol. Process.* **2018**, *7*, 26. [CrossRef]

41. Vohland, M.; Besold, J.; Hill, J.; Fründ, H.-C. Comparing different multivariate calibration methods for the determination of soil organic carbon pools with visible to near infrared spectroscopy. *Geoderma* **2011**, *166*, 198–205. [CrossRef]

42. Peng, X.; Shi, T.; Song, A.; Chen, Y.; Gao, W. Estimating soil organic carbon using VIS/NIR spectroscopy with SVMR and SPA methods. *Remote Sens.* **2014**, *6*, 2699–2717. [CrossRef]

43. Morellos, A.; Pantazi, X.E.; Moshou, D.; Alexandridis, T.; Whetton, R.; Tziotzios, G.; Wiebensohn, J.; Bill, R.; Mouazen, A.M. Machine learning based prediction of soil total nitrogen, organic carbon and moisture content by using VIS-NIR spectroscopy. *Biosyst. Eng.* **2016**, *152*, 104–116. [CrossRef]

44. Stevens, A.; van Wesemael, B.; Bartholomeus, H.; Rosillon, D.; Tychon, B.; Ben-Dor, E. Laboratory, field and airborne spectroscopy for monitoring organic carbon content in agricultural soils. *Geoderma* **2008**, *144*, 395–404. [CrossRef]

45. Sithole, N.J.; Ncama, K.; Magwaza, L.S. Robust Vis-NIRS models for rapid assessment of soil organic carbon and nitrogen in Feralsols Haplic soils from different tillage management practices. *Comput. Electron. Agric.* **2018**, *153*, 295–301. [CrossRef]

46. Heinze, S.; Vohland, M.; Joergensen, R.; Ludwig, B. Usefulness of near-infrared spectroscopy for the prediction of chemical and biological soil properties in different long-term experiments. *J. Plant Nutr. Soil Sci.* **2013**, *176*, 520–528. [CrossRef]

47. Bikindou, F.D.A.; Gomat, H.Y.; Deleporte, P.; Bouillet, J.-P.; Moukini, R.; Mbedi, Y.; Ngouaka, E.; Brunet, D.; Sita, S.; Diazenza, J.-B.; et al. Are NIR spectra useful for predicting site indices in sandy soils under Eucalyptus stands in Republic of Congo? *For. Ecol. Manage.* **2012**, *266*, 126–137. [CrossRef]

48. Liakos, K.G.; Busato, P.; Moshou, D.; Pearson, S.; Bochtis, D. Machine learning in agriculture: A review. *Sensors (Switzerland)* **2018**, *18*, 2674. [CrossRef]

49. Quinlan, J.R. Learning with continuous classes. *Mach. Learn.* **1992**, *92*, 343–348.

50. Rossel, R.A.V.; Webster, R. Predicting soil properties from the Australian soil visible-near infrared spectroscopic database. *Eur. J. Soil Sci.* **2012**, *63*, 848–860. [CrossRef]

51. Xu, S.; Zhao, Y.; Wang, M.; Shi, X. Comparison of multivariate methods for estimating selected soil properties from intact soil cores of paddy fields by Vis–NIR spectroscopy. *Geoderma* **2018**, *310*, 29–43. [CrossRef]

52. Gholizadeh, A.; Saberioon, M.; Carmon, N.; Boruvka, L.; Ben-Dor, E.; Gholizadeh, A.; Saberioon, M.; Carmon, N.; Boruvka, L.; Ben-Dor, E. Examining the Performance of PARACUDA-II Data-Mining Engine versus Selected Techniques to Model Soil Carbon from Reflectance Spectra. *Remote Sens.* **2018**, *10*, 1172. [CrossRef]

53. Carmon, N.; Ben-Dor, E. An Advanced Analytical Approach for Spectral-Based Modelling of Soil Properties. *Int. J. Emerg. Technol. Adv. Eng.* **2017**, *7*.

54. Padarian, J.; Minasny, B.; McBratney, A.B. Using deep learning to predict soil properties from regional spectral data. *Geoderma Reg.* **2019**, *16*, e00198. [CrossRef]

55. Ludwig, B.; Murugan, R.; Parama, V.R.R.; Vohland, M. Use of different chemometric approaches for an estimation of soil properties at field scale with near infrared spectroscopy. *J. Plant Nutr. Soil Sci.* **2018**, *181*, 704–713. [CrossRef]

56. Shi, T.; Wang, J.; Chen, Y.; Wu, G. Improving the prediction of arsenic contents in agricultural soils by combining the reflectance spectroscopy of soils and rice plants. *Int. J. Appl. Earth Obs. Geoinf.* **2016**, *52*, 95–103. [CrossRef]

57. Chen, H.; Liu, X.; Jia, Z.; Liu, Z.; Shi, K.; Cai, K. A combination strategy of random forest and back propagation network for variable selection in spectral calibration. *Chemom. Intell. Lab. Syst.* **2018**, *182*, 101–108. [CrossRef]

58. Xuemei, L.; Jianshe, L. Measurement of soil properties using visible and short wave-near infrared spectroscopy and multivariate calibration. *Meas. J. Int. Meas. Confed.* **2013**, *46*, 3808–3814. [CrossRef]

59. Raj, A.; Chakraborty, S.; Duda, B.M.; Weindorf, D.C.; Li, B.; Roy, S.; Sarathjith, M.C.; Das, B.S.; Paulette, L. Soil mapping via diffuse reflectance spectroscopy based on variable indicators: An ordered predictor selection approach. *Geoderma* **2018**, *314*, 146–159. [CrossRef]

60. Bayer, A.; Bachmann, M.; Müller, A.; Kaufmann, H. A Comparison of feature-based MLR and PLS regression techniques for the prediction of three soil constituents in a degraded South African Ecosystem. *Appl. Environ. Soil Sci.* **2012**, *2012*. [CrossRef]

61. Moura-Bueno, J.M.; Dalmolin, R.S.D.; ten Caten, A.; Dotto, A.C.; Demattê, J.A.M. Stratification of a local VIS-NIR-SWIR spectral library by homogeneity criteria yields more accurate soil organic carbon predictions. *Geoderma* **2019**, *337*, 565–581. [CrossRef]

62. Clingensmith, C.M.; Grunwald, S.; Wani, S.P. Evaluation of calibration subsetting and new chemometric methods on the spectral prediction of key soil properties in a data-limited environment. *Eur. J. Soil Sci.* **2019**, *70*, 107–126. [CrossRef]

63. Debaene, G.; Niedźwiecki, J.; Pecio, A.; Żurek, A. Effect of the number of calibration samples on the prediction of several soil properties at the farm-scale. *Geoderma* **2014**, *214–215*, 114–125. [CrossRef]

64. Lucà, F.; Conforti, M.; Castrignanò, A.; Matteucci, G.; Buttafuoco, G. Effect of calibration set size on prediction at local scale of soil carbon by Vis-NIR spectroscopy. *Geoderma* **2017**, *288*, 175–183. [CrossRef]

65. Vohland, M.; Ludwig, M.; Thiele-Bruhn, S.; Ludwig, B. Determination of soil properties with visible to near- and mid-infrared spectroscopy: Effects of spectral variable selection. *Geoderma* **2014**, *223–225*, 88–96. [CrossRef]

66. St. Luce, M.; Ziadi, N.; Zebarth, B.J.; Grant, C.A.; Tremblay, G.F.; Gregorich, E.G. Rapid determination of soil organic matter quality indicators using visible near infrared reflectance spectroscopy. *Geoderma* **2014**, *232–234*, 449–458. [CrossRef]

67. Rienzi, E.A.; Mijatovic, B.; Mueller, T.G.; Matocha, C.J.; Sikora, F.J.; Castrignanò, A. Prediction of Soil Organic Carbon under Varying Moisture Levels Using Reflectance Spectroscopy. *Soil Sci. Soc. Am. J.* **2014**, *78*, 958. [CrossRef]

68. Wijewardane, N.K.; Ge, Y.; Morgan, C.L.S. Prediction of soil organic and inorganic carbon at different moisture contents with dry ground VNIR: a comparative study of different approaches. *Eur. J. Soil Sci.* **2016**, *67*, 605–615. [CrossRef]

69. Hong, Y.; Yu, L.; Chen, Y.; Liu, Y.; Liu, Y.; Liu, Y.; Cheng, H. Prediction of soil organic matter by VIS-NIR spectroscopy using normalized soil moisture index as a proxy of soil moisture. *Remote Sens.* **2018**, *10*, 28. [CrossRef]

70. Ben-Dor, E. Quantitative remote sensing of soil properties. *Adv. Agron.* **2002**, *75*, 173–244.

71. Marakkala Manage, L.P.; Greve, M.H.; Knadel, M.; Moldrup, P.; de Jonge, L.W.; Katuwal, S. Visible-Near-Infrared Spectroscopy Prediction of Soil Characteristics as Affected by Soil-Water Content. *Soil Sci. Soc. Am. J.* **2018**, *82*, 1333. [CrossRef]

72. Lobell, D.B.; Asner, G.P. Moisture effects on soil reflectance. *Soil Sci. Soc. Am. J.* **2002**, *66*, 722–727. [CrossRef]

73. Nocita, M.; Stevens, A.; Noon, C.; van Wesemael, B. Prediction of soil organic carbon for different levels of soil moisture using Vis-NIR spectroscopy. *Geoderma* **2013**, *199*, 37–42. [CrossRef]

74. Haubrock, S.N.; Chabrillat, S.; Lemmnitz, C.; Kaufmann, H. Surface soil moisture quantification models from reflectance data under field conditions. *Int. J. Remote Sens.* **2008**, *29*, 3–29. [CrossRef]

75. Ji, W.; Viscarra Rossel, R.A.; Shi, Z. Accounting for the effects of water and the environment on proximally sensed vis-NIR soil spectra and their calibrations. *Eur. J. Soil Sci.* **2015**, *66*, 555–565. [CrossRef]

76. Wang, Y.; Veltkamp, D.J.; Kowalski, B.R. Multivariate Instrument Standardization. *J. Electroanal. Chem. Interfaclal Electro-chem.* **1991**, *63*, 2347–2352. [CrossRef]

77. Roger, J.-M.; Chauchard, F.; Bellon-Maurel, V. EPO–PLS external parameter orthogonalisation of PLS application to temperature-independent measurement of sugar content of intact fruits. *Chemom. Intell. Lab. Syst.* **2003**, *66*, 191–204. [CrossRef]

78. Minasny, B.; McBratney, A.B.; Bellon-Maurel, V.; Roger, J.M.; Gobrecht, A.; Ferrand, L.; Joalland, S. Removing the effect of soil moisture from NIR diffuse reflectance spectra for the prediction of soil organic carbon. *Geoderma* **2011**, *167–168*, 118–124. [CrossRef]

79. de Santana, F.B.; de Giuseppe, L.O.; de Souza, A.M.; Poppi, R.J. Removing the moisture effect in soil organic matter determination using NIR spectroscopy and PLSR with external parameter orthogonalization. *Microchem. J.* **2019**, *145*, 1094–1101. [CrossRef]

80. Roudier, P.; Hedley, C.B.; Lobsey, C.R.; Viscarra Rossel, R.A.; Leroux, C. Evaluation of two methods to eliminate the effect of water from soil vis–NIR spectra for predictions of organic carbon. *Geoderma* **2017**, *296*, 98–107. [CrossRef]

81. Rodionov, A.; Pätzold, S.; Welp, G.; Pallares, R.C.; Damerow, L.; Amelung, W. Sensing of Soil Organic Carbon Using Visible and Near-Infrared Spectroscopy at Variable Moisture and Surface Roughness. *Soil Sci. Soc. Am. J.* **2014**, *78*, 949. [CrossRef]

82. Shepherd, K.D.; Walsh, M.G. Development of reflectance spectral libraries for characterization of soil properties. *Soil Sci. Soc. Am. J.* **2002**, *66*, 988–998. [CrossRef]

83. Stevens, A.; Nocita, M.; Tóth, G.; Montanarella, L.; van Wesemael, B. Prediction of Soil Organic Carbon at the European Scale by Visible and Near InfraRed Reflectance Spectroscopy. *PLoS ONE* **2013**, *8*. [CrossRef]

84. Viscarra Rossel, R.A.; Behrens, T.; Ben-Dor, E.; Brown, D.J.; Demattê, J.A.M.; Shepherd, K.D.; Shi, Z.; Stenberg, B.; Stevens, A.; Adamchuk, V.; et al. A global spectral library to characterize the world's soil. *Earth-Science Rev.* **2016**, *155*, 198–230. [CrossRef]

85. Shi, Z.; Wang, Q.L.; Peng, J.; Ji, W.J.; Liu, H.J.; Li, X.; Viscarra Rossel, R.A. Development of a national VNIR soil-spectral library for soil classification and prediction of organic matter concentrations. *Sci. China Earth Sci.* **2014**, *57*, 1671–1680. [CrossRef]

86. Brodský, L.; Klement, A.; Penížek, V.; Kodešová, R.; Boruvka, L. Building soil spectral library of the czech soils for quantitative digital soil mapping. *Soil Water Res.* **2011**, *6*, 165–172. [CrossRef]

87. Cambule, A.H.; Rossiter, D.G.; Stoorvogel, J.J.; Smaling, E.M.A. Building a near infrared spectral library for soil organic carbon estimation in the Limpopo National Park, Mozambique. *Geoderma* **2012**, *183–184*, 41–48. [CrossRef]

88. Tóth, G.; Jones, A.; Montanarella, L. The LUCAS topsoil database and derived information on the regional variability of cropland topsoil properties in the European Union. *Environ. Monit. Assess.* **2013**, *185*, 7409–7425. [CrossRef]

89. Liu, Y.; Shi, Z.; Zhang, G.; Chen, Y.; Li, S.; Hong, Y.; Shi, T.; Wang, J.; Liu, Y.; Liu, Y.; et al. Application of Spectrally Derived Soil Type as Ancillary Data to Improve the Estimation of Soil Organic Carbon by Using the Chinese Soil Vis-NIR Spectral Library. *Remote Sens.* **2018**, *10*, 1747. [CrossRef]

90. Lobsey, C.R.; Viscarra Rossel, R.A.; Roudier, P.; Hedley, C.B. Rs-Local Data-Mines Information From Spectral Libraries To Improve Local Calibrations. *Eur. J. Soil Sci.* **2017**, *68*, 840–852. [CrossRef]

91. Gogé, F.; Gomez, C.; Jolivet, C.; Joffre, R. Which strategy is best to predict soil properties of a local site from a national Vis-NIR database? *Geoderma* **2014**, *213*, 1–9. [CrossRef]

92. Barthès, B.G.; Kouakoua, E.; Clairotte, M.; Lallemand, J.; Chapuis-Lardy, L.; Rabenarivo, M.; Roussel, S. Performance comparison between a miniaturized and a conventional near infrared reflectance (NIR) spectrometer for characterizing soil carbon and nitrogen. *Geoderma* **2019**, *338*, 422–429. [CrossRef]

93. Jiang, Q.; Li, Q.; Wang, X.; Wu, Y.; Yang, X.; Liu, F. Estimation of soil organic carbon and total nitrogen in different soil layers using VNIR spectroscopy: Effects of spiking on model applicability. *Geoderma* **2017**, *293*, 54–63. [CrossRef]

94. Viscarra Rossel, R.A.; Adamchuk, V.I.; Sudduth, K.A.; McKenzie, N.J.; Lobsey, C. Proximal Soil Sensing: An Effective Approach for Soil Measurements in Space and Time. *Adv. Agron.* **2011**, *113*, 243–291.

95. Kuang, B.; Mahmood, H.S.; Quraishi, M.Z.; Hoogmoed, W.B.; Mouazen, A.M.; van Henten, E.J. Sensing Soil Properties in the Laboratory, in situ, and on-line. A review. Elsevier Inc., 2012, 1st ed. Vol. 114. Available online: https://www.sciencedirect.com/science/article/pii/B9780123942753000031 (accessed on 2 September 2019).

96. Christy, C.D. Real-time measurement of soil attributes using on-the-go near infrared reflectance spectroscopy. *Comput. Electron. Agric.* **2008**, *61*, 10–19. [CrossRef]

97. Sinfield, J.V.; Fagerman, D.; Colic, O. Evaluation of sensing technologies for on-the-go detection of macro-nutrients in cultivated soils. *Comput. Electron. Agric.* **2010**, *70*, 1–18. [CrossRef]

98. Gehl, R.J.; Rice, C.W.; Gehl, R.J.; Rice, C.W. Emerging technologies for in situ measurement of soil carbon. *Clim. Change* **2007**, *80*, 43–54. [CrossRef]

99. Waiser, T.H.; Morgan, C.L.S.; Brown, D.J.; Hallmark, C.T. In Situ Characterization of Soil Clay Content with Visible Near-Infrared Diffuse Reflectance Spectroscopy. *Soil Sci. Soc. Am. J.* **2007**, *71*, 389. [CrossRef]

100. Sarkhot, D.V.; Grunwald, S.; Ge, Y.; Morgan, C.L.S. Comparison and detection of total and available soil carbon fractions using visible/near infrared diffuse reflectance spectroscopy. *Geoderma* **2011**, *164*, 22–32. [CrossRef]

101. Bricklemyer, R.S.; Brown, D.J. On-the-go VisNIR: Potential and limitations for mapping soil clay and organic carbon. *Comput. Electron. Agric.* **2010**, *70*, 209–216. [CrossRef]

102. Kweon, G.; Lund, E.; Maxton, C. Soil organic matter and cation-exchange capacity sensing with on-the-go electrical conductivity and optical sensors. *Geoderma* **2013**, *199*, 80–89. [CrossRef]

103. Knadel, M.; Thomsen, A.; Schelde, K.; Greve, M.H. Soil organic carbon and particle sizes mapping using vis-NIR, EC and temperature mobile sensor platform. *Comput. Electron. Agric.* **2015**, *114*, 134–144. [CrossRef]

104. Wetterlind, J.; Piikki, K.; Stenberg, B.; Söderström, M. Exploring the predictability of soil texture and organic matter content with a commercial integrated soil profiling tool. *Eur. J. Soil Sci.* **2015**, *66*, 631–638. [CrossRef]

105. Pei, X.; Sudduth, K.; Veum, K.; Li, M. Improving In-Situ Estimation of Soil Profile Properties Using a Multi-Sensor Probe. *Sensors* **2019**, *19*, 1011. [CrossRef] [PubMed]

106. Veum, K.S.; Parker, P.A.; Sudduth, K.A.; Holan, S.H. Predicting Profile Soil Properties with Reflectance Spectra via Bayesian Covariate-Assisted External Parameter Orthogonalization. *Sensors* **2018**, *18*, 3869. [CrossRef] [PubMed]

107. Kuang, B.; Tekin, Y.; Mouazen, A.M. Comparison between artificial neural network and partial least squares for on-line visible and near infrared spectroscopy measurement of soil organic carbon, pH and clay content. *Soil Tillage Res.* **2015**, *146*, 243–252. [CrossRef]

108. Sorenson, P.T.; Small, C.; Tappert, M.C.; Quideau, S.A.; Drozdowski, B.; Underwood, A.; Janz, A. Monitoring organic carbon, total nitrogen, and pH for reclaimed soils using field reflectance spectroscopy. *Can. J. Soil Sci.* **2017**, *97*, 241–248. [CrossRef]

109. Kodaira, M.; Shibusawa, S. Using a mobile real-time soil visible-near infrared sensor for high resolution soil property mapping. *Geoderma* **2013**, *199*, 64–79. [CrossRef]

110. Kuang, B.; Mouazen, A.M. Non-biased prediction of soil organic carbon and total nitrogen with vis–NIR spectroscopy, as affected by soil moisture content and texture. *Biosyst. Eng.* **2013**, *114*, 249–258. [CrossRef]

111. Mouazen, A.M.; Ramon, H. Development of on-line measurement system of bulk density based on on-line measured draught, depth and soil moisture content. *Soil Tillage Res.* **2006**, *86*, 218–229. [CrossRef]

112. Rodionov, A.; Welp, G.; Damerow, L.; Berg, T.; Amelung, W.; Pätzold, S. Towards on-the-go field assessment of soil organic carbon using Vis-NIR diffuse reflectance spectroscopy: Developing and testing a novel tractor-driven measuring chamber. *Soil Tillage Res.* **2015**, *145*, 93–102. [CrossRef]

113. Franceschini, M.H.D.; Demattê, J.A.M.; Kooistra, L.; Bartholomeus, H.; Rizzo, R.; Fongaro, C.T.; Molin, J.P. Effects of external factors on soil reflectance measured on-the-go and assessment of potential spectral correction through orthogonalisation and standardisation procedures. *Soil Tillage Res.* **2018**, *177*, 19–36. [CrossRef]

114. Wold, S.; Antti, H.; Lindgren, F.; Ohman, J. Orthogonal signal correction of near-infrared spectra. *Chemom. Intell. Lab. Syst.* **1998**, *44*, 175–185. [CrossRef]

115. Ge, Y.; Thomasson, J.A.; Sui, R. Remote sensing of soil properties in precision agriculture: A review. *Front. Earth Sci.* **2011**, *5*, 229–238. [CrossRef]

116. Viscarra Rossel, R.A.; Lobsey, C.R.; Sharman, C.; Flick, P.; McLachlan, G. Novel Proximal Sensing for Monitoring Soil Organic C Stocks and Condition. *Environ. Sci. Technol.* **2017**, *51*, 5630–5641. [CrossRef] [PubMed]

117. Cozzolino, D.; Cynkar, W.U.; Dambergs, R.G.; Shah, N.; Smith, P. In Situ Measurement of Soil Chemical Composition by Near-Infrared Spectroscopy: A Tool Toward Sustainable Vineyard Management. *Commun. Soil Sci. Plant Anal.* **2013**, *44*, 1610–1619. [CrossRef]

118. Gras, J.-P.; Barthès, B.G.; Mahaut, B.; Trupin, S. Best practices for obtaining and processing field visible and near infrared (VNIR) spectra of topsoils. *Geoderma* **2014**, *214–215*, 126–134. [CrossRef]

119. Cambou, A.; Cardinael, R.; Kouakoua, E.; Villeneuve, M.; Durand, C.; Barthès, B.G. Prediction of soil organic carbon stock using visible and near infrared reflectance spectroscopy (VNIRS) in the field. *Geoderma* **2016**, *261*, 151–159. [CrossRef]

120. Bartholomeus, H.; Kooistra, L.; Stevens, A.; van Leeuwen, M.; van Wesemael, B.; Ben-Dor, E.; Tychon, B. Soil Organic Carbon mapping of partially vegetated agricultural fields with imaging spectroscopy. *Int. J. Appl. Earth Obs. Geoinf.* **2011**, *13*, 81–88. [CrossRef]

121. Rodionov, A.; Pätzold, S.; Welp, G.; Pude, R.; Amelung, W. Proximal field Vis-NIR spectroscopy of soil organic carbon: A solution to clear obstacles related to vegetation and straw cover. *Soil Tillage Res.* **2016**, *163*, 89–98. [CrossRef]

122. Guerrero, C.; Zornoza, R.; Gómez, I.; Mataix-Beneyto, J. Spiking of NIR regional models using samples from target sites: Effect of model size on prediction accuracy. *Geoderma* **2010**, *158*, 66–77. [CrossRef]

123. Nawar, S.; Mouazen, A.M. Predictive performance of mobile vis-near infrared spectroscopy for key soil properties at different geographical scales by using spiking and data mining techniques. *CATENA* **2017**, *151*, 118–129. [CrossRef]

124. Nawar, S.; Mouazen, A.M. On-line vis-NIR spectroscopy prediction of soil organic carbon using machine learning. *Soil Tillage Res.* **2019**, *190*, 120–127. [CrossRef]

125. Nawar, S.; Mouazen, A.M. Optimal sample selection for measurement of soil organic carbon using on-line vis-NIR spectroscopy. *Comput. Electron. Agric.* **2018**, *151*, 469–477. [CrossRef]

126. Guerrero, C.; Stenberg, B.; Wetterlind, J.; Viscarra Rossel, R.A.; Maestre, F.T.; Mouazen, A.M.; Zornoza, R.; Ruiz-Sinoga, J.D.; Kuang, B. Assessment of soil organic carbon at local scale with spiked NIR calibrations: Effects of selection and extra-weighting on the spiking subset. *Eur. J. Soil Sci.* **2014**, *65*, 248–263. [CrossRef]

127. Kühnel, A.; Bogner, C. In-situ prediction of soil organic carbon by vis–NIR spectroscopy: An efficient use of limited field data. *Eur. J. Soil Sci.* **2017**, *68*, 689–702. [CrossRef]

128. Chawla, N.V.; Bowyer, K.W.; Hall, L.O.; Kegelmeyer, W.P. SMOTE: Synthetic minority over-sampling technique. *J. Artif. Intell. Res.* **2002**, *16*, 321–357. [CrossRef]

129. Kempen, B.; Dalsgaard, S.; Kaaya, A.K.; Chamuya, N.; Ruipérez-González, M.; Pekkarinen, A.; Walsh, M.G. Mapping topsoil organic carbon concentrations and stocks for Tanzania. *Geoderma* **2019**, *337*, 164–180. [CrossRef]

130. Demattê, J.A.M.; Dotto, A.C.; Bedin, L.G.; Sayão, V.M.; Souza, A.B. Soil analytical quality control by traditional and spectroscopy techniques: Constructing the future of a hybrid laboratory for low environmental impact. *Geoderma* **2019**, *337*, 111–121. [CrossRef]

131. Angelopoulou, T.; Dimitrakos, A.; Terzopoulou, E.; Zalidis, G.; Theocharis, J.; Stafilov, T.; Zouboulis, A. Reflectance Spectroscopy (Vis-NIR) for Assessing Soil Heavy Metals Concentrations Determined by two Different Analytical Protocols, Based on ISO 11466 and ISO 14869-1. *Water. Air. Soil Pollut.* **2017**, *228*. [CrossRef]

132. Ben Dor, E.; Ong, C.; Lau, I.C. Reflectance measurements of soils in the laboratory: Standards and protocols. *Geoderma* **2015**, *245–246*, 112–124. [CrossRef]

133. Viscarra Rossel, R.A.; Brus, D.J.; Lobsey, C.; Shi, Z.; McLachlan, G. Baseline estimates of soil organic carbon by proximal sensing: Comparing design-based, model-assisted and model-based inference. *Geoderma* **2016**, *265*, 152–163. [CrossRef]

134. Adeline, K.R.M.; Gomez, C.; Gorretta, N.; Roger, J.M. Predictive ability of soil properties to spectral degradation from laboratory Vis-NIR spectroscopy data. *Geoderma* **2017**, *288*, 143–153. [CrossRef]

135. Ng, W.; Minasny, B.; Malone, B.P.; Sarathjith, M.C.; Das, B.S. Optimizing wavelength selection by using informative vectors for parsimonious infrared spectra modelling. *Comput. Electron. Agric.* **2019**, *158*, 201–210. [CrossRef]

136. Padarian, J.; Minasny, B.; McBratney, A.B. Transfer learning to localise a continental soil vis-NIR calibration model. *Geoderma* **2019**, *340*, 279–288. [CrossRef]

137. Tsakiridis, N.L.; Theocharis, J.B.; Zalidis, G.C. An evolutionary fuzzy rule-based system applied to real-world Big Data - The GEO-CRADLE and LUCAS soil spectral libraries. *IEEE Int. Conf. Fuzzy Syst.* **2018**, *2018-July*, 1–8.

138. Tsakiridis, N.L.; Theocharis, J.B.; Ben-Dor, E.; Zalidis, G.C. Using interpretable fuzzy rule-based models for the estimation of soil organic carbon from VNIR/SWIR spectra and soil texture. *Chemom. Intell. Lab. Syst.* **2019**, *189*, 39–55. [CrossRef]

139. Hong, Y.; Chen, S.; Liu, Y.; Zhang, Y.; Yu, L.; Chen, Y.; Liu, Y.; Cheng, H.; Liu, Y. Combination of fractional order derivative and memory-based learning algorithm to improve the estimation accuracy of soil organic matter by visible and near-infrared spectroscopy. *CATENA* **2019**, *174*, 104–116. [CrossRef]

140. Zhou, J.; Li, E.; Wei, H.; Li, C.; Qiao, Q.; Armaghani, D.J. Random forests and cubist algorithms for predicting shear strengths of rockfill materials. *Appl. Sci.* **2019**, *9*, 1621. [CrossRef]

141. Gholizadeh, A.; Borůvka, L.; Saberioon, M.; Vašát, R. Visible, Near-Infrared, and Mid-Infrared Spectroscopy Applications for Soil Assessment with Emphasis on Soil Organic Matter Content and Quality: State-of-the-Art and Key Issues. *Appl. Spectrosc.* **2013**, *67*, 1349–1362. [CrossRef] [PubMed]

142. Ackerson, J.P.; Morgan, C.L.S.; Ge, Y. Penetrometer-mounted VisNIR spectroscopy: Application of EPO-PLS to in situ VisNIR spectra. *Geoderma* **2017**, *286*, 131–138. [CrossRef]

143. Yin, Z.; Lei, T.; Yan, Q.; Chen, Z.; Dong, Y. A near-infrared reflectance sensor for soil surface moisture measurement. *Comput. Electron. Agric.* **2013**, *99*, 101–107. [CrossRef]
144. Prudnikova, E.Y.; Savin, I.Y. Study of the optical properties of an exposed soil surface. *J. Opt. Technol.* **2016**, *83*, 642. [CrossRef]
145. Prudnikova, E.; Savin, I.; Vindeker, G.; Grubina, P.; Shishkonakova, E.; Sharychev, D. Influence of Soil Background on Spectral Reflectance of Winter Wheat Crop Canopy. *Remote Sens.* **2019**, *11*, 1932. [CrossRef]
146. England, J.R.; Rossel, R.A.V. Proximal sensing for soil carbon accounting. *SOIL* **2018**, *4*, 101–122. [CrossRef]

Article

Sustainability Assessment of Investments Based on a Multiple Criteria Methodological Framework

Paraskevi Ovezikoglou [1], Dimitrios Aidonis [1,2], Charisios Achillas [2], Christos Vlachokostas [3] and Dionysis Bochtis [4,*]

[1] Department of Quality Management and Technology, School of Science and Technology,
    Hellenic Open University, GR 26335 Patra, Greece; std115418@ac.eap.gr (P.O.); daidonis@ihu.gr (D.A.)
[2] Department of Supply Chain Management, School of Economics and Business Administration,
    International Hellenic University, GR 60100 Katerini, Greece; c.achillas@ihu.edu.gr
[3] Laboratory of Heat Transfer and Environmental Engineering, Aristotle University Thessaloniki,
    GR 54124 Thessaloniki, Greece; vlahoco@auth.gr
[4] Institute for Bio-Economy and Agri-Technology (iBO), Center for Research and Technology–Hellas (CERTH),
    6th km Charilaou-Thermi Rd, Thermi, GR 57001 Thessaloniki, Greece
*   Correspondence: d.bochtis@certh.gr; Tel.: +30-242-109-6740

Received: 26 July 2020; Accepted: 20 August 2020; Published: 21 August 2020

Abstract: The assessment of an investment is currently carried out by using mainly financial tools. This work presents a new model for the assessment of the sustainability of an industrial investment and focuses on the development of a holistic framework with the use of indicators. With the use of multi-criteria decision analysis, the framework evaluates a total of eighteen (18) alternative indicators in order to select the optimal bundle to be used for the assessment of future industrial investments. The proposed indicators are selected based on relevant data from the literature, taking into account the principles of prevention, planning and designing. The alternatives are assessed over four (4) criteria, namely environment, society, economy and technology, which are grounded on the principles of sustainable development. Depending on the special characteristics of the programme that is foreseen to fund the potential investments, the decision-maker is provided with a hierarchized set of indicators over which the alternative investments could be optimally assessed in parallel with widely used indicators that strictly assess economic performance. In the present work, twelve (12) different scenarios are examined, incorporating different values in the coefficients of the criteria. For the majority of the scenarios examined (a sensitivity analysis is also provided), the alternative indicator that is assessed with the highest score is "Resource Savings", followed by "Recycling" and "Research, Innovation, Development".

Keywords: indicators; investments' sustainability; multi-criteria analysis; decision support; ELECTRE III

---

1. Introduction

Environmental performance indicators are considered to assist decision-making processes in managing important environmental, social and financial aspects and perspectives and improving the assessment of the impact caused by business activities [1–5]. Improving environmental performance requires effective control of the activities, products and processes of the business that may trigger a significant burden [6–8]. Improving the business performance can be achieved through a wide spectrum of modifications in corporate activities, products or processes and can range from small fragmentary changes to integrated environmental management.

Changes in this direction include, among others, inter-alias environmental education of workers and stakeholders, informing and/or sensitizing customers and suppliers, investing in environmental protection technologies, adopting optimally available techniques to minimize gaseous pollutants, etc.

A significant process is the so called "Design for Environment" (DfE), i.e., the ecological design of products and services and the creation of environmental reports (corporate environmental reporting), as well as the installation of environmental management systems [9–12].

Environmental performance appraisal is an internal process of business and is essentially a tool designed to provide management with reliable and verifiable information on an ongoing basis to determine whether an enterprise's environmental performance meets the criteria set by the organization's management [13–15]. The literature presents several available environmental indicators that are used at different scales of business activities, namely international, national or local. As evaluation methods vary, environmental performance indicators, as well as the concept of sustainability, also differ and include diverse groups of indicators [16,17]. In particular, indicators related to specific environmental consequences (e.g., climate change) have been developed, using resources (e.g., water footprint) with ecological efficiency measures [18].

There are different approaches to measuring environmental performance, namely production, control, ecological, accounting, economics and quality [19]. These approaches have different guidelines, focus and measurements. It is obvious that performance measurement activities vary in different countries and also among different industries, due to the variety of environmental issues, organizational variables (e.g., the size of the organization or the way it is structured), national conditions and individual corporate strategies [20].

According to the definition provided by the Organization for Economic Co-operation and Development (OECD), an environmental indicator is characterized as "a parameter that describes the state of a phenomenon/environment/area, with significance that extends beyond that directly related to a parameter value" [21]. Therefore, an indicator needs to provide important information about the parameter to be described. If a parameter is complex, such as sustainability, more than one indicator may be required. On the other hand, a group of various indicators may be combined to produce a single indicator, i.e., a Composite Sustainable Development Index (CSDI), if desired. The Environmental Sustainability Index (ESI), the Dow Jones Sustainability Index (DJSI) and the Global Reporting Initiative (GRI) are examples of CSDI. Furthermore, there are various environmental performance indicators, but also specific indicators focused mainly on environmental impacts, such as climate change [22], air pollutants [23] and ad-hoc application of systems of indicators as decision-support tools towards sustainable urban development [24].

The indicators that best describe environmental performance can be divided into the following groups [25]: (a) lagging indicators, which are measures at the end of a process, such as the amount of emitted pollutants; (b) leading indicators, which are performance measurement measures, i.e., they measure the implementation of practices or measures that are expected to lead to improved environmental performance, such as the percentage of facilities that carry out self-monitoring; (c) Environmental Condition Indicators (ECIs) measure the direct impact of an activity on the environment, such as air, water, groundwater and soil concentrations, changes in the size of a population of a particular species in a given area. Each group of indicators has discrete strengths and weaknesses, aiming a different target audience, and this is the reason why many companies use a mix of indicators to measure their environmental performance [25].

In addition to the use of indicators, progress on environmental issues can be assessed by comparative evaluation between companies or by the average performance of the industry to which a company belongs. The International Organization for Standardization (ISO) has compiled a list of conditions that a marker must meet to be useful and relevant to the measurement of environmental performance. According to this list, an indicator must be: (i) in accordance with the environmental policy and the important environmental aspects; (ii) suitable for management, business or environmental activities; (iii) useful and representative of the environmental performance criteria; (iv) understandable to internal and external stakeholders; (v) easily accessible, measurable and informative; (vi) adequate in relation to the quality and quantity of data; (vii) able to respond to changes in environmental performance [26].

This paper focuses on the development of a holistic framework for the assessment of industrial investments' sustainability with the use of indicators. The key research question that is examined in the present work is the identification of the optimal bundle of indicators which could be used for the assessment of alternative industrial investments. It is evident that in most cases, available funds are not adequate to cover all needs, thus decision-makers (either industrial or public) require a concrete methodology for the assessment of alternative investments and the identification of the optimal one, based on sustainability criteria. To date, the evaluation and selection of an investment over competition in most cases is solely realized based on their economic performance and indicators (Return on Investment, Net Present Value, etc.). However, the selection of the optimal criteria for the assessment of investments is, without any doubt, a highly multi-dimensional problem. The proposed framework is the outcome of a research effort that incorporates collection of data with the use of a structured questionnaire. For the determination of the optimal set of indicators that best capture the environmental performance of a given investment, the opinions of a number of different stakeholders, with mutually contradicting priorities (e.g., environmental organizations, companies, universities, public bodies, environmentalists and economists), are considered in the present work. The proposed environmental (non-monetary) indicators are proposed to be used in parallel to the currently widely used economic ones, in order to more efficiently and holistically assess alternative industrial investments.

2. Methodology

A critical point in multidimensional management problems is the evaluation and combination of different types of available information that are ultimately able to lead to an optimal solution [27]. Multi-criteria methods provide the framework for collecting, registering, and ultimately promoting all relevant information, thus making the decision-making process detectable and transparent [28,29]. In this light, the adoption of a decision is based on the result of the analysis of the conflicting parameters and goals of socio-political, economic and environmental nature, thus creating a multidimensional problem that needs special treatment [30]. The nature of decision-making processes makes it difficult to represent them in descriptive models. Furthermore, uncertainty and inaccuracy are inalienable elements of their structure.

Multidisciplinary analysis can be defined as a systematic and mathematically standardized effort to solve problems arising from conflicting goals in an effort to reconcile them [31]. Making a decision is the study of identifying and selecting alternative solutions based on the preferences of the recipient's decision. Decision making also implies that there are alternatives being considered. In this case, the goal is not only to identify as many of the alternatives as possible, but to choose the one that best fits the goals and desires of the decision maker [32]. Decision making with the use of multi-criteria analysis is realized in four discrete steps, as follows: The first step comprises the determination of the alternative scenarios for the selection of environmental indicators. The second step includes the selection of the criteria, the scoring scale of the alternative scenarios and the assessment of the weighting factors by the decision makers. The next step includes the application of the multi-criteria analysis and the extraction of the results, followed by a last step, where the decision is realized for the selection of the appropriate set of indicators, taking into consideration the use results of a sensitivity analysis.

The selection of environmental indicators for investment evaluation is a very complex process. A considerable number of alternatives, often presenting equivalent weightings, need to be evaluated [33]. To efficiently achieve such an assessment, it is necessary to analyze and grade a series of critical parameters, or other criteria. In particular, the formulation of an integrated policy regarding the creation of environmental indicators for investment evaluation is considered a rather complex process, given that the number of environmental indicators (alternatives) can be quite large, and at the same time, each indicator shows advantages and disadvantages on different levels, namely economic, environmental, social or technological (criteria).

The combination of all the parameters that appear makes the selection of an environmental indicator a rather complex problem. The final selection of the most appropriate buddle of indicators between alternative scenarios requires the consideration and evaluation of several parameters, therefore, it is necessary to apply the multi-criteria analysis. In the literature, there are over 50 multi-criteria analysis techniques [34,35], and a different classification can be attempted according to their content and scope.

In the present work, the ELECTRE III technique is selected to process the collected data via the distributed questionnaires. ELECTRE III is a well-known method of multi-criteria analysis, with a long history of successful practical applications internationally [29,36,37]. The method was used often in the past to compare different scenarios in different thematic areas, such as energy, construction, waste management, services, public policy, and transportation. An important advantage of the ELECTRE III method over other methods is its usefulness in examining environmental problems [38]. In addition, ELECTRE III has the ability to include a fairly large number of evaluation criteria for the selection of environmental indicators, combined with the ability of a large number of decision makers [39]. The method requires the determination of three threshold values of the criteria used, i.e., the indifference, the preference and the veto threshold. These thresholds allow the uncertainties of the evaluation criteria to be incorporated into the decision-making process [40].

Determining the recipient's preference data for a decision expressed as a criterion is one of the most important factors of ELECTRE III. It is already noted that the method uses the thresholds of preference and indifference, and includes an additional parameter, the concept of veto [41]. By using these parameters, the method examines not only the two extremes of the problem, strong and weak, but also a whole family of intermediate levels, from the overall strongest to the overall weakest alternative. The process is achieved by assessing, comparing and finalizing the various environmental selection indicators (alternatives) over the criteria considered.

As a ranking technique, ELECTRE III calculates a ranking hierarchy among alternatives. The ranking is based on concordance (c_j) and non-discordance (d_j) binary outranking. In brief, concordance is valid for the cases where alternative X outranks alternative Y when most of the criteria X's performance is better than the alternative's Y. Respectively, non-discordance is valid for the cases where none of the criteria in the minority are strongly opposed to alternative's Y outranking by alternative X. The ELECTRE III methodology calculates a credibility index, which characterizes that X outranks Y. The credibility index shows the real degree of the aforementioned assertion [42].

Following this, alternatives are pairwise compared for every criterion by inserting two more pseudo-criteria, namely the preference (p_j) threshold and the indifference (q_j) threshold. In the case where the difference between the performances of X and Y is lower than the indifference threshold for a specific criterion, then the two alternatives are regarded as similar for that criterion j, and the credibility index $c_j(X,Y)$ equals zero. On the other end, in the case where the difference between the performances of X and Y is larger than the preference threshold for a specific criterion, then X is strongly preferred to Y for that specific criterion j, and the credibility index $c_j(X,Y)$ equals one. In this context, the concordance index $c_j(X,Y)$ for any criterion j is mathematically described with Equation (1).

$$V_j(X) - V_j(Y) \leq q_j \Longleftrightarrow c_j(X,Y) = 0$$
$$q_j < V_j(X) - V_j(Y) < p_j \Longleftrightarrow c_j(X,Y) = \frac{V_j(X) - V_j(Y) - q_j}{p_j - q_j} \tag{1}$$
$$V(X) - V(Y) \geq p_j \Leftrightarrow c_j(X,Y) = 1$$

Taking into account all the concordance indices calculated for each criterion j, and also the weighting factor (relative importance) of each criterion $j(w_j)$, a global concordance index is calculated for every pair of alternatives (X,Y). The global concordance index (C_{XY}) is mathematically described with Equation (2).

$$C_{XY} = \frac{\sum_{j=1}^{n} w_j \times c_j(X,Y)}{\sum_{j=1}^{n} w_j} \tag{2}$$

As a next step in the methodology, a discordance index (d_j) is calculated, with the assistance of preference (p_j) and indifference (q_j) thresholds, as those were described above, and with the use of a third threshold, namely veto (v_j), that gives the acceptable (maximum) difference between the performances of two given alternatives X and Y in criterion j for not rejecting the assertion that X outranks Y, regardless of the alternative's performance in all other criteria. More specifically, in the case that the difference between the performances of X and Y is lower than the preference threshold for a specific criterion, then no discordance exists and $d_j(X,Y)$ equals zero. On the other end, in the case that the difference between the performances of X and Y is larger than the veto threshold for a specific criterion, then Y is globally preferred to X, no matter the performances in all other criteria, and the discordance index $d_j(X,Y)$ equals one. In this context, the discordance index $d_j(X,Y)$ for any criterion j is mathematically described with Equation (3).

$$\begin{cases} V_j(Y) - V_j(X) \leq p_j \Leftrightarrow d_j(X,Y) = 0 \\ p_j < V_j(Y) - V_j(X) < v_j \Leftrightarrow d_j(X,Y) = \frac{V_j(Y)-V_j(X)-p_j}{v_j-p_j} \\ V_j(Y) - V_j(X) \geq v_j \Leftrightarrow d_j(X,Y) = 1 \end{cases} \tag{3}$$

The credibility index δ_{XY} of the assertion "X outranks Y" is mathematically formulated with the use of Equation (4).

$$\delta_{XY} = \Pi_{j\in\overline{F}}\frac{1 - d_j(X,Y)}{1 - C_{XY}}, \text{ where } \overline{F} = \left\{j \in F, d_j(X,Y) > C_{XY}\right\} \tag{4}$$

In order to determine the optimal set of environmental indicators (alternatives) for the assessment of potential investments within the present work, a questionnaire was used, aiming at the collection of the required data that would feed the ELECTRE III methodology. The questionnaire considered a number of indicators that are commonly used to evaluate the environmental performance of a company in operation, the indicators for the evaluation of environmental performance in highly industrialized countries (e.g., USA, UK, EU, Japan) [43], as well as the ISO 14031/2013 standard [44]. The aim was to explore the dominant aspect of tackling the problem, namely the optimal choice of environmental indicators, through the views of experts. A total of 80 experts' responses were collected, representing all different stakeholders involved in the decision-making process, namely NGOs, business, academia, authorities, certified assessors, and scientist/practitioners activated in the field, in order to capture different needs and requirements. More specifically, half of the experts' sample (40 out of 80) represented the private sector (31 practitioners and business consultants activated in industrial investments and nine senior staff in private companies), while the other half represented the public sector (17 representatives from public authorities at local-to-regional level, 9 representatives from the academia sector, 3 certified assessors and 11 representatives from NGOs activated in the field of environmental protection and sustainability). The aim was to evaluate alternative indicators from all different aspects, with the involvement of a balanced sample of experts, since the former (practitioners, business consultants, enterprises) focus mostly on the business success of the proposed investment, while the latter (public authorities, academia, NGOs) place emphasis on the common public interest.

The questionnaire consists of 18 indicators (alternatives) that were coded to facilitate the processing of the results (Table 1). Potential indicators of sustainable development were selected taking into account the pillars of sustainable development and research work in international literature [43–45] and assessed over four discrete criteria by the experts involved in the framework of the present work. The criteria used for the assessment of indicators' suitability are grounded on a set of sustainable development's pillars and principles. The pillars are the environment (Criterion C_1), society (Criterion C_2), and economy (Criterion C_3). To efficiently evaluate an investment, the above criteria are the main priorities for its successful operation and a more sustainable future. In addition, technology (Criterion C_4) was selected as the criterion for evaluation, given that environmental technology and technological infrastructure provide the basis for faster and cost-effective development.

Table 1. Alternative indicators for the environmental assessment of investments.

Code	Thematic Area	Indicator
A_{01}	Recycling	Design for recycling and reuse of materials, and minimization of waste
A_{02}	Gas emissions	Design for minimization of emissions (greenhouse gases, ozone depletion gases, air pollutants, particulate matter, etc.)
A_{03}	Resource savings	Design for energy and other resources (materials, water, etc.) savings, and reduction of non-renewable resources consumption
A_{04}	Biodiversity	Design for biodiversity and habitat conservation
A_{05}	Impact restoration	Design for prevention and remediation of adverse effects on the environment due to the company's operations
A_{06}	Alternative energy forms	Provision for the use of alternative energy sources
A_{07}	Education	Provision for environmental employees' training
A_{08}	Health and safety	Provision for employees' health and safety
A_{09}	Communication & public awareness	Provision for communication with the local community (information and public disclosure of environmental performance, etc.)
A_{10}	Information across supply chain	Provision for information on sustainable development to suppliers and customers
A_{11}	Social actions	Provision for social initiatives (compensations, donations, funding for environmental actions, etc.)
A_{12}	Pollution prevention	Provision for covering the cost of pollution prevention projects
A_{13}	Environmental accounting	Provision for application of environmental accounting
A_{14}	Research, Innovation, Development	Provision for research and development of high-tech and innovative products, development of green products
A_{15}	Environmental policy	Provision for consideration and planning for the implementation of environmental policies and environmental controls, the use of their results in company's operations
A_{16}	Legal framework	Compliance with the legal framework on environmental legislative framework
A_{17}	Environmental standards	Provision for the implementation of environmental management system (e.g., EMAS, ISO 14000)
A_{18}	Corporate governance	Design for implementation of corporate governance rules

3. Results and Discussion

Within the framework of the present study, an assessments' matrix was formed, consisting of the alternative scenarios of environmental indicators, and criteria over which the selected alternative scenarios were assessed by the experts. In Table 2, the evaluations of 80 experts are depicted in a scale from 1 to 9, where 1 represents the worst-case assessment and 9 the best-case one. The assessment of each alternative for each criterion is calculated as follows:

$$V_{ij} = \frac{\sum_1^N v_{ij}}{N}, \; i \in (A_1, A_2, \ldots, A_{18}), \; j \in (C_1, C_2, C_3, C_4) \tag{5}$$

where:

V_{ij}: Assessment of alternative scenario i based on criterion j for all experts
v_{ij}: Performance of an alternative scenario i based on criterion j for each expert
N: Total number of experts

Table 2. Assessment matrix.

	Environment [C_1]	Society [C_2]	Economy [C_3]	Technology [C_4]
Recycling [A_{01}]	8.63	6.69	6.87	7.08
Gas emissions [A_{02}]	8.58	7.23	6.29	7.44
Resources' savings [A_{03}]	8.50	7.23	7.21	7.37
Biodiversity [A_{04}]	8.21	6.42	5.31	5.08
Impact restoration [A_{05}]	8.48	7.23	6.65	6.42
Alternative energy forms [A_{06}]	7.85	6.31	6.56	7.21
Education [A_{07}]	7.12	6.44	5.69	4.75
Health and safety [A_{08}]	5.87	8.02	6.29	4.96
Communication & public awareness [A_{09}]	5.88	8.04	5.25	4.10
Information across supply chain [A_{10}]	6.08	7.63	5.85	4.58
Social actions [A_{11}]	6.67	7.87	6.33	4.31
Pollution prevention [A_{12}]	7.33	6.10	7.83	5.62
Environmental accounting [A_{13}]	6.27	4.56	6.98	4.92
Research, Innovation, Development [A_{14}]	7.98	6.29	7.77	8.67
Environmental policy [A_{15}]	7.85	6.40	6.46	6.19
Legal framework [A_{16}]	7.69	6.94	6.10	4.52
Environmental standards [A_{17}]	7.65	5.69	5.83	5.08
Corporate governance [A_{18}]	5.35	6.06	5.90	4.75
Weighting Factor	*31.9*	*23.6*	*25.1*	*19.4*
Preference threshold	*0.18*	*0.19*	*0.14*	*0.25*
Indifference threshold	*0.05*	*0.06*	*0.04*	*0.08*

In the present work, the weighting factors of each criterion were determined by the experts participating in the research. In particular, experts were required to assign a percentage of importance to each criterion according to their personal opinion. The values of the criterion weighting factors emerged as averages of the views of the various experts, who took part in the qualitative evaluation of the environmental selection scenarios. For the assessment of the importance of the selected criteria (environment, society, economy, technology) a weight scale from 0 to 100% was used.

In respect to the preference and indifference thresholds that are required for the ELECTRE III methodology, the following equations were considered [38,46–49]

$$p_i = \frac{(v_i max - v_i min)}{N} \tag{6}$$

$$q_i = 0.3 \times p_i \tag{7}$$

where

$v_i max$: The maximum value displayed by alternative i for criterion j
$v_i min$: The minimum value displayed by alternative i for criterion j
N: The number of alternative scenarios (here $N = 18$)

According to the ELECTRE III technique, two distillations are calculated (namely, ascending and descending), before the determination of the final order of the available alternatives. In the case under study, the LAMSADE software was used. In Figure 1, the distribution of the ascending and descending distillations is illustrated for the Baseline Scenario (BS), i.e., taking into account: (a) the opinion of the 80 experts with respect to their assessment for the performance of the 18 alternatives over the four selected criteria, (b) the opinion of the 80 experts with respect to the weighting of the four criteria's importance, (c) Equations (6) and (7) for the calculation of the preference and indifference thresholds, in particular, in the *x*-axis, the ascending distillation (i.e., from the worst to the optimal alternative), while in the *y*-axis, the descending distillation (i.e., from the optimal to the worst alternative), are provided. For instance, alternative A_{03} (resources' savings) is the optimal one in both distillations. Correspondingly, alternative A_{01} (recycling) is ranked third (following A_{03} and A_{14}) when considering the optimal hierarchy from the worst to the best one (ascending distillation), while ranked second (only after A_{01}) when considering the optimal hierarchy from the best to the worst one (descending distillation).

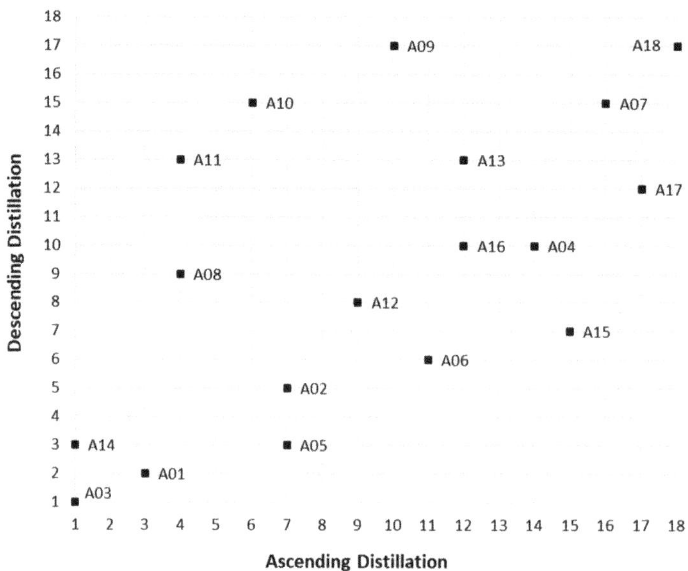

Figure 1. Distribution of ascending and descending preference of alternative environmental indicators for the application of the Baseline Scenario.

The final ranking of the alternatives (indicators) is calculated taking into account the two aforementioned distillations. For the BS, the ranking of the alternative indicators is the following; (i) Resources' savings [A_{03}], (ii) Recycling [A_{01}], (iii) Research, Innovation, Development [A_{14}], (iv) Impact restoration [A_{05}], (v) Health and safety [A_{08}], (vi) Gas emissions [A_{02}], (vii) Social actions [A_{11}], (viii) Pollution prevention [A_{12}], (ix) Alternative energy forms [A_{06}], (x) Information across supply chain [A_{10}], (xi) Legal framework [A_{16}], (xii) Communication and public awareness [A_{09}], (xiii) Environmental policy [A_{15}], (xiv) Environmental accounting [A_{13}], (xv) Biodiversity [A_{04}], (xvi) Education [A_{07}], Environmental standards [A_{17}], Corporate governance [A_{18}].

In addition to the Baseline Scenario (BS), nine (9) additional scenarios were considered for sensitivity analysis purposes. In other words, additional scenarios are examined to study whether changes in the parameters of the problem affect the final solution, with the aim of providing further confidence in decision-making. More specifically, in order to globally assess the alternative indicators (taking into account their performance in the four described criteria, namely environment, society, economy and technology), the following parameters need to be determined: (a) the weighting factor (relative importance) of each criterion, and (b) two pseudo-criteria, namely the preference and the indifference threshold. In the present study, sensitivity analysis is selected to be applied in comparison to the Baseline Scenario. In Table 3, the key parameters (thresholds and weighting factors) of the scenarios (S_x) examined are depicted. The Baseline Scenario (BS) reflects the solution of the mathematical algorithm (Equations (1)–(5)), taking into account the weighting factors (relative importance) of the criteria as averages of the experts' views. In this light, the weighting factor for the environmental criterion [C1], the social criterion [C2], the economic criterion [C3] and the technological criterion [C4], are 31.9%, 23.6%, 25.1% and 19.4%, respectively. Moreover, for the determination of the preference and indifference thresholds, we used the referenced Equations (6) and (7). However, since both the views of the experts are subjective, while the equations for the determination of the pseudo-criteria (preference and indifference thresholds) are based on assumptions, the algorithmic model presented in the methodology is solved with alternative values with respect to weighting factors and thresholds, so as to provide a "what-if" analysis. In this context, the hierarchy of the alternative indicators is re-assessed with the use of different weighting factors (putting more emphasis on different criteria) in each scenario examined. More specifically, in Scenario 1 (S_1), all criteria are equally weighted (25%), compared to the Baseline Scenario (BS), where more emphasis is placed on the environmental pillar of sustainability. In Scenario S_2, more emphasis is put on the economic criterion [C3] which weighs 50%, while each of the environmental [C1] and social [C2] criteria weigh 25% and the technological criterion [C4] is neglected. Similarly, different weight factors are applied in the case of the rest of the different scenarios (S_3–S_9). For the scenarios S_{10} and S_{11}, the weighting factors are based on the experts' views (similarly to the Baseline Scenario), while the preference and indifference thresholds are altered compared to the values calculated with Equations (6) and (7) in order to assess the sensitivity of those thresholds in the optimal hierarchy of the indicators. The analysis, apart from providing robustness to the ranking of the Baseline Scenario, can be used for altering the weights of the criteria based on the individual needs of the funding programmes.

Table 3. Modifications of weighting factors and thresholds for sensitivity analysis purposes.

	Scenario											
	[BS]	[S_1]	[S_2]	[S_3]	[S_4]	[S_5]	[S_6]	[S_7]	[S_8]	[S_9]	[S_{10}]	[S_{11}]
Weighting factor for Environmental criterion [C_1]	31.9	25	25	50	16.7	16.7	16.7	50	0	0	31.9	31.9
Weighting factor for Social criterion [C_2]	23.6	25	25	16.7	50	16.7	16.7	0	50	0	23.6	23.6
Weighting factor for Economic criterion [C_3]	25.1	25	50	16.7	16.7	50	16.7	50	50	50	25.1	25.1
Weighting factor for Technological criterion [C_4]	19.4	25	0	16.7	16.7	16.7	50	0	0	50	19.4	19.4
Preference threshold	p_{BS}	p_{BS}	p_{BS}	p_{BS}	p_{BS}	p_{BS}	p_{BS}	p_{BS}	p_{BS}	p_{BS}	$1.5 \times p_{BS}$	$2 \times p_{BS}$
Indifference threshold	q_{BS}	q_{BS}	q_{BS}	q_{BS}	q_{BS}	q_{BS}	q_{BS}	q_{BS}	q_{BS}	q_{BS}	$1.5 \times q_{BS}$	$2 \times q_{BS}$

In Table 4, the final ranking (hierarchy) of the alternative indicators are illustrated for the scenarios examined. Apparently, the ranking shows significant changes in cases of change in the thresholds of preference and indifference. However, the results provide adequate information in the selection of a bundle of indicators to be used for the assessment of alternative investments. More specifically, for the Baseline Scenario, Alternative A_{03} (Resources' savings) is the highest ranked indicator to be considered in the environmental assessment of potential investments. The same applies for all examined scenarios, except S_9, where the focus is solely on economic and technological criteria, which provides robustness to the optimal solution and confidence to the decision-maker so as to always consider the indicator when assessing the environmental performance of any given investment.

Since, in most programmes, more than one criterion is simultaneously considered in order to select the optimal investment, the bundle of five top indicators in the Baseline Scenario was comprised, apart from Alternative A_{03}, by Alternatives A_{01} (Recycling), A_{14} (Research, Innovation, Development), A_{05} (Impact restoration) and A_{08} (Health and safety). Research, Innovation, Development (A_{14}) is included in the top five criteria for all examined scenarios (being the highest ranked alternative indicator for 3 out of 11 scenarios), while recycling (A_{01}) and impact restoration (A_{05}) are included in the top five criteria for 10 out of 11 examined scenarios (A_{01} is ranked 6th for S_8, while A_{05} is ranked 6th for S_{10}). In this light, the aforementioned criteria should be considered in the top bundle of indicators to be selected when designing a funding programme, accompanied by indicators like A_{14} (Research, Innovation, Development), A_{11} (Gas emissions) and A_{06} (Alternative energy forms), which are placed comparatively highly for most of the examined scenarios. On the contrary, Alternative A_{18} (Corporate governance) is the last indicator to be considered for the assessment of investments, as it is ranked 18th for 8 out of 11 scenarios.

Table 4. Final ranking of environmental indicators (alternatives) for investments' evaluation for each scenario studied.

	1st	2nd	3rd	4th	5th	6th	7th	8th	9th	10th	11th	12th	13th	14th	15th	16th	17th	18th
BS	A_3	A_1	A_{14}	A_5	A_8	A_2	A_{11}	A_{12}	A_6	A_{10}	A_{16}	A_9	A_{15}	A_{13}	A_4	A_7	A_{17}	A_{18}
S₁	A_3	A_1	A_{14}	A_5	A_8	A_2	A_{11}	A_{12}	A_6	A_{10}	A_{16}	A_9	A_{15}	A_{13}	A_4	A_7	A_{17}	A_{18}
S₂	A_3	A_{14}	A_5	A_1	A_{12}	A_2	A_8 & A_{11}		A_{16}	A_{10}	A_6	A_{13}	A_9	A_{15}	A_7	A_4	A_{17}	A_{18}
S₃	A_3	A_1	A_{14}	A_2	A_5 & A_8		A_{11}	A_{12}	A_6	A_{10}	A_{16}	A_9	A_{15}	A_4	A_{13}	A_{17}	A_7	A_{18}
S₄	A_3	A_2	A_1	A_{14}	A_5	A_8	A_{12}	A_{15}	A_{11}	A_6	A_{10}	A_{16}	A_9	A_4	A_7	A_{13}	A_{17}	A_{18}
S₅	A_3 & A_{14}		A_1	A_5	A_{12}	A_2 & A_8		A_6 & A_{11}		A_{13}	A_{16}	A_{10}	A_{15}	A_4	A_{17}	A_9	A_7	A_{18}
S₆	A_3 & A_{14}		A_1	A_2	A_5 & A_8		A_{12}	A_6	A_{10} & A_{11}		A_{16}	A_4	A_9 & A_{15}		A_{13}	A_{17}	A_7	A_{18}
S₇	A_3	A_1	A_{14}	A_5	A_{12}	A_2	A_6 & A_{13}		A_{15}	A_4	A_{16}	A_{11} & A_{17}		A_8	A_7	A_{10}	A_{18}	A_9
S₈	A_3	A_8	A_5	A_{14}	A_{11}	A_1	A_{10} & A_{18}		A_2	A_{12}	A_6	A_{16}	A_9	A_{15}	A_{13}	A_7	A_4	A_{17}
S₉	A_{14}	A_3	A_1	A_{12}	A_5	A_2	A_5 & A_{13}		A_{15}	A_8	A_{17}	A_{16}	A_{18}	A_4 & A_{10} & A_{11}			A_7	A_9
S₁₀	A_3	A_{14}	A_1	A_8	A_2	A_5	A_{11}	A_{12}	A_6	A_{16}	A_{10}	A_{15}	A_4	A_9	A_{13} & A_{17}		A_7	A_{18}
S₁₁	A_3	A_{14}	A_1	A_2 & A_5		A_8	A_{12}	A_6	A_{11}	A_{16}	A_{10}	A_{15}	A_4	A_7	A_9	A_{17}	A_{13}	A_{18}

It should be highlighted from the analysis of the variation of the coefficients that the ranking of the alternative indicators in the Baseline Scenario (BS) and the Scenario S_1 where all criteria (environmental, social, economic, technological) are equally considered (with a weighting factor of 25%), are identical in all ranking positions. Consequently, the experts' opinion on the significance of the criteria does not significantly affect the ranking of the indicators. In this light, all criteria could be equally considered (as realized in S_1) in a real-world case.

Overall, the ranking of environmental indicators for investments' assessment was observed to be significantly influenced by the weighting factors and preference and indifference thresholds. The latter demonstrates the choice of A_{03} as the optimal alternative indicator, but, on the other hand, also reveals that the final choice of the optimal bundle of environmental indicators for the assessment and evaluation of investments is left purely to the decision-maker and to the thresholds used. This provides the funding authority with appropriate "freedom" to apply the most suitable weighting factors and thresholds that best suit the particular needs of the funding programme and the pillar of sustainable development that should be promoted.

4. Conclusions

In the past, the evaluation and selection of an investment over competition was mostly realized based on their economic performance, using mainly financial tools. Nevertheless, the determination of the optimal investment is an interdisciplinary problem and apart from the economic performance, other sustainability criteria need to be considered. The need for an effective bundle of environmental indicators that would lead to best possible investments and cost-efficient use of available funds triggered the need for a multi-criteria analysis methodology, such as the one herein described.

The survey conducted in the framework of the present work, with the involvement of 80 experts closely related to the field, reveals that design for energy and other resources (materials, water, etc.) savings, and reduction in non-renewable resources consumption, are the most appropriate criteria, apart from cost, to introduce as additional indicators in the investments' evaluation. Additionally, the provision for research and development of high-tech and/or innovative products and the development of green products (design for disassembly/recycling/reuse) could be supplemented as a second-best assessment indicator.

The importance of multi-criteria analysis is critical in environmental problems. The ELECTRE III method was preferred over other multi-criteria techniques for selecting environmental indicators to be applied to investment evaluation. In the proposed methodology, 18 scenarios were selected as alternatives, based on relevant data from the literature and taking into account the principles of prevention, planning and design. Moreover, four criteria are considered, namely environmental, social, economic and technological, in order to capture the different pillars of sustainable development. With the use of the ELECTRE III, the optimal bundle of criteria is extracted for twelve (12) scenarios with differentiated weighting factors for the four criteria, and also preference and indifference thresholds.

From the analysis of the scenarios, it is evident that the of environmental indicators is influenced by the selected parameters of the methodology, however, there is a dominant trend demonstrating that specific indicators (resources' savings, recycling, Research–Innovation–Development, impact restoration) should undeniably be considered for the overall assessment of investments. At the same time, the modified rankings, such as those resulting from the realized sensitivity analysis, demonstrate that the final word for the selected bundle of indicators is left to the decision-maker. The latter, on the basis of the particular needs of the funding programme, is responsible for drafting the weights for the criteria and also determines the thresholds for investments' evaluation.

The proposed methodology can be seen as a tool through which decision-makers may select additional indicators that can create a framework of sustainable assessment of potential investments. Based on the results presented herein, the legislative framework could be improved so that sustainable growth indicators can also be incorporated in the decision-making process for the evaluation of an investment. Undeniably, this research can be extended not only to public authorities, but also to businesses in their effort to promote sustainable products and solutions. The present work represents an initial attempt to reach this goal, however, further research is required in terms of sample size (including international bodies and funding agencies) and the criteria considered, so as to demonstrate the optimal indicators that should be incorporated as key performance indicators, alongside financial ones, for the assessment of investments.

Author Contributions: Conceptualization, P.O. and D.A.; methodology, P.O. and D.A.; software, C.A.; validation, C.V. and D.B.; resources, P.O. and C.A.; data curation, P.O., D.A. and C.A.; writing—original draft preparation, P.O.; writing—review and editing, C.A.; visualization, C.A.; supervision, D.A. All authors have read and agreed to the published version of the manuscript.

Funding: This research received no external funding.

Acknowledgments: The authors would like to wholeheartedly thank the experts that participated in the survey.

Conflicts of Interest: The authors declare no conflict of interest

References

1. Medne, A.; Lapina, I. Sustainability and continuous improvement of organization: Review of process-oriented performance indicators. *J. Open Innov. Technol. Mark. Complex.* **2019**, *5*, 49. [CrossRef]
2. Di Vaio, A.; Varriale, L.; Alvino, F. Key performance indicators for developing environmentally sustainable and energy efficient ports: Evidence from Italy. *Energy Policy* **2018**, *122*, 229–240. [CrossRef]
3. Maslesa, E.; Jensen, P.A.; Birkved, M. Indicators for quantifying environmental building performance: A systematic literature review. *J. Build. Eng.* **2018**, *19*, 552–560. [CrossRef]
4. Pilouk, S.; Koottatep, T. Environmental performance indicators as the key for eco-industrial parks in Thailand. *J. Clean. Prod.* **2017**, *156*, 614–623. [CrossRef]
5. Achillas, C.; Vlachokostas, C.; Moussiopoulos, N.; Perkoulidis, G.; Banias, G.; Mastropavlos, M. Electronic waste management cost: A scenario-based analysis for Greece. *Waste Manag. Res.* **2011**, *29*, 963–972. [CrossRef]
6. Feleki, E.; Vlachokostas, C.; Moussiopoulos, N. Characterisation of sustainability in urban areas: An analysis of assessment tools with emphasis on European cities. *Sustain. Cities Soc.* **2018**, *43*, 563–577. [CrossRef]
7. Manrique, S.; Ballester, M.; Pilar, C. Analyzing the effect of corporate environmental performance on corporate financial performance in developed and developing countries. *Sustainability* **2017**, *9*, 1957. [CrossRef]

8. Banias, G.; Achillas, C.; Vlachokostas, C.; Moussiopoulos, N.; Stefanou, M. Environmental impacts in the life cycle of olive oil: A literature review. *J. Sci. Food Agric.* **2017**, *97*, 1686–1697. [CrossRef]

9. Zheng, Y.; Chen, Z.; Pearson, T.; Zhao, J.; Hu, H.; Prosperi, M. Design and methodology challenges of environment-wide association studies: A systematic review. *Environ. Res.* **2020**, *183*. [CrossRef]

10. Saad, E.; Hegab, H.; Kishawy, H.A. Design for sustainable manufacturing: Approach, implementation, and assessment. *Sustainability* **2018**, *10*, 3604.

11. Ariffin, R.; Ghazilla, R.; Sakundarini, N.; Taha, Z.; Abdul-Rashid, S.; Yusoff, S. Design for environment and design for disassembly practices in Malaysia: A practitioner's perspectives. *J. Clean. Prod.* **2015**, *108*, 331–342.

12. Arnette, A.; Brewer, B.; Choal, T. Design for sustainability (DFS): The intersection of supply chain and environment. *J. Clean. Prod.* **2014**, *83*, 374–390. [CrossRef]

13. Tian, J.F.; Pan, C.; Xue, R.; Yang, X.T.; Wang, C.; Ji, X.Z.; Shan, Y.L. Corporate innovation and environmental investment: The moderating role of institutional environment. *Adv. Clim. Chang. Res.* **2020**, *11*, 85–91. [CrossRef]

14. Lee, K.H.; Min, B.; Yook, K.H. The impacts of carbon (CO2) emissions and environmental research and development (R&D) investment on firm performance. *Int. J. Prod. Econ.* **2015**, *167*, 1–11.

15. Theofanidou. *Environmental Performance of Businesses after the Implementation of Environmental Management System*; National Technical University of Athens: Athens, Greece; University of Piraeus: Athens, Greece, 2008; pp. 6–24.

16. Chirico, A.; Hristov, I. The role of sustainability Key Performance Indicators (KPIs) in implementing sustainable strategies. *Sustainability* **2019**, *11*, 5742.

17. Gaidajis, G.; Angelakoglou, K. A conceptual framework to evaluate the environmental sustainability performance of mining industrial facilities. *Sustainability* **2020**, *12*, 2135.

18. Michailidou, A.V.; Vlachokostas, C.; Moussiopoulos, N. A methodology to assess the overall environmental pressure attributed to tourism areas: A combined approach for typical all-sized hotels in Chalkidiki, Greece. *Ecol. Indic.* **2015**, *50*, 108–119. [CrossRef]

19. Berghout, F.; Hertin, J.; Azzone, G.; Carlens, J.; Drunen, M.; Jasch, C.; Noci, G.; Olsthoorn, X.; Tyteca, D.; van der Woerd, F.; et al. Measuring the Environmental Performance of Industry (MEPI). Final Report. 2001. Available online: http://www.sussex.ac.uk/Units/spru/mepi/about/index.php (accessed on 7 July 2020).

20. Dizdaroglu, D. The role of indicator-based sustainability assessment in policy and the decision-making process: A review and outlook. *Sustainability* **2017**, *9*, 1018. [CrossRef]

21. European Commission. *Science for Environment Policy—Indicators for Sustainable Cities. In-Depth Report Produced for the European Commission DG Environment by the Science Communication Unit*; UWE: Bristol, UK, 2018; Available online: http://ec.europa.eu/science-environment-policy (accessed on 7 July 2020).

22. Laurent, A.; Olsen, S.I.; Hauschild, M.Z. Carbon footprint as environmental performance indicator for the manufacturing industry. *CIRP Annals* **2010**, *59*, 37–40. [CrossRef]

23. Vlachokostas, C.; Achillas, C.; Chourdakis, E.; Moussiopoulos, N. Combining regression analysis and air quality modelling to predict benzene concentration levels. *Atmos. Environ.* **2011**, *45*, 2585–2592. [CrossRef]

24. Feleki, E.; Vlachokostas, C.; Moussiopoulos, N. Holistic methodological framework for the characterization of urban sustainability and strategic planning. *J. Clean. Prod.* **2020**, *243*, 118432. [CrossRef]

25. Ethridge, M.A. *Measuring Environmental Performance: A Primer and Survey of Metrics in Use*; Global Environmental Management Initiative (GEMI): Washington, DC, USA, 1998.

26. International Organization for Standardization. *ISO 14031:1999: Environmental Management—Environmental Performance Evaluation—Guidelines*; International Organization for Standardization: Geneva, Switzerland, 1999.

27. Watróbski, J.; Jankowski, J.; Ziemba, P.; Karczmarczyk, A.; Zioło, M. Generalised framework for multi-criteria method selection. *Omega* **2019**, *86*, 107–124. [CrossRef]

28. Cinelli, M.; Coles, S.; Kirwan, K. Analysis of the potentials of multi criteria decision analysis methods to conduct sustainability assessment. *Ecol. Indic.* **2014**, *46*, 138–148. [CrossRef]

29. Hokkanen, J.; Salminen, P. Choosing a solid waste management system using multicriteria decision analysis. *Eur. J. Oper. Res.* **1997**, *98*, 19–36. [CrossRef]

30. Achillas, C.; Aidonis, D.; Vlachokostas, C.; Folinas, D.; Moussiopoulos, N. Re-designing industrial products on a multi-objective basis: A case study. *J. Oper. Res. Soc.* **2013**, *64*, 1336–1346. [CrossRef]

31. Hashemi, S.S.; Hajiagha, S.H.R.; Zavadskas, E.K.; Mahdiraji, H.A. Multicriteria group decision making with ELECTRE III method based on interval-valued intuitionistic fuzzy information. *Appl. Math. Model.* **2016**, *40*, 1554–1564. [CrossRef]

32. Zopounidis, C.; Pardalos, P. (Eds.) *Handbook of Multicriteria Analysis*; Applied Optimization Series; Springer: Berlin, Germany, 2010.

33. Hass, J.L.; Palm, V. *Using the Right Environmental Indicators: Tracking Progress, Raising Awareness and Supporting Analysis*; Nordic Council of Ministers: Copenhagen, Denmark, 2012.

34. Arcibugi, F.; Nijkamp, P. *Economy and Ecology: Towards Sustainable Development*; Klumer Academic Publishers: Dordrecht, The Netherlands, 1989.

35. Wu, Z.; Abdul-Nour, G. Comparison of multi-criteria group decision-making methods for urban sewer network plan selection. *CivilEng* **2020**, *1*, 26-48. [CrossRef]

36. Azziz, M.B.-H. *Multiple Criteria Outranking Algorithm: Implementation and Computational Tests. ELECTRE III Method*; Department of Engineering and Management, Instituto Superior Técnico: Lisbon, Portugal, 2015.

37. Vlachokostas, C.; Michailidou, A.V.; Matziris, E.; Achillas, C.; Moussiopoulos, N. A multiple criteria decision-making approach to put forward tree species in urban environment. *Urban Clim.* **2014**, *10*, 105–118. [CrossRef]

38. Rogers, M.; Bruen, M. Choosing realistic values of indifference, preference and veto thresholds for use with environmental criteria within ELECTRE. *Eur. J. Oper. Res.* **1998**, *107*, 542–551. [CrossRef]

39. Xiaoting, W. Study of Ranking Irregularities When Evaluating Alternatives by Using Some ELECTRE Methods and A Proposed New MCDM Method Based on Regret and Rejoicing. Master's Thesis, Louisiana State University, Baton Rouge, LA, USA, 2007.

40. Roy, B.; Electre, A. III lgorithme de classement base sur une representation floue des preferences en presence de criteres multiples. *Cahiers de CERO* **1978**, *20*, 3–24.

41. Roy, B.; Bouyssou, D. *Aide Multicritere a la Decision: Methods et Cas*; Economica: Paris, France, 1993.

42. Roussat, N.; Dujet, C.; Mehu, J. Choosing a sustainable demolition waste management strategy using multicriteria decision analysis. *Waste Manag.* **2009**, *29*, 2–20. [CrossRef] [PubMed]

43. Moussiopoulos, N.; Achillas, C.; Vlachokostas, C.; Spyridi, D.; Nikolaou, K. Environmental, social and economic information management for the evaluation of sustainability in urban areas: A system of indicators for Thessaloniki, Greece. *Cities* **2010**, *27*, 377–384. [CrossRef]

44. Falqi, I.; Alsulamy, S.; Mansour, M. Environmental performance evaluation and analysis using ISO 14031 guidelines in construction sector industries. *Sustainability* **2020**, *12*, 1774. [CrossRef]

45. Ocampo, A.L.; Estanislao-Clark, E. Developing a framework for sustainable manufacturing strategies selection. *DLSU Bus. Econ. Rev.* **2014**, *23*, 115–131.

46. Michailidou, A.V.; Vlachokostas, C.; Moussiopoulos, N. Interactions between climate change and the tourism sector: Multiple-criteria decision analysis to assess mitigation and adaptation options in tourism areas. *Tour. Manag.* **2016**, *55*, 1–12. [CrossRef]

47. Spyridi, D.; Vlachokostas, C.; Michailidou, A.V.; Sioutas, C.; Moussiopoulos, N. Strategic planning for climate change mitigation and adaptation: The case of Greece. *Int. J. Clim. Chang. Strateg. Manag.* **2015**, *7*, 272–289. [CrossRef]

48. Kourmpanis, B.; Papadopoulos, A.; Moustakas, K.; Kourmoussis, F.; Stylianou, M.; Loizidou, M. An integrated approach for the management of demolition waste in Cyprus. *Waste Manag. Res.* **2008**, *26*, 573–581. [CrossRef]

49. Haralambopoulos, D.; Polatidis, H. Renewable energy projects: Structuring a multi-criteria group decision-making framework. *Renew. Energy* **2003**, *28*, 961–973. [CrossRef]

Article

Application of Artificial Neural Networks for Natural Gas Consumption Forecasting

Athanasios Anagnostis [1,2,*], Elpiniki Papageorgiou [1,3] and Dionysis Bochtis [1]

[1] Institute for Bio-economy and Agri-technology (iBO), Center for Research and Technology—Hellas (CERTH), Thessaloniki GR57001, Greece; e.papageorgiou@certh.gr or elpinikipapageorgiou@uth.gr (E.P.); d.bochtis@certh.gr (D.B.)

[2] Department of Computer Science, University of Thessaly, Lamia GR35131, Greece

[3] Department of Energy Systems, Faculty of Technology, University of Thessaly, Geopolis Campus Ring Road of Larissa-Trikala, Larissa GR41500, Greece

* Correspondence: a.anagnostis@certh.gr or athananagno@uth.gr; Tel.: +30-6947072707

Received: 3 July 2020; Accepted: 6 August 2020; Published: 9 August 2020

Abstract: The present research study explores three types of neural network approaches for forecasting natural gas consumption in fifteen cities throughout Greece; a simple perceptron artificial neural network (ANN), a state-of-the-art Long Short-Term Memory (LSTM), and the proposed deep neural network (DNN). In this research paper, a DNN implementation is proposed where variables related to social aspects are introduced as inputs. These qualitative factors along with a deeper, more complex architecture are utilized for improving the forecasting ability of the proposed approach. A comparative analysis is conducted between the proposed DNN, the simple ANN, and the advantageous LSTM, with the results offering a deeper understanding the characteristics of Greek cities and the habitual patterns of their residents. The proposed implementation shows efficacy on forecasting daily values of energy consumption for up to four years. For the evaluation of the proposed approach, a real-life dataset for natural gas prediction was used. A detailed discussion is provided on the performance of the implemented approaches, the ANN and the LSTM, that are characterized as particularly accurate and effective in the literature, and the proposed DNN with the inclusion of the qualitative variables that govern human behavior, which outperforms them.

Keywords: machine learning; artificial neural networks; natural gas; demand forecasting

---

1. Introduction

The consumption of natural gas has seen a substantial increase during recent years, as it presents a reliable and economical energy and heating solution for businesses as well as households. Its wide acceptance from large-scale infrastructures to small houses has created diverse consumption patterns, especially during high-demand occasions. Inevitably, this has perplexed any attempt of forecasting its demand, especially when one considers the diversity of the consumers and the finite restrictions of the natural gas infrastructure, i.e., low accumulation ability within the grid.

Analytical modelling of such complicated systems would require substantial effort in order to design the grid architecture and each of its consumers, apply correct heat losses throughout the pipes, and in general, include a variety of intricate parameters into the whole system before running the simulation computations. On the other hand, data-driven models are invariant of such parameter tuning and can properly model a system by learning valuable patterns from its collected data. Machine learning algorithms create models by recurrently learning from data, until they can model a process in the best way possible. Being dependent on data alone, alternative scenarios based on different energy resources like fossil fuels, oil, or electricity may as well utilize these methods for their own forecasting purposes.

State-of-the-art published studies which focus on natural gas forecasting of production, consumption, demand, market volatility and fluctuation in prices, and income elasticity have been surveyed and are presented in [1]. Efficient energy supply planning is essential for any country's socio-economic state since it is crucial, especially for building successful development plans [2]. There is a large number of papers found in the relevant literature that tackle the problem of accurate forecasting of natural gas consumption, mostly focusing in hourly intervals [3]. Short-term forecasting is based on the pattern analysis of time series in order to predict accurate values of consumption or demand [4]. Artificial intelligence, machine learning, and other statistical methods are typically used in short-, medium-, and long-term forecasts of energy demand [5]. Based on research studies from the literature, there are notable findings that utilized artificial neural network (ANN) algorithms on forecasting natural gas demand, and whose day-ahead predictions had high accuracy [6–15]. Multiple variants of neural networks, especially deep neural networks, have been extensively used to tackle the problem of short-term demand forecasting of natural gas. Deep learning was firstly used by Merkel et al. for forecasting short-term load of natural gas [16,17], and then to be compared to traditional ANN and linear regression models on 62 different areas with at least 10 years of data [18].

Other data-driven approaches, such as neuro-fuzzy methods or genetic algorithms, have tackled the problem of natural gas demand [19–21]. Hybrid approaches including Wavelet Transform (WT), Genetic Algorithm (GA), Adaptive Neuro-Fuzzy Inference System (ANFIS), and Feed-Forward Neural Network (FFNN) have been used by Panapakidis and Dagoumas in order to forecast natural gas demand in the Greek natural gas grid [22]. Moreover, other soft computing techniques, like fuzzy cognitive maps (FCMs), enhanced by evolutionary algorithms, have been applied for modeling time series problems [23–28]. In [29], Poczeta and Papageorgiou conducted a preliminary study on implementing FCMs with ANNs for natural gas prediction, showing for first time the capabilities of evolutionary FCMs in this domain. Furthermore, the research team in [30] recently conducted a study for time series analysis devoted to natural gas demand prediction in three Greek cities, implementing an efficient ensemble forecasting approach through combining ANN, RCGAFCM, SOGA-FCM, and hybrid FCM-ANN. In this research study, the advantageous features of intelligent methods, through an ensemble to multivariate time series prediction, have emerged.

Many works can be found in the literature that address the accurate forecasting of natural gas demand with a methodology that was based solely on an artificial neural network, or was used in combination with other methods in hybrid forecasting systems; however, in the present work, an innovative approach that includes vital social factors in deep neural network (DNN) models was studied exclusively, contributing to the novelty of the current study. The main aim of this study is the development of a non-linear time series model that can accurately predict future energy demand and estimate how the introduction of important social factors can improve the accuracy of its predictions. As a case study for the demonstration of the approach's applicability, natural gas energy data from various cities in Greece, which present socio-economic aspects and thus different consumption attributes, have been implemented.

Contrary to most studies that focus on quantitative-only inputs, there are some studies that take into consideration the impact of social or socio-economic factors with machine learning based approaches [31]. The behavioral habits and characteristics of consumers are strong indicators in forecasting electricity load in households [32–34]. Social factors were taken into consideration in the prediction of total energy demand in several cases such as Spain [35], China [36], and Turkey [37,38]. The application of social components alongside meteorological and past consumption data was also studied in district heating networks [39,40]. In all relevant studies, the results showed that the inclusion of social parameters in the modelling can increase the model's overall accuracy [41,42].

Our effort focuses on investigating three types of approaches. The first approach relies on a simple Artificial Neural Network (ANN), namely a single-layer perceptron, that takes into account only quantitative variables. The second approach is based on the state-of-the-art recurrent neural network (RNN), namely Long Short-Term Memory (LSTM) network, which uses single-variable time series and can predict a variable's value for the next point in time by "memorizing" past variations. The third approach is the proposed Deep Neural Network (DNN) that takes as input not only quantitative, but also qualitative variables. The DNN consists of more nodes and layers than the ANN since it needs to process a larger and more diverse amount of inputs, both numerical and categorical, in a more appropriate fashion. The qualitative variables that are being used in the proposed DNN approach are social factors that fit the characteristics of the country of Greece and will be described in detail in paragraph 2.2. For the case of natural gas demand forecasting, the consumption of energy is bound to the behavior of the human population, which is dependent on social habits, a factor whose impact is investigated extensively in this study.

In the present study, the aim is to build a robust forecasting model based on a proposed deep neural network (DNN) and compare it with an artificial neural network (ANN) and a recurrent neural network (RNN), both of which are able to accurately forecast energy demand [43]. This comparative analysis aims to investigate whether the factors that dictate human behavior can offer crucial information and increase the accuracy of our forecasts. The results clearly demonstrate that the proposed DNN approach, with the inclusion of social factors, has attained better accuracies than other state of the art intelligent models for natural gas consumption forecasting.

2. Materials and Methods

The Hellenic Gas Transmission System Operator S.A. (DESFA) (www.desfa.gr) is the operator that manages and develops the Greek natural gas infrastructure. DESFA handles all natural gas off-takes, deliveries, and general distribution, as well as the collection of useful data. They have provided with the dataset that was used in this study. Details on the dataset and its features are provided below.

2.1. Dataset

The dataset contains historical data of time series from the natural gas consumption of multiple cities all over Greece, as well as the average daily temperature of each city's surrounding area. The data spans from 1 March 2010, or later on for some cities, until 31 October 2018. Specifically for some centralized larger cities of Greece such as Athens and Larissa, where the natural gas distribution system was installed early on, there is data since 1 March 2010, as seen in Figure 1, while in some other large cities, like Thessaloniki, also seen in Figure 1, or smaller ones, like Alexandroupoli, seen in Figure 2, data collection started later on. The exact starting dates of data collection for each city are given later on in Section 3.

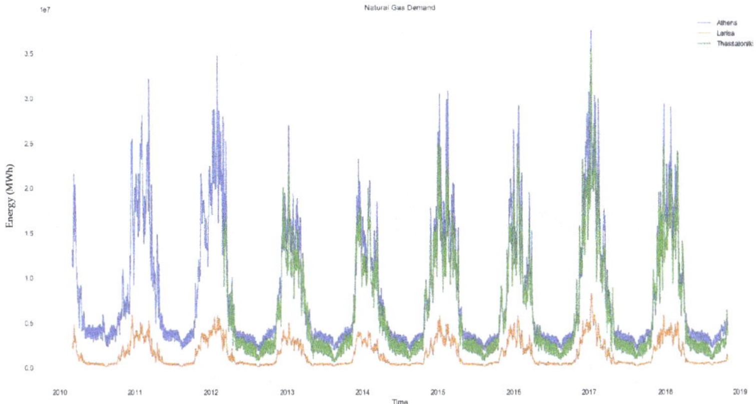

Figure 1. Natural gas energy consumption [MWh] for Athens, Thessaloniki, and Larissa over the years.

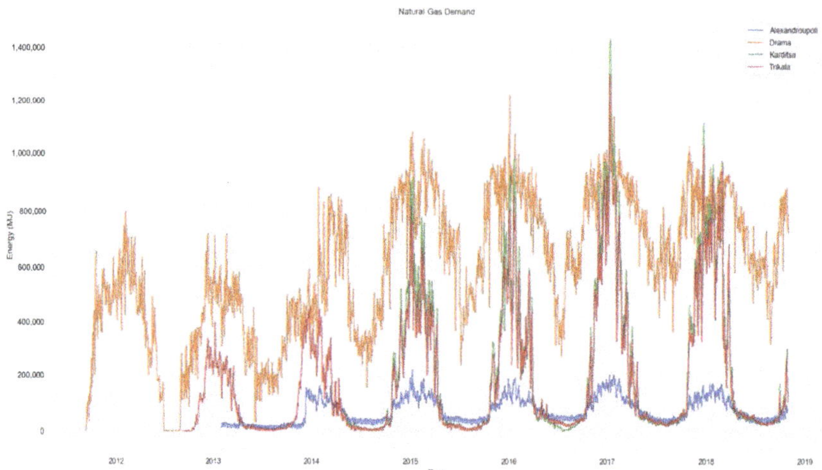

Figure 2. Natural gas energy consumption [MWh] for Alexandroupoli, Drama, Karditsa, and Trikala over the years.

As provided, the dataset contained time and dates, natural gas consumption of each city's distribution point, and the daily average temperature of the area in Celsius degrees. On top of the existing data, social indicators have been added to the dataset such as a month indicator, a day indicator, a weekday/weekend indicator, and a bank holiday indicator. The proper addition of social variables is a key factor to the study since the aim is to see if qualitative social traits can improve the performance of a forecasting model, and by how much compared to other methods.

2.2. Feature Engineering

A certain amount of feature engineering is required for the qualitative data to take proper form, in order to be readable by the machine learning algorithms. This takes place during the preprocessing phase and is conducted in the following way: Months and days are described by a name, e.g., September, Tuesday, etc., and need to be transformed into categorical values, e.g., 9, 2, etc., in a serial way. Therefore, the following association is considered: January-1, February-2, etc., and Monday-1, Tuesday-2, etc. Each of these values are then transformed into vectors with the size of the value range of the variable. In detail, the "month" variable contains 12 different values, one for each month, therefore the size of the vector is 1×12. Respectively, the "day" variable contains 7 values, one for each day, therefore the size for this vector is 1×7. Consecutively, the "month" variable is transformed into 12 variables, one for each month, and the "day" variable is transformed into 7 variables, one for each day of the week. The "bank holiday" variable denotes a public or religious holiday that affects social behavior (businesses are closed, people are out celebrating, etc.) and is binary, therefore there is no need for any kind of further transformation.

Time and date data are transformed into a single timedate variable which is then used as an index, thus leading to the total amount of 22 variables that are being taken as inputs for the modeling of the energy forecast of the proposed methodology. The desirable variable for the forecast is the natural gas energy consumption from the specific distribution point, which is used as output. The correlation of the consumption energy with the mean daily temperature for the city of Athens is shown in Figure 3.

The correlation plots of consumption energy and mean temperature for all the investigated cities are given in the Appendix A, where it is obvious that not all cities have the same pattern of correlation between the mean temperature and the consumption of natural gas. This variation in patterns is one of the reasons that the implemented models achieve different accuracies for each different city, as will be shown later.

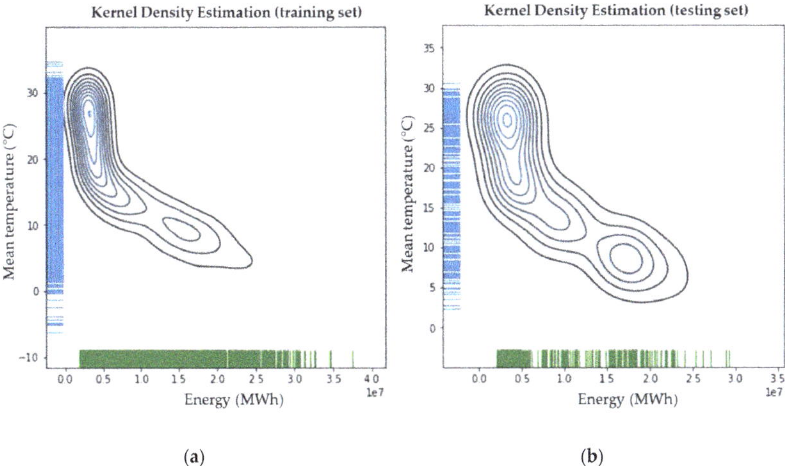

(a)   (b)

Figure 3. Correlation of the energy demand and the mean temperature for the training set (**a**) and the test set (**b**) for the city of Athens.

3. Methodology

Three types of neural network variants were modeled and tested for accurately forecasting natural gas energy demand. Their implementations differ even though they all belong to the neural network family of algorithms. Each approach has different input variables, however, the general approach remains the same; historical data train a model that is able to produce accurate forecasts of natural gas energy consumption. The approaches, as well as the general process flow, are described in the following sections.

3.1. Artificial Neural Networks (ANN)

Artificial Neural Networks (ANN) [44] were hypothesized as a method to imitate the human brain and its functions while it performs cognitive tasks, or when it learns. For the mathematical structure of such networks, nodes are used to represent the neurons, and layers are used to represent their interactive synapses in the same fashion as it is within the brain. They received wide acceptance ever since they were conceptualized, however, they started gaining more and more fame ever since the computational power, especially in graphic processing units, has become cheaper and thus easily attainable. Another important reason for their wide acceptance is the vast availability of immense amounts of data that have been collected throughout the past years. This adoption has enhanced the scientific progress of the ANN algorithms, which started from the single-layer perceptron [45], moved to multi-layer perceptron [46], introduced the back-propagation algorithm [47], and led to many new derivatives of ANNs, the most well-known being the deep neural networks that are described in the next paragraph.

3.2. Long Short-Term Memory Networks (LSTM)

Long Short-Term Memory are also neural networks which are built upon a recurrent fashion by introducing memory cells and their in-between connections, in order to construct a graph directed over a sequence. In general, recurrent networks process sequences by using said memory cells in a fashion that is different than that of the simple ANNs, and even though they are well suited for problems with time dependency, they often face the problem of vanishing gradients, or not being able to "memorize" large portions of data. LSTMs [48] solved this problem because of the specific cell structure they have, which allows the network the ability to variate the amount of retained information. These cell structures are called gates, and they control which information is stored in the long memory and which is discarded, thus optimizing the memorizing process. Dynamic temporal behavior problems, i.e., time sequences, were suited for such approaches.

3.3. Deep Neural Networks (DNN)

Deep neural networks [49] are the basis of deep learning, one of the most influential areas of the artificial intelligence for the past decade. Based on the ANN, the DNN is comprised by more layers and nodes in the same sequential fashion. For very deep implementations, the problem of "vanishing" or "exploding" gradients would not allow the network to learn properly, therefore new techniques were introduced, such as "identity shortcut connections" as seen in Resnet [50], as well as others, in order to solve these obstacles. The "deep" approach has been used in all derivatives of neural networks. Deep convolutional neural networks (CNN) have been used for image classification and object detection [51,52], and deep recurrent neural networks (RNN) have been used for word prediction [53] and time series forecasting [54].

3.4. Process Flow

In order to accurately model the natural gas energy forecast, a specific process flow has been designed. The steps that have been followed, are described below.

3.4.1. Preprocessing

The initial preprocessing is focused on the organization of the original unstructured data. Same-variable columns have been aligned and set in correct time and date order, and columns that were empty or contained plenty of NaN (Not a Number) values were removed completely. The columns "date" and "time" were merged into a TIMEDATE column, which was consecutively used as index. The MONTH and DAY variables were manually added and later transformed into multiple categorical variables as described in Section 2.

For the ANN implementation, only the daily mean temperature is considered as input, for the LSTM, only the energy consumption is used as both input and output (past and present values respectively), and for the DNN implementation, all the aforementioned variables are used with the addition of vectorial representations of qualitative variables. The inputs that are used in each neural network variant, as well as the output variable, that are being used in this study, are shown in Table 1.

Table 1. Variables that are used as output and input for each implementation of the neural network variants. ANN: Artificial neural network; LSTM: Long Short-Term Memory; DNN: Deep neural network.

	Output		Inputs				
	Energy Current Day (MJ)	**Daily Mean Temperature** (°C)	**Energy 1 Previous Day** (MJ)	**Energy 2 Previous Days** (MJ)	**Month** (Jan, Feb, Mar, Apr, May, Jun, Jul, Aug, Sep, Oct, Nov, Dec)	**Day of the Week** (Mon, Tue, Wen, Thu, Fri, Sat, Sun)	**Bank Holiday** (Yes, No)
ANN	×	×					
LSTM	×						
DNN	×	×	×	×	×	×	×

3.4.2. Data Split

The data was split in the following way: The last year, ranging from 1 November 2017 till 31 October 2018, was used as the testing period for all models, approaches, and for all cities. The starting date of data collection for each city, as well as the ratio of training/testing portions of each dataset, is shown in Table 2.

During the training phase, 20% of the training dataset is used as a validation set, in order to identify whether our model tends to under- or overfit, and to be able to measure its loss and accuracy.

Table 2. Starting dates and ratio of training/testing portions of the dataset per city.

City	Start Date	Ratio of Training/Testing
Agioi Theodoroi	07/06/2014	3.41
Alexandroupoli	02/02/2013	4.78
Athens	01/03/2010	7.68
Drama	07/09/2011	6.15
Karditsa	01/05/2014	3.51
Kilkis	01/03/2010	7.68
Lamia	01/02/2013	4.75
Larissa	01/03/2010	7.68
Laurio	01/03/2010	7.68
Markopoulo	01/03/2010	7.68
Serres	01/06/2013	4.42
Thessaloniki	01/03/2012	5.67
Trikala	12/09/2012	5.14
Volos	01/03/2010	7.68
Xanthi	01/03/2010	7.68

For the ANN implementation, only the numerical variables were used as input, i.e., the energy consumption of 2 prior days and the mean temperature. For the LSTM implementation, the natural gas energy consumption is used both as input and output, so the previous 365 values are used to find the future trend, i.e., the energy demand. For the DNN implementation, all the variables described in paragraph 2.2 are used as inputs, and the natural gas energy consumption is used as output, as it is with all implementations.

3.4.3. Standardization

All data was normalized with Python's SciKit Learn MinMaxScaler, between 0 and 1 values [55]. This way, the performance metrics are common for all cities, therefore direct comparisons can be conducted, but more importantly because during training it allows the non-convex cost function to converge to the global minimum faster and in a more appropriate fashion.

3.4.4. Processing

In all ANN, LSTM, and DNN approaches, we used the "ReLU" activation function, the "adam" optimizer for the cost function, mean squared error for measuring the loss of training and validation, and an early stopping function (EarlyStopping callback in Keras [56] with 5 epochs patience) in order to avoid overfitting [57]. For the LSTM approach, the previous 365 energy values were used as input, and the consumption for the next 365 days was forecasted. After the model training, all models are being evaluated on the testing portion of the dataset. Their performance is measured based on certain metrics that are described in the following section.

4. Evaluation metrics

The performance and robustness of each studied natural gas forecasting model is based on four of the most common evaluation metrics. Mean square error (MSE), absolute error (MAE), mean absolute percentage error (MAPE), and coefficient of determination (R^2) are all being used in order to determine the best performing model [5,7,43].

All the modelling, tests, and evaluations were performed with the use of Python 3.7 and the Tensorflow 1.14, Keras 2.3, SciKit Learn 0.21, Pandas 0.25, Numpy 1.17, Matplotlib 3.1, Seaborn 0.9 libraries. The mathematical equations of these evaluation metrics are described below:

Mean Squared Error:

$$\text{MSE} = \frac{1}{T} \sum_{t=1}^{T} (Z(t) - X(t))^2, \tag{1}$$

Mean Absolute Error:

$$\text{MAE} = \frac{1}{T} \sum_{t=1}^{T} |Z(t) - X(t)|, \tag{2}$$

Mean Absolute Percentage Error:

$$\text{MAPE} = \frac{1}{T} \sum_{t=1}^{T} \left| \frac{Z(t) - X(t)}{Z(t)} \right|, \tag{3}$$

Coefficient of Correlation:

$$R = \frac{T \sum_{t=1}^{T} Z(t) \cdot X(t) - \left(\sum_{t=1}^{T} Z(t) \right)\left(\sum_{t=1}^{T} X(t) \right)}{\sqrt{T \sum_{t=1}^{T} (Z(t))^2 - \left(\sum_{t=1}^{T} Z(t) \right)^2} \cdot \sqrt{T \sum_{t=1}^{T} (X(t))^2 - \left(\sum_{t=1}^{T} X(t) \right)^2}}, \tag{4}$$

where $X(t)$ is the predicted value, $Z(t)$ is the real value, t is the iteration at each point ($t = 1, \ldots, T$), and T is the number of testing records.

Low MSE, MAE, and MAPE values signify small error, therefore higher accuracy. On the contrary, R^2 value close to 1 is preferred, signifying better performance for the model and that the regression curve is well fit on the data. A coefficient of determination value of 1 would signify that the regression line fits the data perfectly; however, this could also denote overfitting on the data.

To summarize, the whole process so far can be visually represented into an algorithmic flowchart. Starting from the data preprocessing, to the training of the algorithm and the prediction of the results, all the consecutive steps are shown in Figure 4.

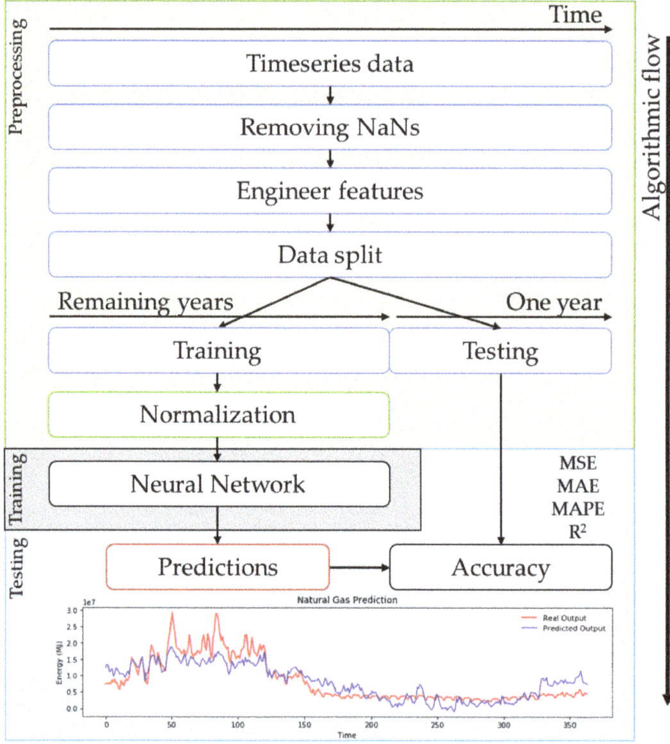

Figure 4. Process flow for the algorithm that has been used for the study.

5. Results

Athens was chosen as a reference point for searching the best fitting parameters of each method. The outcomes of each parameter selection for the ANN, the LSTM, and the DNN architectures are presented in paragraphs 5.1, 5.2, and 5.3. The parameters that resulted to the best performing model for all 15 cities are presented in paragraph 5.4. For the evaluation of all tests, MSE, MAE, MAPE, and R^2 were used.

5.1. Results from ANN

For the ANN implementation, an architecture of a single-layer perceptron with 8 nodes in the hidden layer was selected, after a concise exploratory analysis. The simple ANN was selected for benchmarking purposes. Having a simplistic model as a baseline, we can investigate the performance improvement of the other approaches. The initial architecture, seen in Figure 5, was tested and evaluated without any dropout function, and the performance metrics are shown in Table 3.

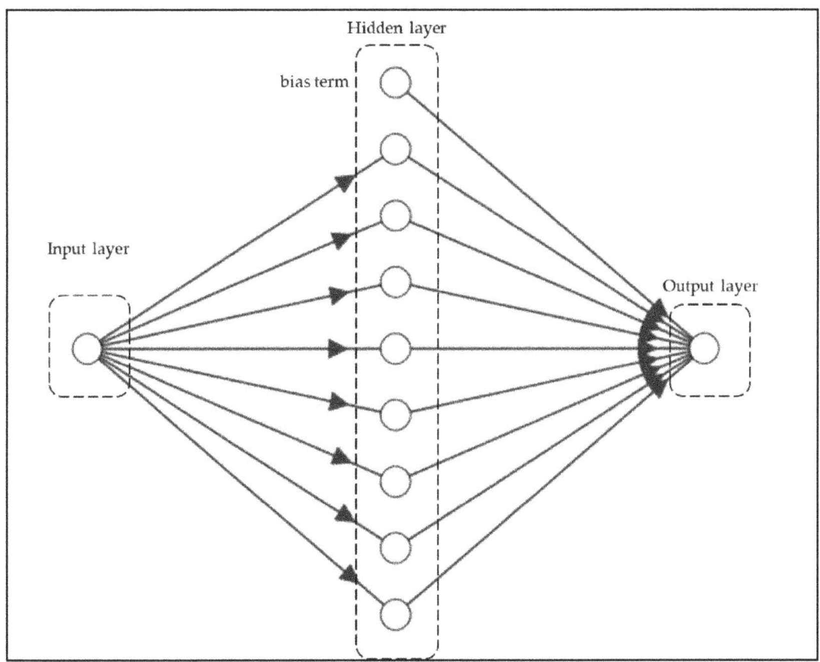

Figure 5. Architecture of the ANN approach.

Table 3. Performance metrics of the selected ANN architecture. MSE: Mean square error; MAE: Mean absolute error; MAPE: Mean absolute percentage error; R^2: Coefficient of determination.

ANN		Architecture Selection			
Layers	**Nodes**	**MSE (MJ^2)**	**MAE (MJ^2)**	**MAPE (%)**	**R^2**
1	8	7.70×10^{-3}	7.01×10^{-2}	16.41	0.87

Next, the effect of the dropout rate is investigated, and in order to understand how it affects the model's performance, four distinct percentages have been tested. The results are presented in Table 4.

Table 4. Comparison of dropout rate in ANN.

1 Layer/ 8 Nodes	Dropout Comparison			
Dropout	MSE (MJ^2)	MAE (MJ^2)	MAPE (%)	R^2
0	7.70×10^{-3}	7.01×10^{-2}	16.41	0.87
0.25	8.20×10^{-3}	6.94×10^{-2}	30.30	0.86
0.50	9.00×10^{-3}	7.41×10^{-2}	52.74	0.84
0.75	1.25×10^{-2}	8.66×10^{-2}	73.02	0.78

The ability of forecasting the energy demand in yearly intervals was also investigated. Even though this study is focused in one-year ahead forecasts, a timeframe of yearly depths up to four years ahead was investigated. This investigation was conducted in order to see the magnitude of the forecasts' accuracy through time, and the results are presented in Table 5.

Table 5. Comparison of year-ahead forecasting in ANN.

1 Layer/ 8 Nodes	Forecasting Comparison			
Years ahead	MSE (MJ^2)	MAE (MJ^2)	MAPE (%)	R^2
1	7.70×10^{-3}	7.01×10^{-2}	16.41	0.87
2	1.08×10^{-1}	3.07×10^{-1}	48.61	−1.39
3	9.28×10^{-2}	2.83×10^{-1}	52.46	−1.31
4	1.42×10^{-1}	3.53×10^{-1}	49.73	−2.57

The ANN approach is able to capture the general trend, however, it deviates significantly from the real consumption values, something that could signify that the model cannot give better forecasts for longer time ahead. Figure 6 shows the plots of the ANN implementation for (a) one-, (b) two-, (c) three-, and (d) four-year ahead forecasting. The prediction of the energy demand in MWh is depicted in blue, and the real output is depicted in red.

It is evident that the ANN model, even though it can follow the trend of the fluctuation, fails to forecast accurately the consumption of the natural gas. Furthermore, the more the forecasting time increases, the greater this deviation gets. Even though ANNs are powerful algorithms for forecasting, in this particular problem, the single-layer perceptron is not enough to model the problem accurately.

5.2. Results from LSTM

An investigatory analysis was conducted also for the LSTM implementation. The number of layers and memory units were explored in order to find the best combination, which was comprised of one LSTM layer with 200 memory units. The architecture of this implementation is seen in Figure 7, and its performance is shown in Table 6.

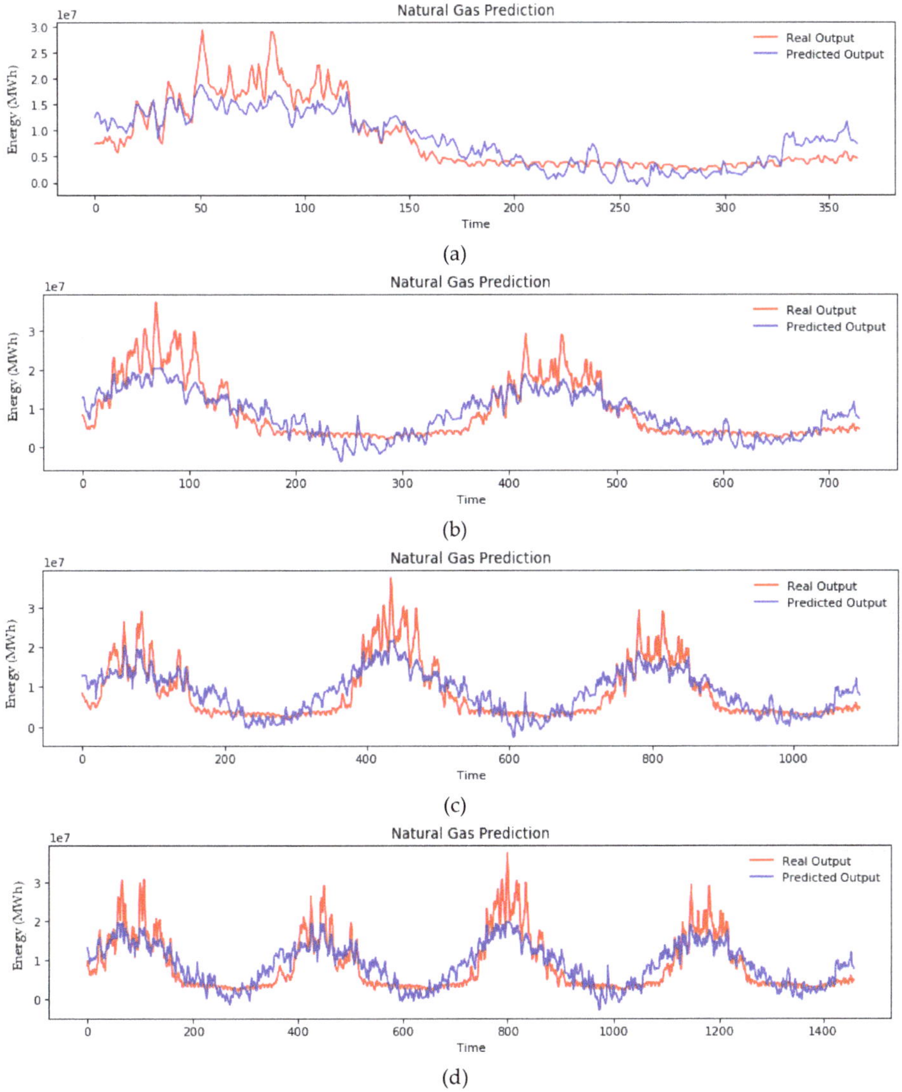

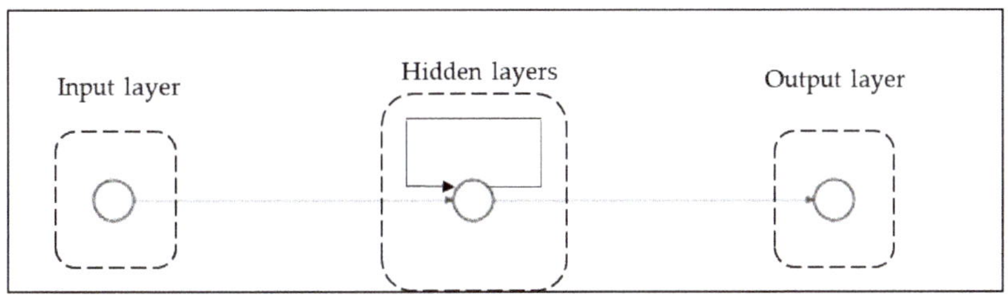

Figure 6. Forecasts of (**a**) one-, (**b**) two-, (**c**) three-, and (**d**) four-year ahead of natural gas demand with the use of ANN.

Figure 7. The architecture of the LSTM approach.

Table 6. Performance metrics of the selected LSTM architecture.

LSTM		Architecture Selection			
Layers	**Units**	**MSE (MJ^2)**	**MAE (MJ^2)**	**MAPE (%)**	**R^2**
1	200	2.20×10^{-3}	2.95×10^{-2}	11.07	0.96

The dropout's effect on the LSTM implementation was also investigated, and its effect on the performance of the model is seen in Table 7.

Table 7. Comparison of dropout rate in LSTM.

1 Layer/ 200 Units		Dropout Comparison			
Dropout	**MSE (MJ^2)**	**MAE (MJ^2)**	**MAPE (%)**	**R^2**
0	2.20×10^{-3}	2.95×10^{-2}	11.07	0.96
0.25	2.40×10^{-3}	3.31×10^{-2}	20.94	0.96
0.50	2.20×10^{-3}	3.10×10^{-2}	18.50	0.96
0.75	2.10×10^{-3}	2.92×10^{-2}	9.77	0.96

Dropout application seems to increase performance over a non-dropout approach, and in fact the highest rate selected has given the best results.

Again, forecasts of up to four years ahead were evaluated in order to investigate how the predictions are affected. The results are presented in Table 8.

Table 8. Comparison of year-ahead forecasting in LSTM.

1 Layer/ 8 Nodes		Forecasting Comparison			
Years ahead	**MSE (MJ^2)**	**MAE (MJ^2)**	**MAPE (%)**	**R^2**
1	2.10×10^{-3}	2.92×10^{-2}	9.77	0.96
2	6.30×10^{-3}	5.90×10^{-2}	15.00	0.86
3	5.26×10^{-2}	2.06×10^{-1}	59.06	−0.31
4	8.73×10^{-2}	2.44×10^{-1}	80.87	−1.20

The plots of the LSTM setup are shown in Figure 8 for (a) one-, (b) two-, (c) three-, and (d) four-year ahead forecasting. The prediction of the energy demand in MWh is depicted in blue, and the real output is depicted in red.

In the case of LSTM implementation, it is clear that the forecasts for the one- and two-year ahead demands are more accurate than that of the ANN implementation. However, it is evident that anything beyond the two-year ahead forecast is tremendously inaccurate, resulting even in negative R^2 values. LSTMs can offer excellent accuracy for single-variable time series; however, it is evident that they are highly susceptible to the depth of the forecasting period, as well as to the data that are required for proper training.

5.3. Results from DNN

For the DNN implementation, a deeper, more complex network was constructed that is comprised of 4 hidden layers with 32 nodes in each layer. The proposed architecture is structured in such way so that it can take as input the vectorial representations of categorical values, the ones mentioned above, the quantitative values from the current time (in each step), as well as the energy values from past inputs. The architecture of the ANN approach is seen in Figure 9. No dropout was initially set for the exploratory analysis, and the performance metrics for the selected setup is shown in Table 9.

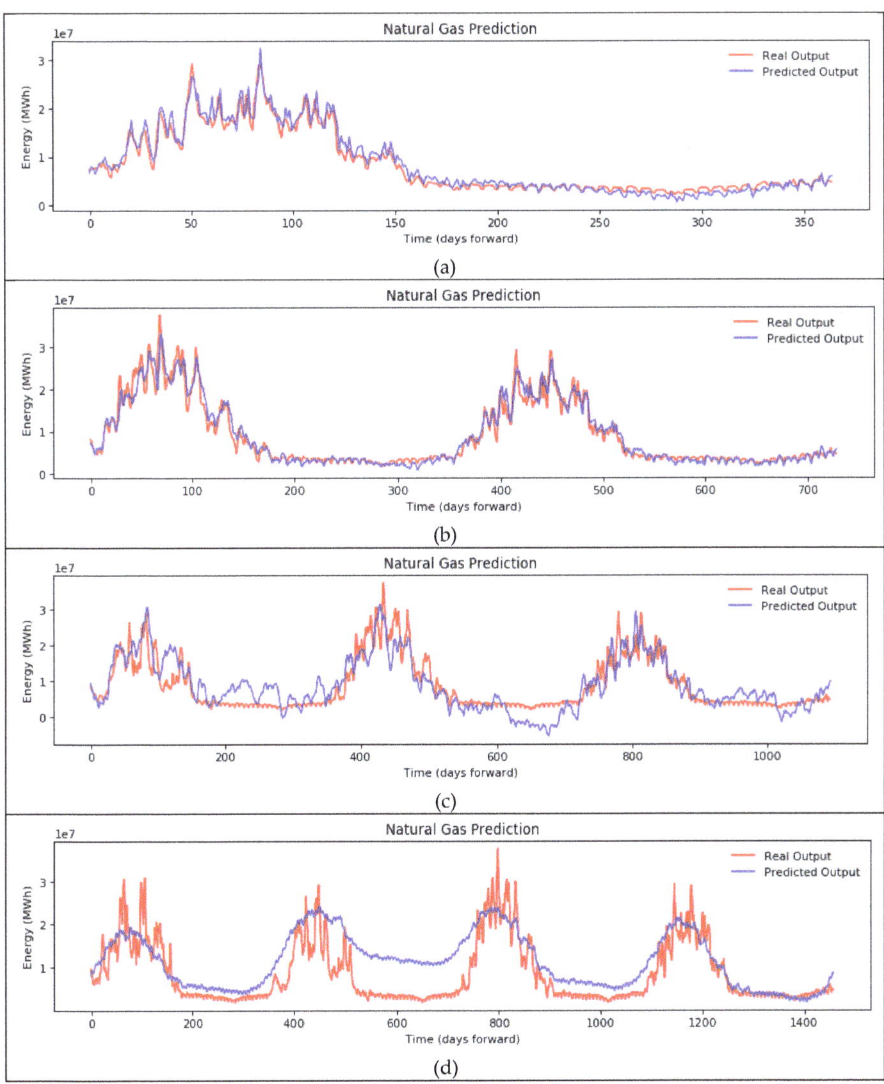

Figure 8. One year ahead forecast of natural gas demand with the use of LSTM.

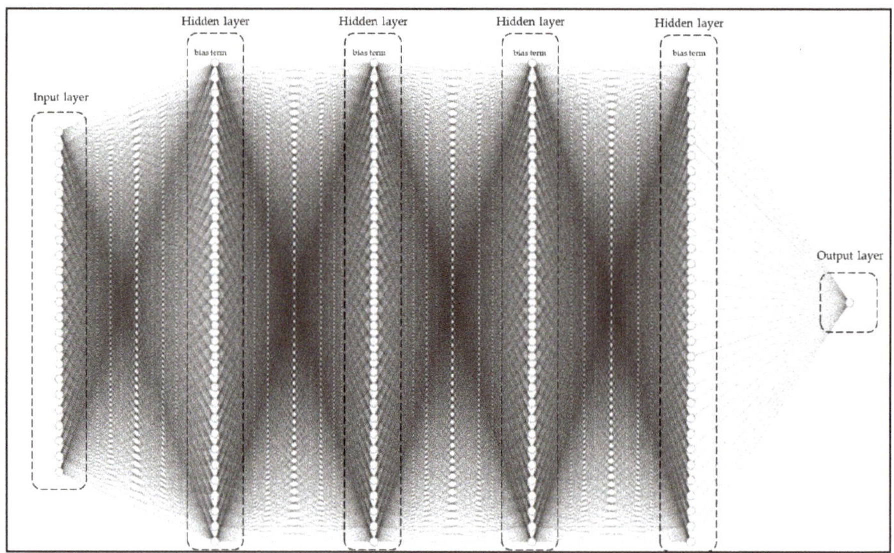

Figure 9. Architecture of the DNN approach.

Table 9. Performance metrics of the selected DNN architecture.

DNN		Architecture Selection			
Layers	**Nodes**	**MSE (MJ^2)**	**MAE (MJ^2)**	**MAPE (%)**	**R^2**
4	32	1.70×10^{-3}	2.75×10^{-2}	7.53	0.97

The effect of the dropout rate is investigated once again, and the same four percentages have been tested and evaluated, the results of which are presented in Table 10.

Table 10. Comparison of dropout rate in DNN.

4 Layer/ 32 Nodes	Dropout Comparison			
Dropout	**MSE (MJ^2)**	**MAE (MJ^2)**	**MAPE (%)**	**R^2**
0	1.70×10^{-3}	2.75×10^{-2}	7.53	0.97
0.25	2.20×10^{-3}	3.11×10^{-2}	10.98	0.96
0.50	2.90×10^{-3}	4.32×10^{-2}	26.45	0.95
0.75	3.00×10^{-3}	3.82×10^{-2}	48.61	0.95

It is evident that the proposed DNN model performs better than the ANN and the LSTM. Testing its forecasting capabilities for up to four years ahead will show its ability to generalize well and properly model the consumption pattern. The results are presented in Table 11.

Table 11. Comparison of year-ahead forecasting in DNN.

4 Layer/ 32 Nodes	Forecasting Comparison			
Years ahead	**MSE (MJ^2)**	**MAE (MJ^2)**	**MAPE (%)**	**R^2**
1	1.70×10^{-3}	2.75×10^{-2}	7.53	0.97
2	1.50×10^{-3}	2.76×10^{-2}	8.98	0.97
3	1.30×10^{-3}	2.06×10^{-2}	8.68	0.97
4	1.70×10^{-3}	3.22×10^{-2}	9.40	0.96

At this point, it is interesting to mention that the accuracy of the predictions is hardly affected for up to four years ahead. This is due to the yearly periodicity of the energy demand that is caused not only by the general temperature trends, but also by the social aspects that govern human behavior in certain periods of time. Figure 10 shows the plots of the best performing setup for the DNN implementation for (a) one-, (b) two-, (c) three-, and (d) four-year ahead forecasting. The prediction of the energy demand in MWh is depicted in blue, and the real output is depicted in red.

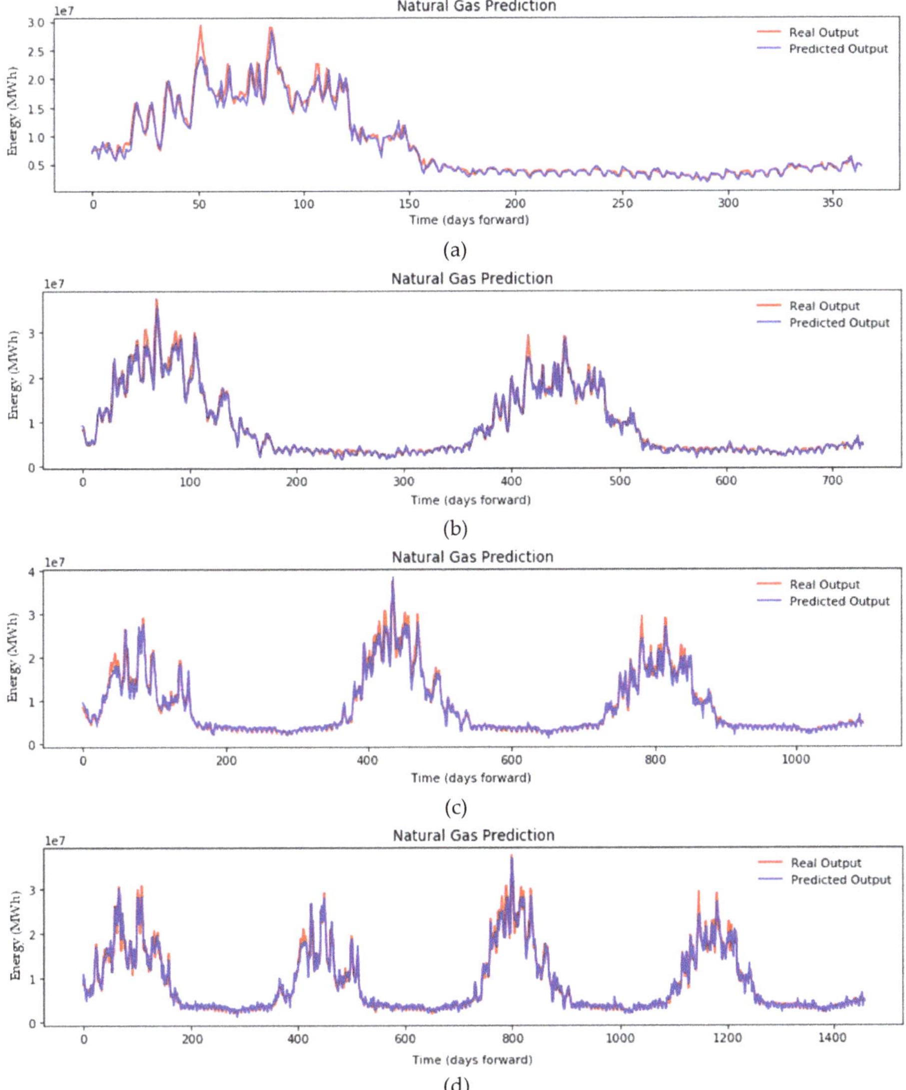

Figure 10. Forecasts of (**a**) one-, (**b**) two-, (**c**) three-, and (**d**) four-year ahead of natural gas demand with the use of DNN.

For the proposed DNN, it is clear that its forecasting capabilities surpass by far the ANN and the state-of-the-art LSTM models. The inclusion of qualitative social variables alongside measurable quantities has improved not only the accuracy of the forecasts, but also the depth of forecasting time into the future. This is an indicator that the deeper network, alongside with behavioral knowledge, has offered a generalized understanding of the energy consumption trend.

5.4. Comparison (Cities)

The trained models of all approaches were applied on a range of fifteen cities all around Greece for the sake of comparison. Each city had different energy distributions during the documented years depending on its size, population, and specific natural gas network characteristics. Testing all implementations on different cities, offers insight on whether each model can provide accurate one-year ahead forecasts in cities that are in different geographical locations, but also with different behavioral patterns. The confidence interval (CI) [58] is also included in the following analysis, to demonstrate the range of energy values, in which 95% of the predictions fall within, for each city. The performance metrics for all cities are shown in Table 12 for the ANN, Table 13 for the LSTM, and Table 14 for the proposed DNN.

Table 12. Comparison of cities for the ANN implementation.

ANN	Cities Comparison				
Cities	**MSE (MJ^2)**	**MAE (MJ^2)**	**MAPE (%)**	**R^2**	**CI**
Agioi Theodoroi	4.96×10^{-2}	1.70×10^{-1}	58.29	0.14	[92,246–101,257]
Alexandroupoli	8.30×10^{-3}	6.72×10^{-2}	13.43	0.89	[80,527–88,566]
Athens	2.40×10^{-3}	3.53×10^{-2}	10.94	0.96	[7,710,057–9,006,556]
Drama	8.50×10^{-3}	6.77×10^{-2}	6.42	0.78	[736,302–764,636]
Karditsa	2.60×10^{-3}	3.54×10^{-2}	28.66	0.97	[234,893–293,622]
Kilkis	2.87×10^{-2}	1.23×10^{-1}	20.16	0.43	[1,023,806–1,088,807]
Lamia	3.96×10^{-2}	1.62×10^{-1}	33.76	0.20	[125,129–135,568]
Larissa	2.60×10^{-3}	3.43×10^{-2}	14.18	0.96	[1,329,088–1,609,373]
Laurio	1.32×10^{-2}	7.59×10^{-2}	86,605.00	0.74	[6,004,693–7,432,331]
Markopoulo	3.06×10^{-2}	1.38×10^{-1}	19.38	0.27	[250,400–264,171]
Serres	3.60×10^{-3}	4.52×10^{-2}	11.69	0.96	[400,111–467,284]
Thessaloniki	3.40×10^{-3}	4.52×10^{-2}	16.80	0.95	[6,283,167–7,449,531]
Trikala	2.40×10^{-3}	2.76×10^{-2}	25.62	0.97	[213,058–266,020]
Volos	3.90×10^{-3}	4.92×10^{-2}	11.41	0.91	[1,662,841–1,845,579]
Xanthi	2.94×10^{-2}	1.33×10^{-1}	36.12	0.17	[153,935–168,218]

Table 13. Comparison of cities for LSTM implementation.

LSTM	Cities Comparison				
Cities	**MSE (MJ^2)**	**MAE (MJ^2)**	**MAPE (%)**	**R^2**	**CI**
Agioi Theodoroi	3.56×10^{-2}	1.50×10^{-1}	59.68	0.38	[86,688–94,424]
Alexandroupoli	4.20×10^{-3}	4.60×10^{-2}	8.29	0.94	[86,103–95,178]
Athens	2.10×10^{-3}	2.92×10^{-2}	9.77	0.96	[7,634,555–8,986,641]
Drama	1.06×10^{-2}	7.82×10^{-2}	9.86	0.72	[765,187–790,777]
Karditsa	8.40×10^{-3}	7.61×10^{-2}	72.99	0.89	[290,871–357,268]
Kilkis	1.15×10^{-2}	8.17×10^{-2}	10.97	0.77	[1,008,757–1,071,517]
Lamia	1.86×10^{-2}	1.11×10^{-1}	32.19	0.62	[157,729–167,902]
Larissa	3.30×10^{-3}	4.43×10^{-2}	17.39	0.95	[1,377,457–1,661,699]
Laurio	2.45×10^{-2}	1.30×10^{-1}	78,544.00	0.53	[8,712,884–10,054,310]
Markopoulo	8.00×10^{-3}	6.79×10^{-2}	9.70	0.81	[258,990–272,945]
Serres	4.00×10^{-3}	4.85×10^{-2}	15.78	0.96	[375,137–443,421]
Thessaloniki	3.90×10^{-3}	4.79×10^{-2}	19.54	0.94	[6,268,458–7,571,520]
Trikala	4.60×10^{-3}	5.41×10^{-2}	32.77	0.94	[256,379–318,363]
Volos	4.80×10^{-3}	5.21×10^{-2}	13.18	0.89	[1,480,596–1,649,836]
Xanthi	9.80×10^{-3}	7.60×10^{-2}	24.43	0.72	[169,605–180,582]

Table 14. Comparison of cities for the proposed DNN approach.

DNN	Cities Comparison				
Cities	**MSE (MJ^2)**	**MAE (MJ^2)**	**MAPE (%)**	**R^2**	**CI**
Agioi Theodoroi	2.46×10^{-2}	1.19×10^{-1}	38.70	0.57	[89,946–98,778]
Alexandroupoli	3.00×10^{-3}	4.01×10^{-2}	7.57	0.96	[80,268–88,430]
Athens	1.70×10^{-3}	2.75×10^{-2}	7.53	0.97	[7,571,538–8,863,422]
Drama	1.03×10^{-2}	7.69×10^{-2}	7.32	0.73	[729,057–757,898]
Karditsa	3.50×10^{-3}	4.84×10^{-2}	25.10	0.95	[230,724–290,272]
Kilkis	8.20×10^{-3}	6.73×10^{-2}	10.36	0.84	[984,497–1,049,268]
Lamia	1.41×10^{-2}	9.71×10^{-2}	21.23	0.71	[126,988–137,111]
Larissa	1.10×10^{-3}	2.40×10^{-2}	9.95	0.98	[1,413,437–1,698,290]
Laurio	2.03×10^{-2}	1.17×10^{-1}	78,518.00	0.61	[6,822,570–8,233,887]
Markopoulo	6.80×10^{-3}	6.29×10^{-2}	10.10	0.84	[245,857–259,800]
Serres	4.50×10^{-3}	5.21×10^{-2}	17.46	0.95	[396,910–466,370]
Thessaloniki	1.60×10^{-3}	2.72×10^{-2}	10.25	0.98	[6,289,857–7,456,553]
Trikala	4.40×10^{-3}	5.04×10^{-2}	22.98	0.95	[222,492–276,310]
Volos	5.10×10^{-3}	5.70×10^{-2}	11.73	0.89	[1,644,087–1,825,076]
Xanthi	1.02×10^{-2}	7.44×10^{-2}	26.04	0.71	[165,997–179,896]

For the ANN implementation, the performance of the model ranges from ~14% for Agioi Theodoroi till ~97% for Trikala considering R^2. Seven out of fourteen cities achieved an accuracy of >90%, however, for the other seven cities, the performance of the model is disappointing.

For the LSTM implementation, the prediction accuracies are better than the ANN, ranging from ~39% for Agioi Theodoroi to ~96% for Athens, using R^2 as the primary metric. Here, six cities achieved an accuracy of >94%, with the rest achieving higher accuracy when compared to the ANN.

For the proposed DNN implementation, the performance of the model ranges from ~58% for Agioi Theodoroi till ~99% for Larissa considering R^2. For seven out of fourteen cities, the proposed methodology achieved an accuracy of >94%, which is considered very satisfactory for prediction, considering that the MSE of these models is also very low.

5.5. Sensitivity Analysis

A sensitivity analysis is conducted on the dataset of Athens. The selected method for the sensitivity analysis is the Partial Dependence Plots (PDP) [59,60], where target variables (features) of the dataset are investigated through their range of values in order to visualize their dependence to the target outcome. The numerical variables used in the datasets, i.e., daily mean temperature, 1-day and 2-days prior consumption energy are used as the target features and the dependence of the target outcome, i.e., the present-day consumption energy, is shown in Figure 11.

Both for the ANN and the DNN, the mean daily temperature is inversely proportional to the daily energy consumption, which is expected since heating needs are lower when the external temperature is high. For the DNN, the 1-day prior consumption is directly proportional to the target outcome, the same applying for the 2-days prior consumption as well. We notice that for the 2-days prior, the scale is two orders of magnitude less than for the 1-day prior and one order less than the mean daily temperature. This signifies that the model interprets a weaker relationship with this variable and the outcome, meaning that the dependence of the target outcome from this feature is less significant than the others.

Qualitative values cannot be included in the sensitivity analysis because they don't span over a range of values. Also, for the LSTM model, there can be no sensitivity analysis because only one variable is used as time series, therefore only past values of energy consumption are used for future predictions.

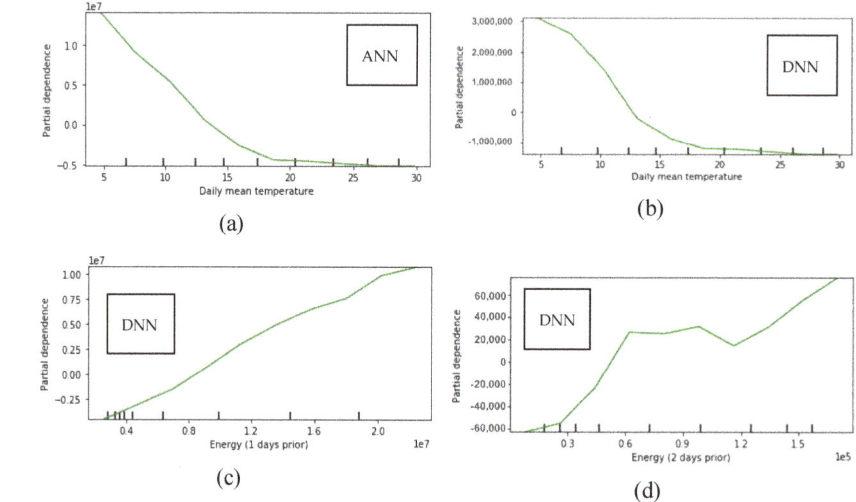

Figure 11. Partial Dependence Plots for the dataset of Athens on the mean daily temperature for the ANN (**a**) and the DNN (**b**) models, and the energy consumption values of 1-day (**c**) and 2-days (**d**) prior.

6. Discussion

Three methods, focused on tackling the problem of accurate forecasting of natural gas energy consumption in fifteen cities all over Greece, were investigated and applied in this study. The first method is an ANN that takes into consideration quantitative variables only, like the energy consumption and the external daily temperature. The second method is a LSTM that takes into consideration 365 previous values of only energy consumption of each city. The third method is a DNN that takes into consideration not only the quantitative variables used in the ANN, but also qualitative variables that govern human behavior such as weekdays, weekends, and bank holidays. Comparison analyses were conducted for each method in order to find the optimal architecture for each one.

All models perform adequately in most cases. The value of artificial neural networks and their derivatives is well known, however, the purpose of this study is to increase the accuracy and the time-depth of the forecasting capabilities. For the larger cities, high accuracies in forecasting energy consumption is achieved. The proposed DNN implementation achieved the highest R^2 for the city of Larissa (0.9846) while the LSTM implementation for the city of Athens (0.9644) and the ANN for the city of Trikala (0.699). For the worst-case scenario, the city of Agioi Theodoroi, has consistently obtained the worst accuracies, with the DNN (0.5748) achieving significant higher accuracy, even though still not so good, compared to the LSTM (0.3848) and ANN (0.1440) implementation. The dataset of Agioi Theodoroi is the smallest compared to the rest, being one reason for achieving these low accuracies. It ca be argued that the size of the city (<5000 habitants) is another important reason, since the consumption trends are sparser due its low population.

For the city of Agioi Theodoroi, the DNN increased the accuracy of its forecasts by 49.362% compared to the LSTM, and 299.195% compared to the ANN. Regarding the city of Athens, the DNN increased the accuracy of its forecasts by 0.682% compared to the LSTM, and 1.292% compared to the ANN. The proposed DNN increased the accuracy of the forecasts in almost all cases, however, its main impact was on small-scale cities such as Kilkis (+94.698%), Lamia (+259.457%), Markopoulo (+207.667%), and Xanthi (+330.273%), which have small populations, energy consumption, and less amount of gathered data. In a previous study [43] where only the ANN and LSTM approaches were applied on quantitative-only datasets, the LSTM approach offered the best results. The particular problem of forecasting energy values, is time-dependent, thus allowing the LSTM approach to excel. However, since there are other factors that affect the behavior of the consumers, and consequently the consumption of energy, the DNN was considered as an approach that could improve the accuracy of the forecasts.

For the implementation of the LSTM, the application of dropout has improved performance for the one-year-ahead forecast. By selecting 200 units in the layer of the selected implementation, the LSTM is able to capture a measurable part of the input (365 days); however, in order to generalize well, the model should randomly drop a percentage of the weights it has "learned". This way it has the ability to "memorize" large inputs, however, these inputs are generalized and do not overfit on the past data. This conflicts with the other implementations of ANN and DNN; however, LSTM utilizes information in a different way than the ANN and DNN. The reason why the LSTM performs better with a high dropout rate is because it tends to overfit soon during training, and even if it could reach high training accuracy, its validation (and therefore testing) accuracy would be weak. In this study, there is a trend based on seasonality, and in order to have an LSTM model that is not overly simplistic (therefore needing at least 200 units), and to train as long as possible, generalization was achieved via high dropout [61].

For the long-term predictions, the ANN and LSTM models fail to produce accurate predictions, resulting in negative R^2 values. This can be derived from several facts. The more important is that since the dataset is finite, the further ahead in time the prediction is, the less training data the model is left with to "learn" from. Machine learning models are highly dependent on data, and their performance is highly correlated to the data quality and quantity. Particularly for the ANN approach, it's simplistic implementation cannot capture the complexity that is required for the long-term forecasting, even if in general ANNs are powerful. Another reason is that the scale of the energy prediction units is large (in absolute numbers), thus the worse the prediction is, the larger is the penalty for it. Additionally, since the forecasting timescale increases for additionally 1, 2, and 3 years, the ill-fitted models produce large errors in predictions which are accumulated, because the forecasting time is 1, 2, and 3 times larger, respectively. The R^2 metric is based on the MSE, and is scale-dependent, while MAPE is not, therefore it is useful for understanding the performance of the models. It is considered that R^2 is still probably the best metric for forecasts [62], however, MAPE can still be used because the percentage of error makes sense and there are no zero values in our dataset.

In our proposed architecture, social behavior variables were added as inputs and the number of layers and nodes in our neural network was increased, in order to investigate the effect of these additions on the forecasting accuracy. These variables are strong indicators of social behavior and habits of the majority of the Greek population, which can affect the energy consumption in specific days/occasions. Overfitting was avoided by monitoring loss and accuracy throughout the training phase.

Furthermore, in order to show the effectiveness of the proposed DNN forecasting methodology, a comparative analysis was conducted with a SOGA-FCM, which was applied in one year ahead of natural gas consumption predictions concerning the same dataset of the three Greek cities (Athens, Thessaloniki, and Larissa) in [29], and a recent soft computing technique for time series forecasting using evolutionary fuzzy cognitive maps and their ensemble combination [30]. This comparison is shown in Table 15, where the MSE and MAE are used as performance metrics.

Table 15. Comparison of results between machine learning and soft computing methods for three benchmark cities.

	Cities					
	Athens		Thessaloniki		Larissa	
Methods	MSE (MJ^2)	MAE (MJ^2)	MSE (MJ^2)	MAE (MJ^2)	MSE (MJ^2)	MAE (MJ^2)
ANN	7.70×10^{-3}	7.01×10^{-2}	3.40×10^{-3}	4.52×10^{-2}	2.60×10^{-3}	3.43×10^{-2}
LSTM	2.10×10^{-3}	2.92×10^{-2}	3.90×10^{-3}	4.79×10^{-2}	3.30×10^{-3}	4.43×10^{-2}
Hybrid FCM	3.20×10^{-3}	3.28×10^{-2}	3.30×10^{-3}	3.81×10^{-2}	4.10×10^{-3}	4.17×10^{-2}
Ensemble (EB)	3.10×10^{-3}	3.28×10^{-2}	3.10×10^{-3}	3.69×10^{-2}	4.00×10^{-3}	4.17×10^{-2}
Proposed DNN	$\mathbf{1.70 \times 10^{-3}}$	$\mathbf{2.75 \times 10^{-2}}$	$\mathbf{1.60 \times 10^{-3}}$	$\mathbf{2.72 \times 10^{-2}}$	$\mathbf{1.10 \times 10^{-3}}$	$\mathbf{2.40 \times 10^{-2}}$

It is evident that all methods achieve high accuracy in the predictions of the energy consumption patterns in their relative timescales. The ensemble and hybrid methods achieve the same accuracy as the ANN, with the LSTM performing slightly better. The proposed DNN, by utilizing inputs of social variables into its learning patterns and having a deeper architecture, outperforms all other methods with significant difference. The significance of this outcome lies on the fact that qualitative variables that dictate human behavior can be learned by computational algorithms and be utilized to improve forecasting accuracy furthermore.

The case of Greece has some sensitive aspects, since multiple dynamics in natural gas consumption were introduced due to the financial crisis of the previous years. This instability created an additional obstacle to the accurate forecasting of energy demand, thus increasing the need for efficient forecasting models that can accurately offer in-depth insight on the demand trends of each city and adapt to their different conditions. Since the proposed method offers high accuracy and long forecasting capabilities, it can be used by any utilities and distribution operators, as a solution to upgrade operational long-term planning, as well as to provide insight on policy making from the side of the state.

7. Conclusions

Summing up, three different forecasting approaches have been implemented in order to develop models for predicting energy demand of natural gas. Investigative analysis took place for an ANN, a LSTM, and the proposed DNN implementations in order to find a desired architecture for each method. Fifteen cities all around Greece were tested, each one with a dataset of measurements that spanned from 3 to 7 years. The investigated cities differ both in size as well as in geographical location, amplifying as much as possible the variability of each use case examined. Despite the fact that this study is focused on cities that are only in Greece, the proposed methodology is highly generalizable for any other city that can provide sufficient amount of data, both measurable and behavioral.

The goal of this study was to propose an efficient neural network implementation that utilizes a variety of quantitative and qualitative inputs, as well as a deep architecture with many layers and nodes, to demonstrate how social factors can improve the performance of the model and increase the accuracy of its forecasts. The proposed methodology has outperformed both the simple ANN approach as well as the state-of-the-art LSTM approach even though both still offer good accuracy in most cases. The inclusion of social factors in the proposed DNN approach offered consistently more generalized, high-accuracy results. This derives from the fact that by exploring longer forecasts, the four-year ahead forecast was achieved only with the proposed DNN implementation, while the LSTM could only provide accurate results up to two years ahead, and the ANN was deviating systematically.

Applying a combination of multi-parametric social factors, by also taking advantage of the memory cells structure of the LSTM implementation will be the base of the future work that will aim to outperform the DNN implementation. Additional Fuzzy Cognitive Maps structures will be also considered for increasing the interpretability of the models and how the inputs affect the performance.

Author Contributions: Conceptualization, A.A., E.P. and D.B.; methodology, A.A.; software, A.A.; validation, A.A., E.P. and D.B.; formal analysis, A.A.; investigation, A.A.; resources, E.P.; data curation, A.A.; writing—original draft preparation, A.A.; writing—review and editing, A.A., E.P., and D.B.; visualization, A.A; supervision, E.P. and D.B.; project administration, D.B. All authors have read and agree to the published version of the manuscript.

Funding: This research received no external funding.

Conflicts of Interest: The authors declare no conflict of interest.

Appendix A Appendix

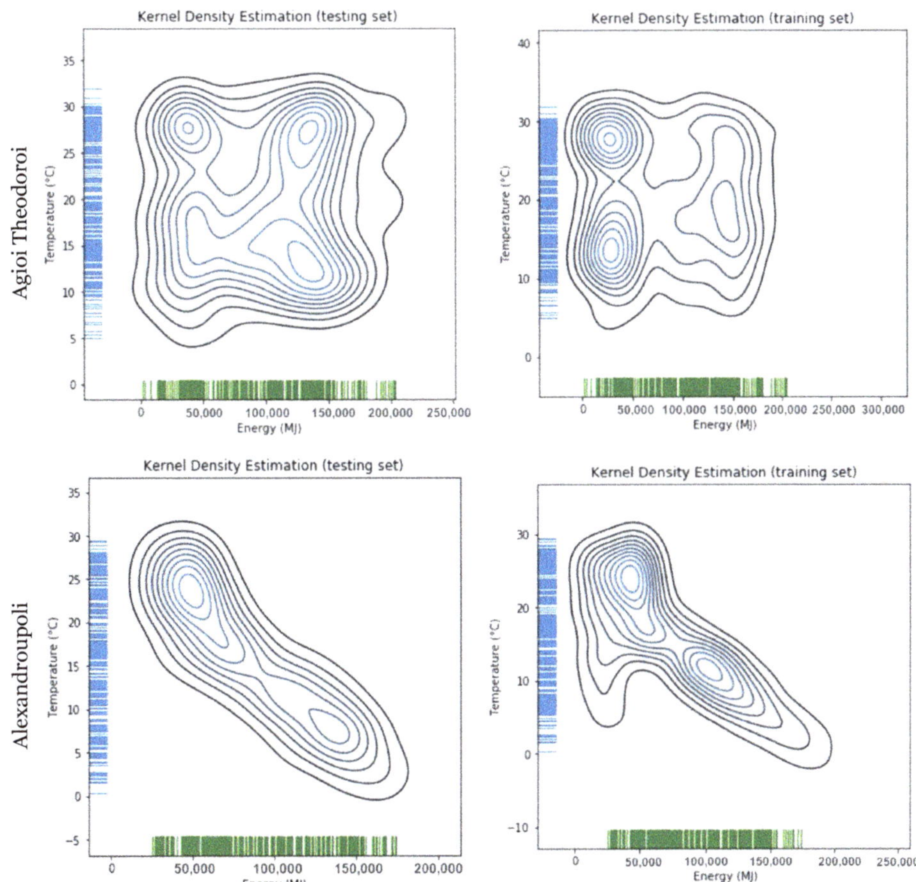

Figure A1. *Cont.*

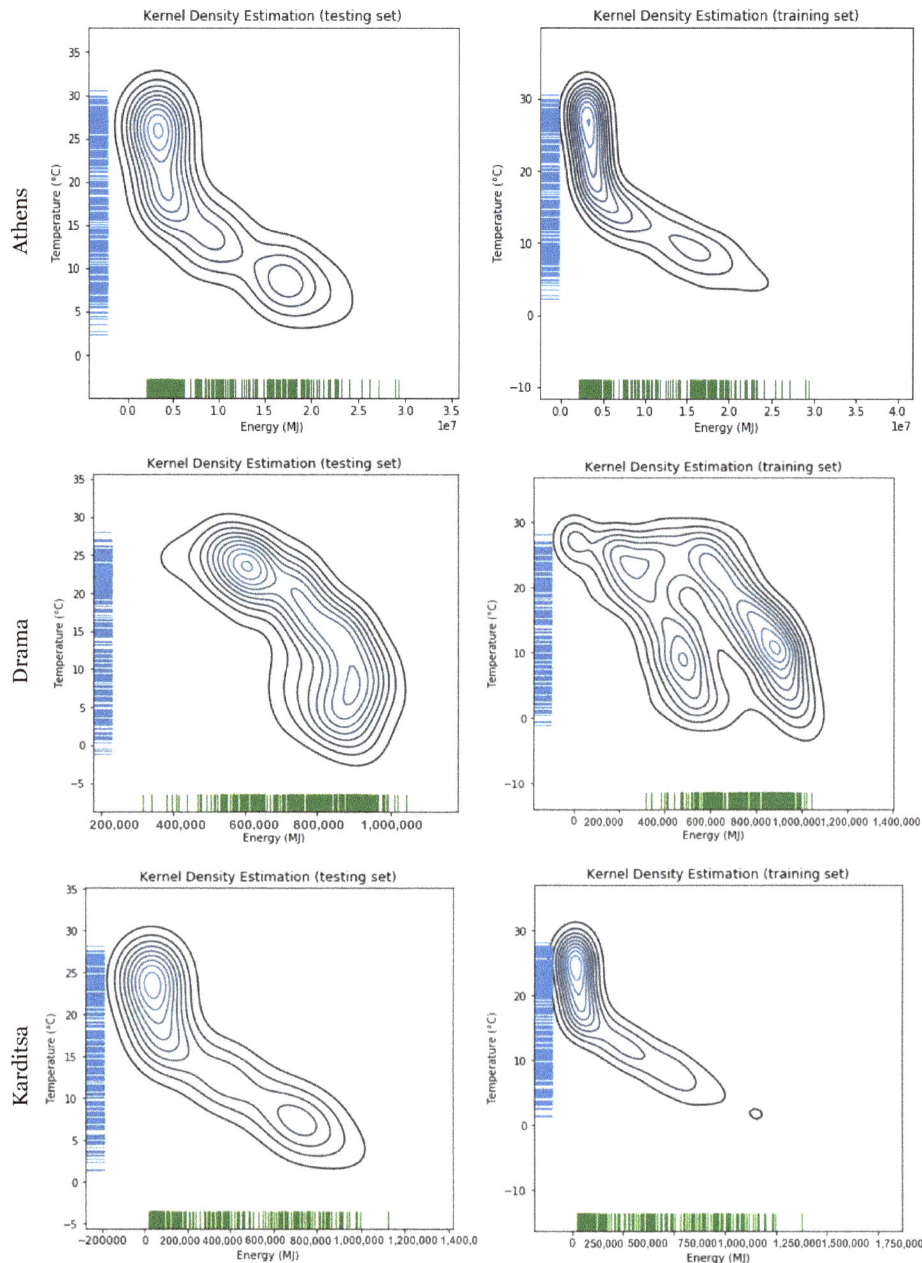

Figure A1. *Cont.*

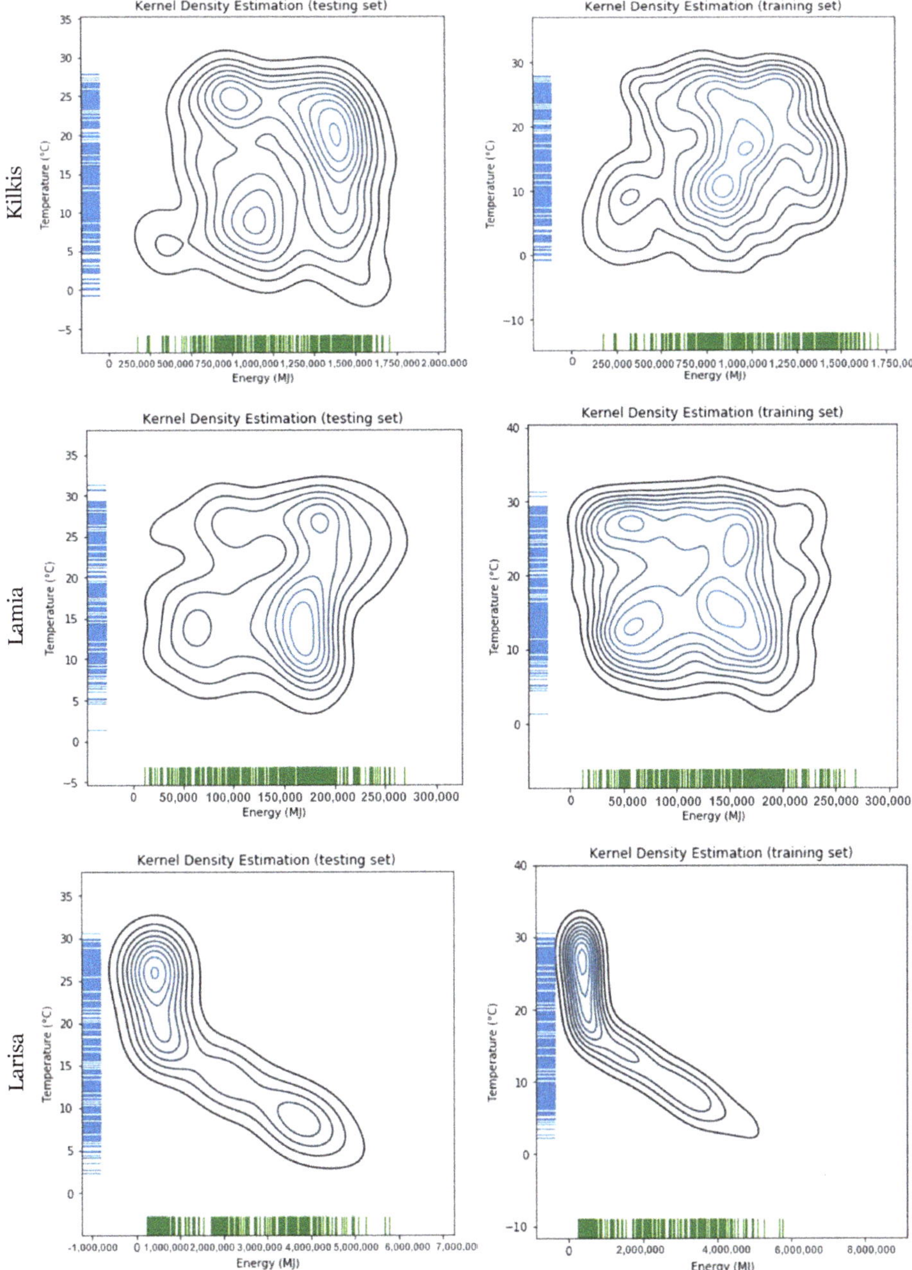

Figure A1. *Cont.*

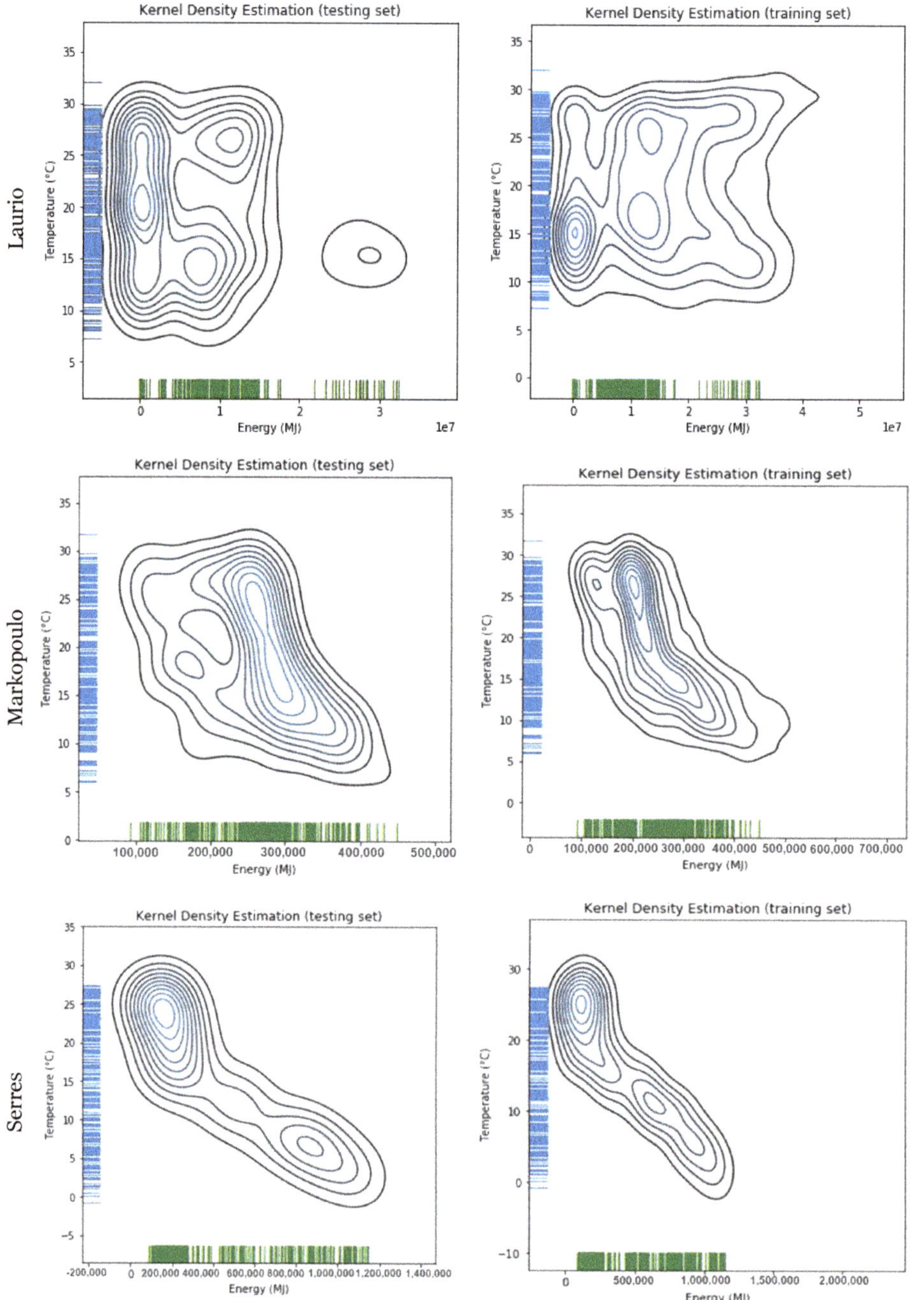

Figure A1. *Cont.*

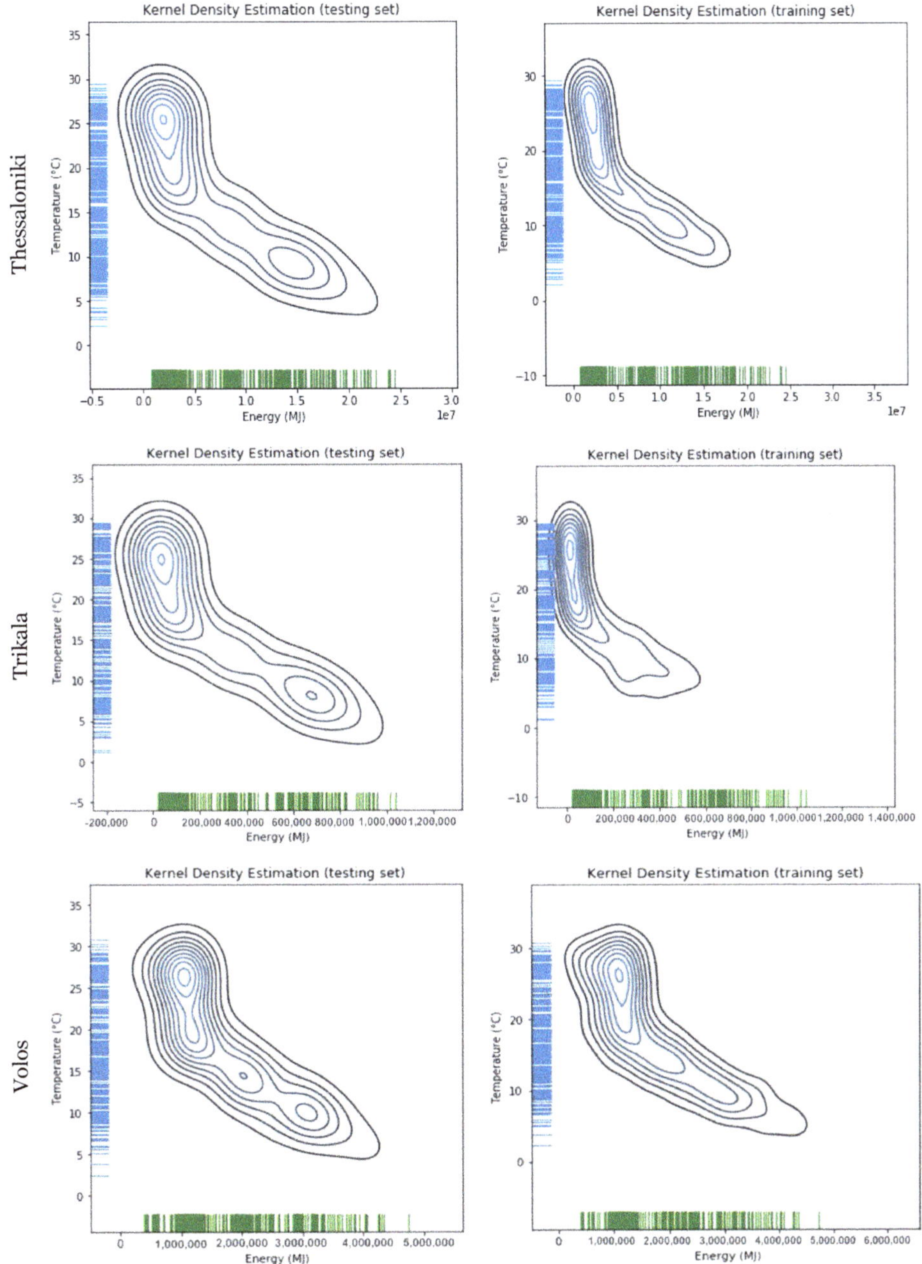

Figure A1. *Cont.*

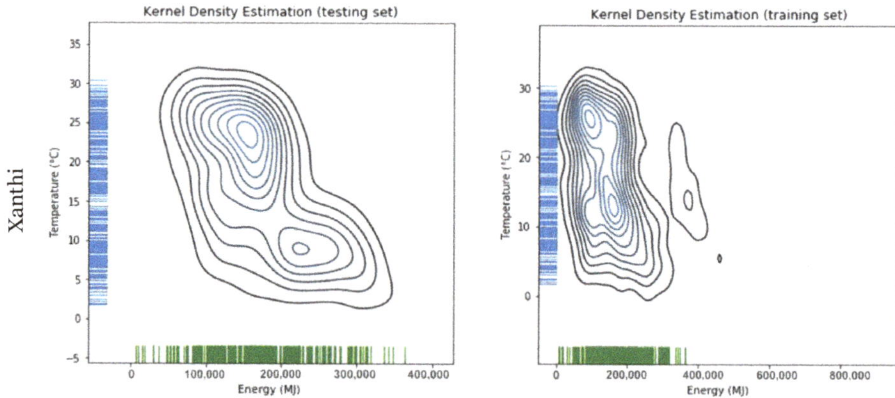

Figure A1. Correlation of the energy demand and the mean temperature for the training set and the test set for all the examined cities.

References

1. Tamba, J.G.; Essiane, S.N.; Sapnken, E.F.; Koffi, F.D.; Nsouandélé, J.L.; Soldo, B.; Njomo, D. Forecasting natural gas: A literature survey. *Int. J. Energy Econ. Policy* **2018**, *8*, 216–249.
2. Ivezić, D. Short-Term Natural Gas Consumption Forecast. *FME Trans.* **2006**, *34*, 165–169.
3. Ghalehkhondabi, I.; Ardjmand, E.; Weckman, G.R.; Young, W.A. An overview of energy demand forecasting methods published in 2005–2015. *Energy Syst.* **2017**, *8*, 411–447. [CrossRef]
4. Motlagh, O.; Grozev, G.; Papageorgiou, E.I. A Neural Approach to Electricity Demand Forecasting. In *Artificial Neural Network Modelling*; Shanmuganathan, S., Samarasinghe, S., Eds.; Springer International Publishing: Cham, Switzerland, 2016; pp. 281–306.
5. Raza, M.Q.; Khosravi, A. A review on artificial intelligence based load demand forecasting techniques for smart grid and buildings. *Renew. Sustain. Energy Rev.* **2015**, *50*, 1352–1372. [CrossRef]
6. Gorucu, F.B.; Geriş, P.U.; Gumrah, F. Artificial Neural Network Modeling for Forecasting Gas Consumption. *Energy Sources* **2004**, *26*, 299–307. [CrossRef]
7. Papageorgiou, E.I.; Poczęta, K. A two-stage model for time series prediction based on fuzzy cognitive maps and neural networks. *Neurocomputing* **2017**, *232*, 113–121. [CrossRef]
8. Karimi, H.; Dastranj, J. Artificial neural network-based genetic algorithm to predict natural gas consumption. *Energy Syst.* **2014**, *5*, 571–581. [CrossRef]
9. Khotanzad, A.; Elragal, H. Natural gas load forecasting with combination of adaptive neural networks. In Proceedings of the International Joint Conference on Neural Networks, Washington, DC, USA, 10–16 July 1999; IEEE: Piscataway Township, NJ, USA, 2002.
10. Khotanzad, A.; Elragal, H.; Lu, T.L. Combination of artificial neural-network forecasters for prediction of natural gas consumption. *IEEE Trans. Neural Netw.* **2000**, *11*, 464–473. [CrossRef]
11. Kizilaslan, R.; Karlik, B. Comparison neural networks models for short term forecasting of natural gas consumption in Istanbul. In Proceedings of the 1st International Conference on the Applications of Digital Information and Web Technologies, ICADIWT 2008, Ostrava, Czech Republic, 4–6 August 2008.
12. Kizilaslan, R.; Karlik, B. Combination of neural networks forecasters for monthly natural gas consumption prediction. *Neural Netw. World* **2009**, *19*, 191–199.
13. Musilek, P.; Pelikan, E.; Brabec, T.; Simunek, M. Recurrent Neural Network Based Gating for Natural Gas Load Prediction System. In Proceedings of the 2006 IEEE International Joint Conference on Neural Network Proceedings, Vancouver, BC, Canada, 16–21 July 2006.
14. Soldo, B. Forecasting natural gas consumption. *Appl. Energy* **2012**, *92*, 26–37. [CrossRef]
15. Szoplik, J. Forecasting of natural gas consumption with artificial neural networks. *Energy* **2015**, *85*, 208–220. [CrossRef]

16. Merkel, G.D.; Povinelli, R.J.; Brown, R.H. Deep neural network regression as a component of a forecast ensemble. In Proceedings of the International Symposium on Forecasting, Cairns, Australia, 25–28 June 2017.

17. Merkel, G.D. Deep Neural Networks as Time Series Forecasters of Energy Demand. Master's Thesis, Marquette University, Milwaukee, WI, US, 2017. Available online: https://epublications.marquette.edu/theses_open/434/ (accessed on 5 May 2020).

18. Merkel, G.D.; Povinelli, R.J.; Brown, R.H. Short-term load forecasting of natural gas with deep neural network regression. *Energies* **2018**, *11*, 2008. [CrossRef]

19. Azadeh, A.; Asadzadeh, S.M.; Ghanbari, A. An adaptive network-based fuzzy inference system for short-term natural gas demand estimation: Uncertain and complex environments. *Energy Policy* **2010**, *38*, 1529–1536. [CrossRef]

20. Behrouznia, A.; Saberi, M.; Azadeh, A.; Asadzadeh, S.M.; Pazhoheshfar, P. An adaptive network based fuzzy inference system-fuzzy data envelopment analysis for gas consumption forecasting and analysis: The case of South America. In Proceedings of the 2010 International Conference on Intelligent and Advanced Systems, ICIAS 2010, Kuala Lumpur, Malaysia, 15–17 June 2010.

21. Yu, F.; Xu, X. A short-term load forecasting model of natural gas based on optimized genetic algorithm and improved BP neural network. *Appl. Energy* **2014**, *134*, 102–113. [CrossRef]

22. Panapakidis, I.P.; Dagoumas, A.S. Day-ahead natural gas demand forecasting based on the combination of wavelet transform and ANFIS/genetic algorithm/neural network model. *Energy* **2017**, *118*, 231–245. [CrossRef]

23. Homenda, W.; Jastrzebska, A.; Pedrycz, W. Modeling time series with fuzzy cognitive maps. In Proceedings of the IEEE International Conference on Fuzzy Systems, Beijing, China, 6–11 July 2014.

24. Homenda, W.; Jastrzębska, A.; Pedrycz, W.; Pedrycz, W. Nodes selection criteria for fuzzy cognitive maps designed to model time series. *Adv. Intell. Syst. Comput.* **2015**, *323*, 859–870.

25. Salmeron, J.L.; Froelich, W. Dynamic optimization of fuzzy cognitive maps for time series forecasting. *Knowl. Based Syst.* **2016**, *105*, 29–37. [CrossRef]

26. Froelich, W.; Salmeron, J.L. Evolutionary learning of fuzzy grey cognitive maps for the forecasting of multivariate, interval-valued time series. *Int. J. Approx. Reason.* **2014**, *55*, 1319–1335. [CrossRef]

27. Papageorgiou, E.I.; Poczeta, K.; Laspidou, C. Application of Fuzzy Cognitive Maps to water demand prediction. In Proceedings of the IEEE International Conference on Fuzzy Systems, Istanbul, Turkey, 2–5 August 2015.

28. Poczęta, K.; Yastrebov, A.; Papageorgiou, E.I. Learning fuzzy cognitive maps using structure optimization genetic algorithm. In Proceedings of the 2015 Federated Conference on Computer Science and Information Systems, FedCSIS 2015, Lodz, Poland, 13–16 September 2015.

29. Poczeta, K.; Papageorgiou, E.I. Implementing fuzzy cognitive maps with neural networks for natural gas prediction. In Proceedings of the Proceedings—International Conference on Tools with Artificial Intelligence, ICTAI, Volos, Greece, 5–7 November 2018.

30. Papageorgiou, K.I.; Poczeta, K.; Papageorgiou, E.; Gerogiannis, V.C.; Stamoulis, G. Exploring an ensemble of methods that combines fuzzy cognitive maps and neural networks in solving the time series prediction problem of gas consumption in Greece. *Algorithms* **2019**, *12*, 235. [CrossRef]

31. Bhattacharyya, S.C.; Timilsina, G.R. Energy Demand Models for Policy Formulation: A Comparative Study of Energy Demand Models. 2009. Available online: https://core.ac.uk/reader/6521667 (accessed on 24 November 2019).

32. Han, Y.; Sha, X.; Grover-Silva, E.; Michiardi, P. On the impact of socio-economic factors on power load forecasting. In Proceedings of the 2014 IEEE International Conference on Big Data, IEEE Big Data 2014, Washington, DC, USA, 27–30 October 2014; IEEE: Piscataway Township, NJ, USA, 2015.

33. McLoughlin, F.; Duffy, A.; Conlon, M. Characterising domestic electricity consumption patterns by dwelling and occupant socio-economic variables: An Irish case study. *Energy Build.* **2012**, *48*, 240–248. [CrossRef]

34. Beckel, C.; Sadamori, L.; Santini, S. Automatic socio-economic classification of households using electricity consumption data. In Proceedings of the e-Energy 2013—Proceedings of the 4th ACM International Conference on Future Energy Systems, Berkeley, CA, USA, 22–24 May 2013.

35. Sánchez-Oro, J.; Duarte, A.; Salcedo-Sanz, S. Robust total energy demand estimation with a hybrid Variable Neighborhood Search—Extreme Learning Machine algorithm. *Energy Convers. Manag.* **2016**, *123*, 445–452. [CrossRef]

36. Liu, B.; Fu, C.; Bielefield, A.; Liu, Y.Q. Forecasting of Chinese Primary Energy Consumption in 2021 with GRU artificial neural network. *Energies* **2017**, *10*, 1453. [CrossRef]
37. Oğcu, G.; Demirel, O.F.; Zaim, S. Forecasting Electricity Consumption with Neural Networks and Support Vector Regression. *Procedia Soc. Behav. Sci.* **2012**, *58*, 1576–1585. [CrossRef]
38. Günay, M.E. Forecasting annual gross electricity demand by artificial neural networks using predicted values of socio-economic indicators and climatic conditions: Case of Turkey. *Energy Policy* **2016**, *90*, 92–101. [CrossRef]
39. Parfenenko, Y.; Shendryk, V.; Vashchenko, S.; Fedotova, N. The forecasting of the daily heat demand of the public sector buildings with district heating. In *Information and Software Technologies*; Springer: Cham, Switzerland, 2015.
40. Benalcazar, P.; Kamiński, J. Short-term heat load forecasting in district heating systems using artificial neural networks. In Proceedings of the IOP Conference Series: Earth and Environmental Science, Krakow, Poland, 14–17 November 2017; IOP: London, UK, 2019.
41. Dolinay, V.; Vasek, L.; Novak, J.; Chalupa, P.; Kral, E. Heat demand model for district heating simulation. In Proceedings of the MATEC Web of Conferences, Majorca, Spain, 14–17 July 2018.
42. Powell, K.M.; Sriprasad, A.; Cole, W.J.; Edgar, T.F. Heating, cooling, and electrical load forecasting for a large-scale district energy system. *Energy* **2014**, *74*, 877–885. [CrossRef]
43. Anagnostis, A.; Papageorgiou, E.; Dafopoulos, V.; Bochtis, D. Applying Long Short-Term Memory Networks for natural gas demand prediction. In Proceedings of the 2019 10th International Conference on Information, Intelligence, Systems and Applications, Patras, Greece, 15–17 July 2019; pp. 1–7.
44. McCulloch, W.S.; Pitts, W. A logical calculus of the ideas immanent in nervous activity. *Bull. Math. Biophys.* **1943**, *5*, 115–133. [CrossRef]
45. Rosenblatt, F. The perceptron: A probabilistic model for information storage and organization in the brain. *Psychol. Rev.* **1958**, *65*, 386–408. [CrossRef]
46. Pal, S.K.; Mitra, S. Multilayer Perceptron, Fuzzy Sets, and Classification. *IEEE Trans. Neural Netw.* **1992**, *3*, 683–697. [CrossRef]
47. Linnainmaa, S. Taylor expansion of the accumulated rounding error. *BIT* **1976**, *16*, 146–160. [CrossRef]
48. Hochreiter, S.; Schmidhuber, J. Long Short-Term Memory. *Neural Comput.* **1997**, *9*, 1735–1780. [CrossRef]
49. Lecun, Y.; Bengio, Y.; Hinton, G. Deep learning. *Nature* **2015**, *521*, 436–444. [CrossRef] [PubMed]
50. He, K.; Zhang, X.; Ren, S.; Sun, J. Deep residual learning for image recognition. In Proceedings of the Proceedings of the IEEE Computer Society Conference on Computer Vision and Pattern Recognition, Las Vegas, NV, USA, 27–30 June 2016.
51. Huang, G.; Liu, Z.; Van Der Maaten, L.; Weinberger, K.Q. Densely connected convolutional networks. In Proceedings of the 30th IEEE Conference on Computer Vision and Pattern Recognition, CVPR 2017, Honolulu, HI, USA, 21–26 July 2017.
52. Szegedy, C.; Liu, W.; Jia, Y.; Sermanet, P.; Reed, S.; Anguelov, D.; Erhan, D.; Vanhoucke, V.; Rabinovich, A. Going deeper with convolutions. In Proceedings of the Proceedings of the IEEE Computer Society Conference on Computer Vision and Pattern Recognition, Boston, MA, USA, 7–12 June 2015.
53. Kenton, L.; Kristina, T.; Devlin, J.; Chang, M.-W. BERT: Pre-training of Deep Bidirectional Transformers for Language Understanding. *arXiv* **2017**, arXiv:1810.04805.
54. Chung, J.; Gulcehre, C.; Cho, K.; Bengio, Y. Gated feedback recurrent neural networks. In Proceedings of the 32nd International Conference on Machine Learning, ICML 2015, Lille, France, 6–11 July 2015.
55. Anagnostis, A.; Asiminari, G.; Papageorgiou, E.; Bochtis, D. A Convolutional Neural Networks Based Method for Anthracnose Infected Walnut Tree Leaves Identification. *Appl. Sci.* **2020**, *10*, 469. [CrossRef]
56. Chollet, F.; Allaire, J.J. Keras 2015. Available online: https://keras.io/getting_started/faq/#how-should-i-cite-keras (accessed on 27 March 2015).
57. Srivastava, N.; Hinton, G.; Krizhevsky, A.; Sutskever, I.; Salakhutdinov, R. Dropout: A simple way to prevent neural networks from overfitting. *J. Mach. Learn. Res.* **2014**, *15*, 1929–1958.
58. Hazra, A. Using the confidence interval confidently. *J. Thorac. Dis.* **2017**, *9*, 4125–4130. [CrossRef]
59. Ruppert, D. The Elements of Statistical Learning: Data Mining, Inference, and Prediction. *J. Am. Stat. Assoc.* **2004**, *99*, 567. [CrossRef]
60. Molnar, C. *Interpretable Machine Learning: A Guide for Making Black Box Models Explainable*; Lulu: Morrisville, NC, USA, 2019.

61. Graham, B.; Reizenstein, J.; Robinson, L. Efficient batchwise dropout training using submatrices. *arXiv* **2015**, arXiv:1502.02478.
62. Cameron, A.C.; Windmeijer, F.A.G. An R-squared measure of goodness of fit for some common nonlinear regression models. *J. Econom.* **1997**, *77*, 329–342. [CrossRef]

Article

Proposing a Paradigm Shift in Rural Electrification Investments in Sub-Saharan Africa through Agriculture

George Kyriakarakos *, Athanasios T. Balafoutis and Dionysis Bochtis *

Institute for Bio-Economy and Agri-Technology (iBO), Centre for Research and Technology–Hellas (CERTH), 6th km Charilaou-Thermi Rd, GR 57001 Thermi, Thessaloniki, Greece; a.balafoutis@certh.gr
* Correspondence: g.kyriakarakos@certh.gr (G.K.); d.bochtis@certh.gr (D.B.); Tel.: +30-2421-096-740 (D.B.)

Received: 20 March 2020; Accepted: 8 April 2020; Published: 12 April 2020

Abstract: Almost one billion people in the world still do not have access to electricity. Most of them live in rural areas of the developing world. Access to electricity in the rural areas of Sub-Saharan Africa is only 28%, roughly 600 million people. The financing of rural electrification is challenging and, in order to accomplish higher private sector investments, new innovative business models have to be developed. In this paper, a new approach in the financing of microgrid electrification activities is proposed and investigated. In this approach, agriculture related businesses take the lead in the electrification activities of the surrounding communities. It is shown that the high cost of rural electrification can be met through the increased value of locally produced products, and cross-subsidization can take place in order to decrease the cost of household electrification. The approach is implemented in a case study in Rwanda, through which the possibility of local agricultural cooperatives leading electrification activities is demonstrated.

Keywords: developing world; rural electrification; Sub-Saharan Africa; energy; agriculture

---

1. Introduction

Almost one billion people in the world still do not have access to electricity [1]. Most of them live in rural areas of the developing world. Access to electricity in the rural areas of Sub-Saharan Africa is only 28%, roughly 600 million people. In recent years, poverty has been officially recognized to be at the core of development. The Sustainable Development Goal (SDG) No. 1 of the United Nations (UN) is "No poverty", and it is acknowledged that economic growth must be inclusive to provide sustainable jobs and promote equality. Infrastructure is recognized as having a high importance on a country's economic development. At the same time, two key factors are needed for the move to higher productivity, economies of scale, and specialization [2], which have considerable benefits in large urban centers [3]. Urbanization has been considered to be the key to economic development, but recent studies have shown that enforcing policies to simply pursue accelerated urbanization will not necessarily lead to increased economic growth [4]. At the same time, urbanization has been increasing Greenhouse Gas (GHG) emissions [5], contributing to climate change [6], and affecting water resources [7] and biodiversity [8], as well as human health [9]. Based on these drawbacks of urban economic development, recent studies have shown that, especially for the least-developed countries, rural development is an alternative way of increasing income and quality of live, with agriculture being the starting point [10].

Rural development can be defined as "the process of improving the quality of life and economic well-being of people living in rural areas, often relatively isolated and sparsely populated" [11]. The main sectors of activity that have been traditionally regarded as the core of rural development are agriculture, forestry, and natural resources extraction. In recent times, the importance of social infrastructure related to health and education [12] has been acknowledged, along with tourism, recreation, and decentralized manufacturing [13]. However, it has to be highlighted that agriculture is still considered to be the most important sector for the rural areas of the developing world [10]. Sustainable development, in turn, has been defined, according to the Brundtland Commission's "Our Common Future" document [14], as "development that meets the needs of the present without compromising the ability of future generations to meet their own needs".

Rural electrification and the sustainable development of rural communities has been investigated extensively, and it has been shown that there is an evident relationship between them; rural electrification can act as an enabler, a facilitator, and a driver for sustainable development [15]. Indeed, rural electrification has provided extended benefits to rural populations in health [16], education [17], and income [18,19]. Agricultural electrification has been found to produce long-lasting positive effects on communities. A review of rural electrification that took place in the US between 1930 and 1960 showed that, in the short term, electrification mostly impacted the agriculture related economy, increasing employment and rural farm population. The fact that benefits exceeded historical costs even in low density rural areas is highlighted. Another important and interesting fact is that the areas that got electrified first showed increased economic growth that was maintained for decades, even after the whole of the US was electrified [20]. Newer rural electrification programs show comparable results. In a 2018 review of the Chinese rural electrification program, one of the main outcomes was that it considerably increases the farmer's agricultural income [21]. Another important result is that a proper matching of electrification needs can considerably increase its effectiveness [21]. It has been acknowledged that it is beneficial to utilize an approach where an investigation of each area's needs is performed first and the electrification activities are subsequently tailored to those needs [22].

How to measure access to electricity is also an important aspect when considering electrification activities. The United Nations Global Tracking Network for Sustainable Energy for All has proposed a multi-tier framework, which is a comprehensive approach in measuring access [23]. The main attributes of this framework are that it has five tiers, it is based on six attributes of electricity supply, and that, with electricity supply improvement, there is an increase in the possible electricity services availability.

The Index of access to electricity is defined as:

$$i = \sum P_T \times T,$$

where P_T refers to the proportion of households at tier $T \in \{0, 1, \ldots, 5\}$.

Table 1 presents the main attributes of each tier, while Table 2 presents the main services provided. As is understandable, high energy efficiency devices and appliances need to be used in order to achieve each tier access cost effectively. It should be pointed out that Tier 1 is considered as having access to electricity and therefore people with, for example, a solar lantern with the ability to also charge a mobile phone are not included in the ~1 billion people without access to electricity.

Rural electrification can be accomplished using three main approaches; grid extension, microgrids, and solar home systems [24]. The electrification costs for grid extension is estimated at between 2000 and 3000 USD, for microgrids at between 500 and 1200 USD, and for solar home systems at between 150 and 500 USD [25]. Based on technology availability and reliability nowadays, grid extension does not seem to be the optimum solution in remote locations due to the high cost and extensive infrastructure requirements; therefore, for most people in rural areas, electrification will come with systems that produce and distribute electricity locally, through solar home systems or microgrids [26]. The choice between an investment in a microgrid or solar home systems is ultimately based on a techno-economic evaluation for any given project, with population density playing an important role [27]. In Sub-Saharan Africa there are 602 million people living in the dark [1]. If all these people were to get access to electricity with microgrids, more than 480 billion USD would be needed. It is interesting to mention here that the total official financing received by African countries in 2018, according to the OECD, was ~64 billion USD [28], and this includes the financing for all activity sectors, not just energy. It is understandable that it cannot be expected that development financing alone will make the world succeed in fulfilling Sustainable Development Goal No. 7 of ensuring access to affordable, reliable, sustainable, and modern energy for all. This, in turn, means that the available financing tools should be utilized in the most effective way possible in order to stimulate sustainable economic development, which in turn will contribute to the required financing.

Table 1. Access Tier attributes.

Attributes	Tier 0	Tier 1	Tier 2	Tier 3	Tier 4	Tier 5
Peak available capacity (W)	-	>3	>50	>200	>800	>2000
Duration (h)	-	≥4	≥4	≥8	≥16	≥23
Evening supply (h)	-	≥1	≥2	≥3	≥4	≥4
Reliability	-	-	-	-	Max 14 disruptions per week	Max 3 disruptions per week of total duration <2 h
Quality	-	-	-	-	Voltage problems do not affect the use of desired appliances	
Affordability	-	-	-	Cost of a standard consumption package of 365 kWh y^{-1} <5% of the household income		
Legality	-	-	-	-	Bill paid to the utility, pre-paid card seller, or authorized representative	
Health and Safety	-	-	-	-	Absence of past accidents and perception of high risk in the future	

Source: Bhatia M., Angelou N., "Beyond Connections—Energy Access Redefined", 2015, ESMAP, World Bank.

Table 2. Services provided in each Tier.

Tier 0	Tier 1	Tier 2	Tier 3	Tier 4	Tier 5
-	Task lighting AND Phone charging	General Lighting AND Phone charging AND Television AND Fan (if needed)	Tier 2 AND Any medium-power appliances	Tier 3 AND Any high-power appliances	Tier 2 AND Any very high-power appliances

Source: Bhatia M., Angelou N., "Beyond Connections—Energy Access Redefined", 2015, ESMAP, World Bank.

One of the most extended and current studies of installed microgrids in Sub-Saharan Africa was performed by the International Finance Corporation of the World Bank Group [29]. The main findings are summarized in Table 3. According to the study, households can spend very low amounts of money per month and the average payback period is more than seven years. The average access provided is Tier 2. Furthermore, most microgrids presented in this report rely heavily on grant funding and equity investment. If an increase of microgrid investments is to take place, commercial lending will also have to play an important role. Finally, it is important to note the 50-50 split between generation (which includes storage and power electronics) and distribution (poles and cables). This is due to the declining costs of photovoltaics and batteries, whereas the cost of poles and cables has remained practically unchanged.

Table 3. Summary of findings of Benchmarking microgrids in Sub-Saharan Africa.

Indicator	Value
Monthly Average Revenue per user	7 USD
Average Investment per user	920 USD
Tier 2 Average Residential Consumption	11 kWh m^{-1}
Average Generation Capacity	34 kW
Average Number of Connections	~100
A/C vs. D/C	85% vs. 15%
Operational Expenditure (OPEX) as a % of revenue	58%
Capital Expenditure (CAPEX) payback period	>7 years
Split of CAPEX spending on distribution vs. generation	50% vs. 50%
Average Distance from National Grid	23 km

Source: International Finance Corporation World Bank Group (WBG). Benchmarking Mini-grid Distribution Companies (DESCOs) 2017 update–Summary of Findings. (2018).

It is evident that new business models need to be developed and utilized if an increase in electricity access is sought. Digital technologies offer various tools for cost decrease and their use by investors is increasing. The use of pay-as-you-go models, coupled with the internet of things, blockchain technologies, and mobile money is increasing [30]. There is a need, though, to target electrification activities that can also increase economic activity in the area [31]. This is the only path to enabling the electricity consumers to pay for more electricity and, as such, to increase the income for the microgrid operators, because it is acknowledged that revenue collection is one of the biggest challenges in the operation of microgrids [32]. In order to achieve this, there is a need to power appliances, devices, and equipment directly associated with economic development. These include water pumping, water desalination, refrigeration, space heating and cooling, incubators for poultry farming, milking machines, rice and maize hullers, polishers, threshers, graters, grain mills, oil presses, tailoring, workshop machinery (e.g., drills, chainsaws, rotary hammers, grinders, jigsaws, routers, etc.), and hairdresser equipment, among others [33]. Many studies have demonstrated that one of the short-term gains of rural electrification is related to increased agricultural income [20,21,34,35], showcasing that agriculture related load electrification needs to be a priority. Furthermore, experience from Asia (Bangladesh, Nepal, and Myanmar) and the USA has shown that the involvement of socially oriented cooperatives can also play an important role in rural electrification, acknowledging the fact that they can provide electricity with lower profit margins than the private sector, as long as the policy and regulatory frameworks are facilitating this involvement [20,35,36].

In Sub-Saharan Africa there are some good examples, initiated by the private sector, of linking electrification activities with agriculture. JUMEME (www.jumeme.com) is a private sector driven microgrid electrification program based on strong community engagement in the increased productive uses of electricity [37]. JUMEME aims to create value chains that can increase local income and in turn allow the local population to afford electrification, both for productive and household use. In one location, rice is grown, and JUMEME aided the local farmers to access financing in order to install water pumps to allow for irrigation [38]. An important aspect of this approach is that it is initiated by the private sector whose aim is to sell electricity [38].

In this paper, a new approach to the financing of microgrid electrification activities in Sub-Saharan Africa is proposed and investigated. In this approach, agriculture related businesses and cooperatives take the lead in the electrification activities of the surrounding communities. The aim of this work is to demonstrate that the high cost of electrification can be met through the increased value of locally produced products and, in parallel, internal cross-subsidization can take place in order to decrease the cost of household electrification. The proposed paradigm shift and the possibility of local agricultural cooperatives leading the electrification activities are investigated and presented through a case study implementing Rwanda conditions. The study is deployed according to the following steps:

Step 1. A microgrid is designed that meets the electrification needs of the households of the typical village considered. Through this investigation, the levelized cost of electricity (LCOE) is determined. A sensitivity analysis takes place in terms of the maximum allowed annual capacity shortage, the loads not met, and the electricity lost.

Step 2. A microgrid is designed to meet the needs of both the households and also a deferrable load. In the focal case study, the deferrable load considered is maize milling to produce flour.

Step 3. The possibility of an agricultural cooperative investing in a mill powered by a photovoltaics (PV)/battery system in order to be able to sell maize flour instead of dry maize is considered.

Step 4. An investigation takes place on whether the cooperative system can be extended to a microgrid to power the local households.

2. Materials and Methods

2.1. Case Study Assumptions

For the case study purposes, a typical village in Rwanda with 100 households was considered. This translates to roughly 550 inhabitants [39]. The meteorological data of the capital city of Kigali were used. It was assumed that the main agricultural activity in the area was maize production. In fact, over 65% of the farmers grow maize in Rwanda for both household consumption and commercial sale to traders and millers. Moreover, Rwanda is a net importer of maize grain and net exporter of maize meal [40]. Based on data derived from [39] and personal communication information, the maize value chain map in Rwanda is presented in Figure 1. As is visible, farmers usually sell fresh or dry maize. The annual cultivated area and the total and unit area yield for the past decade, according to FAO/FAOSTAT data, are listed in Table 4. The average maize farm size in Rwanda is 0.6 ha [41]. The average annual maize flour consumption in Rwanda is between 50 and 129 kg per capita [42]. Based on the above data, for the considered case study the consumption per capita of maize flour is considered as 100 kg y^{-1}.

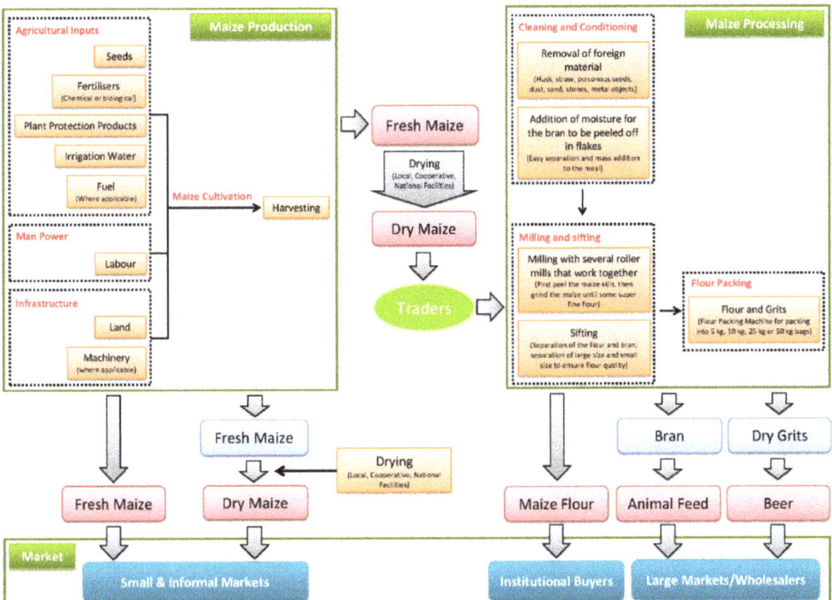

Figure 1. Maize value chain map in Rwanda.

Table 4. Maize yield in Rwanda for the past decade.

Year	Cultivated Area (ha)	Production (t)	Yield (t ha^{-1})
2008	144,896	166,853	1.15
2009	147,129	286,946	1.95
2010	184,658	432,404	2.34
2011	223,414	525,679	2.35
2012	253,698	573,038	2.26
2013	292,326	667,833	2.28
2014	233,150	583,096	2.50
2015	241,713	370,140	1.53
2016	237,658	374,267	1.57
2017	297,447	358,417	1.20
Average	225,609	433,867	1.92

Source: FAO/FAOSTAT.

2.2. Simulation Software, Assumptions and Parameters

The HOMER Energy Legacy v.2.68 software (HOMER Energy, Boulder, CO, USA) was used to perform the simulation and optimization process and the economic analysis in the cases investigated. This particular software was chosen as it is a free-to-use tool.

The consumption load of the 100 households is considered in line with Tier 2 access. Figure 2 presents the created hourly synthetic profile and Table 5 presents key power and energy data for the appliances used and the total needs of the 100 households. It has to be noted that super energy efficient appliances were considered. In order to add realism to the simulation, and because the predicted load profile, even based on surveys, is often not accurate [43,44], random variability was applied to this profile under a 5% random variability from day-to-day and a 5% variability from hour-to-hour. The operating reserve variables present in HOMER were set to 5% of the hourly load, 0% of the peak load, and 25% of the solar power output. This choice of variables offers a degree of balance between having a low-cost system and reliability of supply. It has to be highlighted that, for Tier 2 access, there are no restrictions in terms of reliability (see Table 1). The system topology chosen is that of an AC low voltage, single phase microgrid. This is presented in Figure 3. High quality power electronics, photovoltaic panels, and LiFePO$_4$ batteries were used. It has to be noted that the battery was not allowed to deep discharge below a 10% state of charge in order to prevent premature aging and need for replacement. When reaching a 10% state of charge, all loads were disconnected until the battery was charged again. The related cost elements used in the simulations, as presented in Table 6, were in line with real market prices in Rwanda for 2019. Furthermore, a sensitivity analysis was carried out for different levels of annual capacity shortage to evaluate its effect on the levelized cost of electricity and load not met. The techno-economic investigation horizon was set to a 20-year period. Finally, the real interest rate for Rwanda for the year 2018 according to World Bank data was adopted, which was recorded as 17.9%.

Table 5. Key consumption data.

Parameter	Value	Notes
Lighting	2 W	Three LED lamps were considered for each household
Cell phone charging	5 W	Typical USB charger
Television	15 W	In line with the consumption of 19–22 inch TVs that won the Global LEAP awards [45]
Radio	2 W	Typical energy efficient radio
Min Power	0 W	
Max Power	2100 W	For 100 households
Average Power	291.67 W	
Energy per day	7000 Wh	

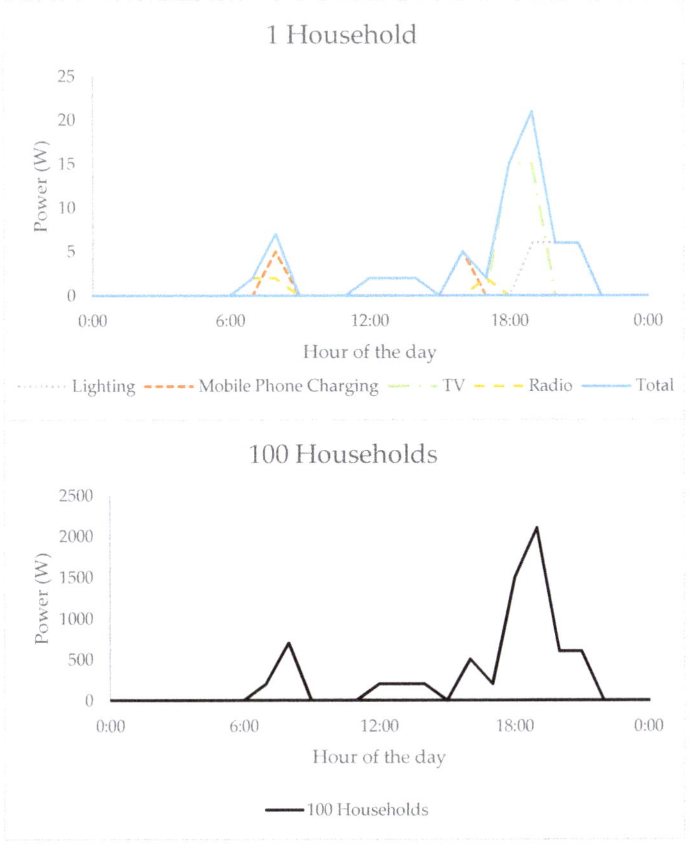

Figure 2. Load profiles.

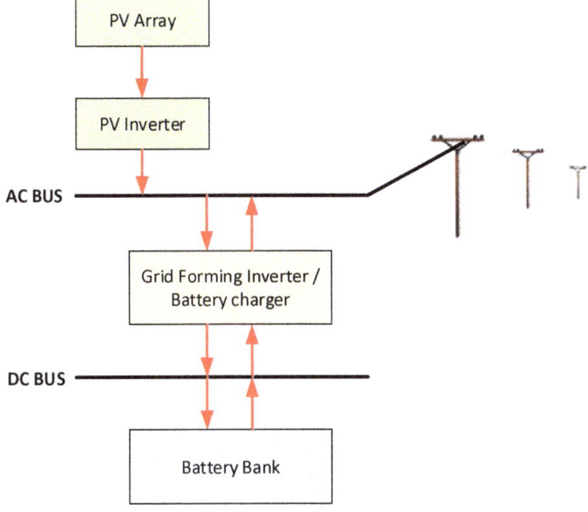

Figure 3. Microgrid topology.

Table 6. Cost data.

Cost Category	Cost
Photovoltaic panels, including inverter cost for AC microgrid topology.	0.750 € Wp^{-1}
Grid forming inverter cost	3500 €
Transportation and installation cost	5000 €
AC and DC equipment including cabling, equipment, appliances, consumables etc.	8000 €
Supplementary costs (e.g., fencing)	2000 €
Smart meters/monitoring system	5000 €
LiFePO$_4$ batteries	600 € kWh
Operation and Maintenance cost	1% of Capital Expenditure (CAPEX) [46]
Grid infrastructure cost	10,000 €

3. Systems Sizing and Economic Investigation of Rural Electrification

3.1. Step 1: Sizing of a System to Meet the Household Needs

A microgrid is designed to meet the electrification needs of the households of the typical village considered. Through this investigation, the levelized cost of electricity is determined. A sensitivity analysis takes place in terms of the maximum allowed annual capacity shortage, the loads not met, and the electricity lost. The results of the optimizations are presented in Table 7.

Table 7. Household electrification results.

Case No	Annual Capacity Shortage Allowed	PV (kWp)	Batteries (kWh)	CAPEX (€)	OPEX (€)	Net Present Cost (€)	Levelized Cost of Electricity (€ kWh^{-1})	Unmet Load kWh·y^{-1}	%	Excess Electricity kWh·y^{-1}	%
1.1	0%	3.75	7.68	40,921	449	43,335	3.164	1.8	0.1	2670	48.2
1.2	5%	2.75	5.12	38,635	440	41,000	3.131	113	4.5	1321	32.5
1.3	10%	2.25	5.12	38,260	433	40,587	3.196	187	7.3	664	20
1.4	20%	3	2.56	37,286	441	39,659	3.526	457	17.9	2093	47.3
1.5	30%	1.75	2.56	36,349	421	38,611	3.862	689	27	504	19.5

Intermediate conclusions from the results are listed below:

The lowest LCOE is observed for Case 1.2. This is understandable because, subsequently, the system operator sells less kWhs. In Case 1.1, the LCOE is higher because, even if more kWhs are sold, the cost of the system is considerably higher.

An investor would most probably choose Case 1.2 because it has almost the same CAPEX as Case 1.3, but lower LCOE than all the other cases.

In all cases, electricity is lost because it is not cost-effective to store it. Figure 4 graphically presents the cumulative served load, the unmet load, and the excess load throughout the year. The excess load is simply discarded and, thus, wasted. The most important highlight of this is that the wasted energy from the system is much higher than the utilized electricity.

- Figure 5 presents two typical days of the year for Case 1.2, which further showcase how the system operates under excess power and unmet load. On July 1st, the solar irradiation is low and, as such, the PV produced power is low. The battery bank starts at 15% state of charge. The morning load is able to be met by the PV production and the battery, and even though the PV production is low throughout the day, the load is met until the evening. From then on, the battery reaches 10% state of charge and all loads are disconnected. The load is going to be served again when the battery gets charged by the PV array. On the 22nd of October, the PV array produces power from early in the morning until late in the evening. The battery gets full and is able to meet the load in the evening hours when the sun has set. For almost seven hours, the battery remains full, and during this time the produced power is lost because there is no extra storage capacity.

- The LCOE is extremely high in all cases and it is very difficult to sell electricity at that price. For reference, grid kWh for Rwanda is sold between 0.12 € and 0.18 €.
- Such an investment would need a high amount of grant money to cover the CAPEX in order to become viable.

The most important outcome of this analysis is the high cost of electricity and the high amounts of electricity wasted. The most common solution used to address this is to be able to connect a deferrable load to the system, such as a water pump, an ice making machine, etc. In the next step of this study, a deferrable load in the form of a maize mill will be considered.

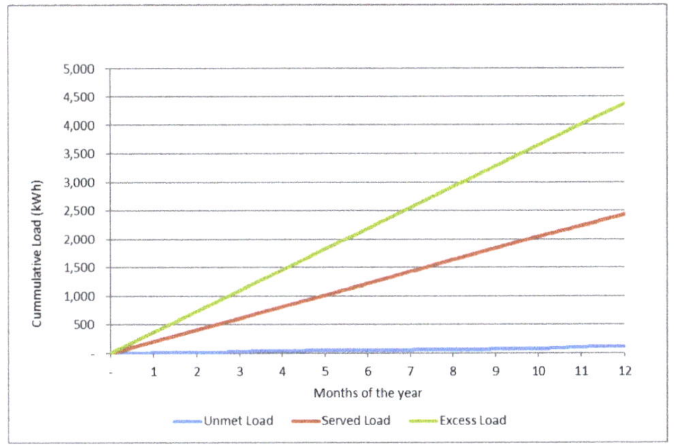

Figure 4. Step 1 cumulative served, unmet, and excess load.

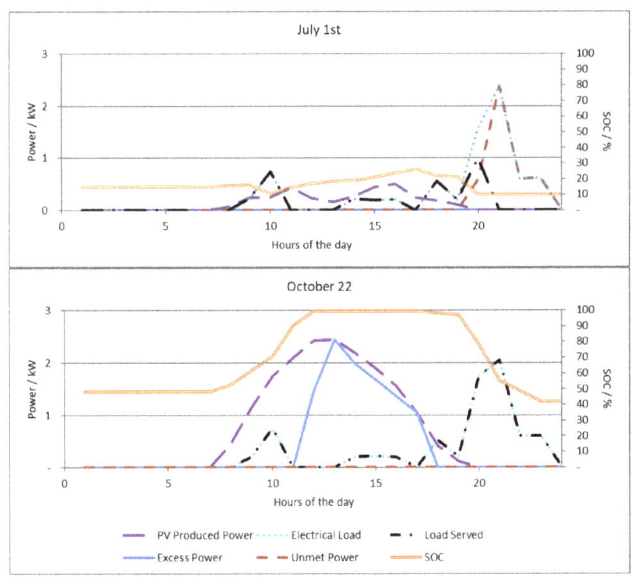

Figure 5. Hourly power figures of two typical days for Step 1.

3.2. Step 2: Addition of a Deferrable Load

A commercial high efficiency mill targeted for rural areas of the developing world was considered for the case study. This mill can process 30 kg of maize per hour to produce flour with a consumption equal to 1.1 kW when in operation. The cost of this mill, along with the supporting structures to allow easy supply of maize and storage of the produced flour, is considered to be 10,000 € and is in line with market costs for off-grid high efficiency mills.

Based on the data and assumptions made in Section 2.1, a 55 t consumption of maize flour takes place at the village. The mill cost is assumed to be an investment of an agricultural cooperative or a private sector business, which will buy electricity from the microgrid operator. The same analysis as in Section 3.1 takes place, and the results are presented in Table 8.

Table 8. Household electrification with deferrable load results.

Case No	Annual Capacity Shortage Allowed	PV (kWp)	Batteries (kWh)	CAPEX (€)	OPEX (€)	Net Present Cost (€)	Levelized Cost of Electricity (€ kWh^{-1})	Unmet Load kWh·y^{-1}	%	Excess Electricity kWh·y^{-1}	%
2.1	0%	5	7.68	41,858	458	44,320	1.811	3.7	0.1	2518	34.1
2.2	5%	3.25	5.12	39,010	439	41,369	1.809	156	3.54	254	5.29
2.3	10%	4	2.56	38,036	450	40,454	1.824	416	9.2	1536	26
2.4	20%	2.25	2.56	36,724	421	38,986	2.377	683	18.3	61.1	1.8
2.5	30%	1.75	2.56	36,349	413	38,571	3.087	778	25.1	56.9	2.2

Based on the results, the following can be extracted:

- The LCOE is able to drop to ~1.8 € kWh^{-1}, which is considerably lower than the systems that supplied electricity only to households in Step 1.
- For Case 2.2, the excess electricity is 5.29%, so a significant improvement is observed in relation to the systems that targeted only households.
- For Cases 2.4 and 2.5, almost all of the produced electricity is consumed; however, there is a high percentage of unmet load. As such, the LCOE increases.
- An investor would most probably choose Case 2.2 because it has the lowest LCOE than all the other cases.
- Considerably less electricity is wasted from the system using a deferrable load. Figure 6 graphically presents the cumulative served load, the unmet load, the excess load, and the deferrable load throughout the year for Case 2.2. As can be seen, much less energy is wasted in comparison with the system from Step 1.
- Figure 7 presents two typical days of the year for Case 2.2, which further showcases how the system operates under excess power and unmet load. July the 1st is a day with low PV production, and the performance observed is comparable with the system in Step 1. As is expected, the deferrable load is not activated at all during this day. On the 22nd of October, the PV array produces power from early in the morning until late in the evening. The battery gets full and is able to meet the load in the evening hours when the sun has set. The deferrable load is activated for 9 h and, as such, much less electricity is wasted in comparison with the system in Step 1.

It is understandable that, with further optimization (e.g., demand side management, variable tariffs, and different tariffs to different customers), a microgrid investor can push the LCOE down even further to approximately 1.5 € kWh^{-1}. Given that access to grant funding is available, these results are in line with the results presented in Table 3. The business models, as in the JUMEME case presented in Section 1, aim to ensuring that enough productive use and deferrable loads are present in remote locations, along with the ability to pay for the electricity consumed, which, in the end, make the rural electrification investment viable.

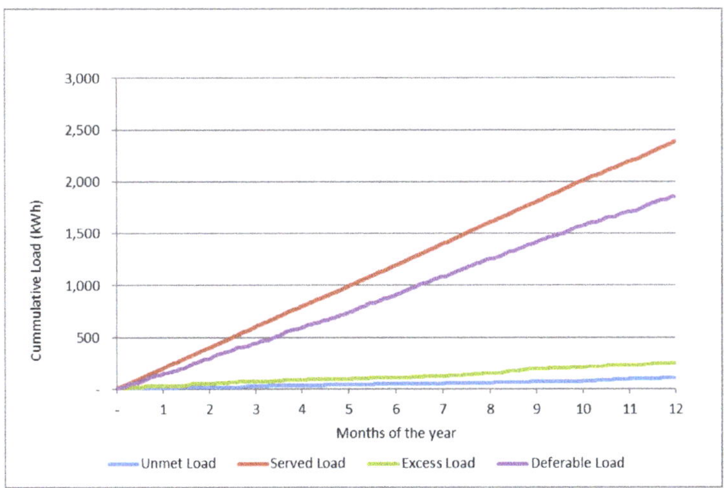

Figure 6. Step 2 cumulative served, unmet, excess, and deferrable load.

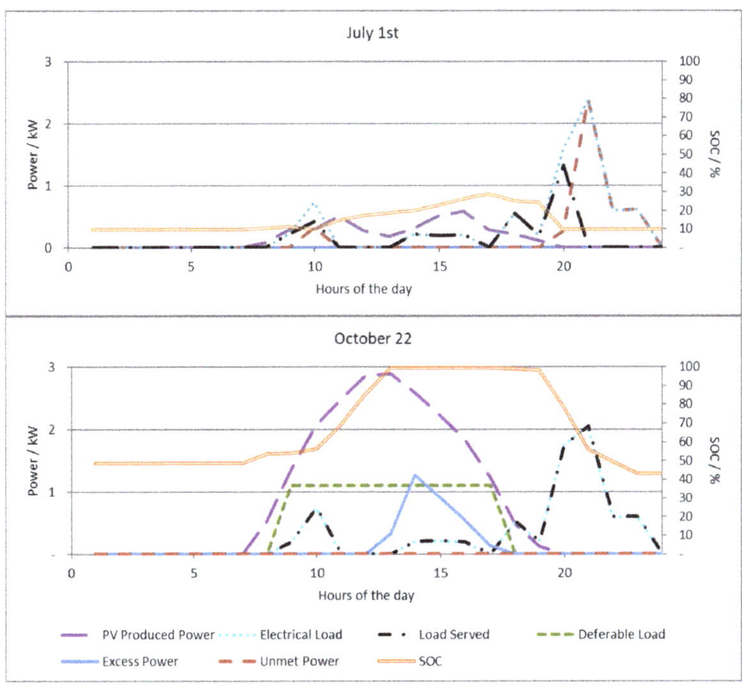

Figure 7. Hourly power figures of two typical days for Step 2.

3.3. Step 3: Agricultural Cooperative Business Expansion

In many areas of the developing world, cooperatives are looking to expand their activities and to produce higher value goods in rural areas in order to increase their income. For example, as presented in Figure 1, drying takes place in rural areas to prepare dry maize before selling it to traders. This section investigates the possibility of the cooperative producing flour on-site. In 2019, the price of maize, fixed by the Rwandan Ministry of Agriculture, was 200 RWF per kilo, which is ~0.20 €. At the same time, the price of maize flour was at 550 RWF per kilo, which is ~0.54 €. A cooperative could invest in a mill to produce flour.

An autonomous system is designed to produce the electricity needed by the mill, and the overall investment is evaluated. The production of maize in the village is considered to be 115 t based on the assumptions in Section 2.1. An assumption is made that flour is produced throughout the year, as dry maize can be stored on the cooperative premises. Since the system is designed for commercial purposes, the load has to be continuously met. The optimal system design, which includes the cost of the mill, as calculated by HOMER, is presented in Table 9. Figure 8 presents the cumulative served, unmet, and wasted load throughout the year.

Table 9. Results of the Mill system.

PV (kWp)	Batteries (kWh)	CAPEX (€)	OPEX (€)	Net Present Cost (€)	Levelized Cost of Electricity (€ kWh⁻¹)	Unmet Load		Excess Electricity	
						kWh y⁻¹	%	kWh y⁻¹	%
6.5	10.24	44,519	490	47,153	1.821	3.15	~0	4610	48

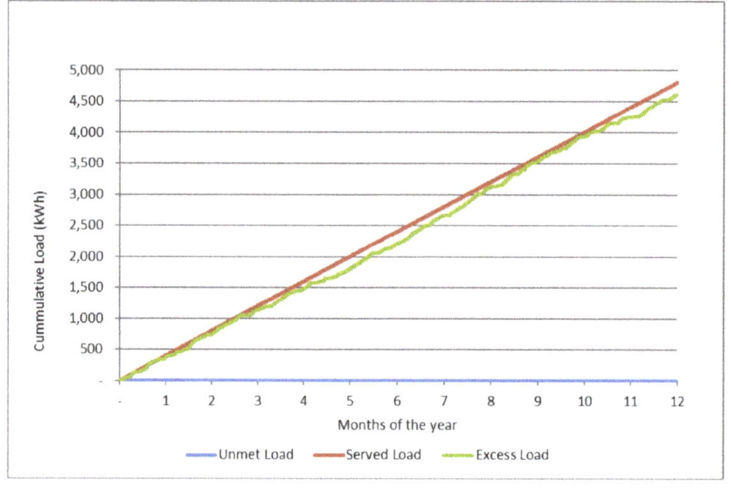

Figure 8. Step 3 cumulative served, unmet, and excess load.

With the system above, the cooperative, instead of selling the dry maize for 23,000 €, would sell maize flour for 62,100 €, so the profit would increase by 70%. As the CAPEX of the system is 44,519 €, the simple payback period for this investment is less than two years. As is understandable, such an investment is very profitable and could boost the economic activity of rural areas considerably, improving the livelihood of the local population. Because of the considerable socio-economic benefits of such investments, in recent years, the interest in using renewable energy for decentralized solutions in the agri-food chain has been rising [47].

Still, as is clearly visible, such a system is characterized by very high excess electricity throughout the year. Figure 8 clearly showcases this, as the wasted power is roughly the same as the utilized electricity. At the same time, as was expected by the design constraints, there is practically zero unmet load throughout the year.

In reality, the cooperative does not care about the actual cost of the kWh because the system on a whole is built to process maize and produce flour. The increased value of flour in reality can allow high electricity costs because electricity, in this case, is just an intermediate good.

The next and final step investigates household electrification activity by the agricultural cooperative.

3.4. Step 4: The Local Agricultural Cooperative as the Village Household Electrification Investor

In this step, the option to include village household electrification to the system used for milling by the agricultural cooperative is investigated.

Firstly, a technical investigation takes place to see if the electricity system designed in Section 3.3 can also provide electrification to households, as it presents very high excess electricity. In this example, the only extra cost the cooperative would have to invest in is the grid infrastructure of the village, which is estimated for this case study at 10,000 € [29]. The results of this simulation are presented in Table 10.

Table 10. Results of the Mill system also serving households.

PV (kWp)	Batteries (kWh)	CAPEX (€)	OPEX (€)	Net Present Cost (€)	Levelized Cost of Electricity (€ kWh^{-1})	Unmet load		Excess Electricity	
						kWh y^{-1}	%	kWh y^{-1}	%
6.5	10.24	54,519	600	54,519	1.555	462	6.3	2247	23.4

The simulation shows that the unmet load to the households will be 462 kWh per year, which translates to ~18% of the household consumptions (it is comparable to Case 1.4 in Section 3.1).

This is in line with Tier 2 access, as presented in Table 1. This is further showcased in Figure 9. However, there is a high excess of electricity observed in this example, which could be utilized further, decreasing the unmet load to the households through the implementation of an advanced energy management system.

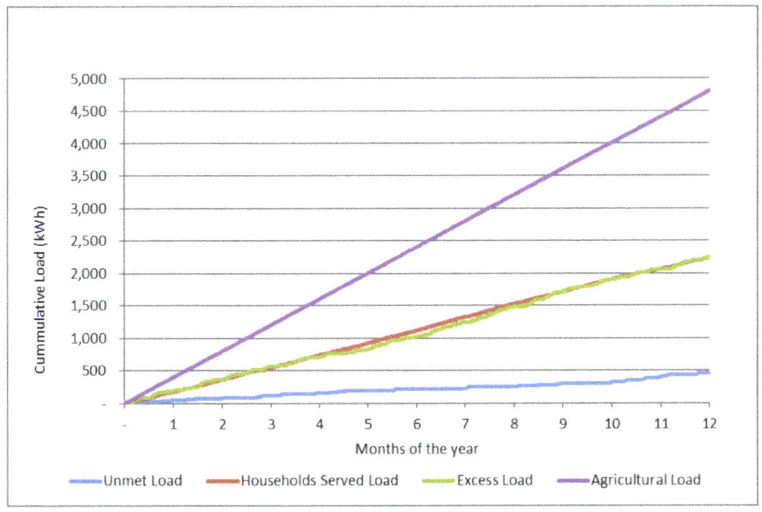

Figure 9. Step 3 cumulative agricultural, households served, unmet, and excess load.

Moreover, the LCOE is the lowest observed in all the systems analyzed in this study, at 1.555 €/kWh. Given the very high profitability of the mill investment, as presented in Section 3.3, and the fact that a cooperative is essentially a union of the local population, it is relatively easy to develop a business model for the provision of electricity to local households at very low cost. Essentially, a cross-subsidy between the agricultural activity and the household electrification can take place. Furthermore, the increased income of the farmers due to the mill can lead to further investments in the area, such as, for example, an expansion of the microgrid to increase the tier access provided.

4. Environmental Benefits of Decentralized Agricultural Activity

Maize flour production on a local basis would result in increased income for the farmers/cooperative members, as was described in Section 3.3. This, in turn, would give them the opportunity to invest in technology and advanced techniques of production resulting in higher yields for the same acreage. This would make it possible for the cooperative to either strengthen its relationship with large mills, thus gaining higher grain prices, or increase the milling capacity by adding small scale mills on a modular basis (keeping the same mill type and capacity in each upgrade and adding units). The energy producing system would also increase accordingly, providing the village with the ability to increase its tier in electricity access.

Furthermore, such an activity would also reduce the amount of maize grain that centralized large mills are currently processing and result in less environmental impact through reduced energy consumption and GHG emissions. Such facilities are based on grid electricity that, 47.2% of the time in the case of Rwanda, comes from fossil fuels [48]. According to Energy Experts International, conventional mills of European standard, based on their capacity and size, require 361–1186 MJ t^{-1} for flour production [49]. If the solar mill considered in Section 3.2 is used for local flour milling, then the energy required will reach 132 MJ t^{-1}, reducing it by 63–89%, not taking into account the possible energy needs involved in grain transportation to the centralized mill. In addition, if we take into account the UN standardized baseline Rwanda grid emission factor of 0.767 kg CO_2eq per kWh [50], then the centralized mill produces directly 76.7–253 kg CO_2eq per ton of produced flour, while the solar decentralized system has zero direct GHG emissions. Therefore, for the 115 t that the village presented above produces, the energy use reduction would reach 26.3–121.2 GJ, and the environmental benefit would be 8.82–29.1 tCO_2eq (equivalent to 1.9–6.3 regular passenger cars on an annual use basis [51]). This energy surplus could be used for several purposes, either for other industrial needs or to provide electrification to urban and peri-urban populations; the urban access to electricity figure for Rwanda in 2018 was 69%. In the hypothetical, but technically possible, scenario of replacing 10% of industrial maize processing with local solar mills, then 6.760 t y^{-1} of processed maize [52] would be processed locally, and about 515–1.710 tCO_2eq in total would not be emitted (equivalent to 112–372 regular passenger cars on an annual use basis).

5. Discussion

It has been demonstrated that rural electrification is a key enabler, facilitator, and driver for sustainable development of rural areas of the developing world. It is also acknowledged that the most important economic sector for these areas is agriculture because, as a first step, they can provide the food needed by the local population and, as a next step, can stimulate economic activity and development. Rural electrification is challenging, and it is difficult to have viable profitable investments.

It is widely acknowledged at this point in time that private investment is needed in order to provide electricity to people living in rural areas of the developing world [53]. Development partners like the World Bank [54], the European Union (EU) [55] and the USA [56] are providing financing schemes to innovative business models related to rural electrification. Moreover, support for developing new policy models has also been heavily supported. The EU, through its Technical Assistance Facility for Sustainable Energy, has developed the "Guidelines for Institutional and Policy Model for Micro-/Mini-grids" for the benefit of the African Union Commission [57], which includes provisions that facilitate innovative business models for rural electrification, including the model proposed in this paper. Furthermore, community based models have also been investigated because they can increase community engagement, address some of the challenges faced by other business models, and be able to offer more socio-economic benefits to the local community [58]. Technological advancements can indeed decrease overall costs, contributing to the viability of rural electrification investments. This has been acknowledged, and upcoming innovative financing schemes like the Digital Energy Facility by Agence Française de Développement (AFD) and the EU have this at their core [59]. The most advanced business models in rural electrification, such as the JUMEME example that was presented in the introduction, have rural electrification companies at their center and are trying to develop the electricity needs of the local population in order to be able to sell more electricity and ultimately ensure the viability of the electrification investment.

Going beyond the state-of-the-art and building upon the lessons learned, the case study of this work shows that there is another way possible, which can address some of the challenges faced and eventually ensure that two targets are met at the same time; stimulation of economic activity through the electrification of productive loads mainly related to agriculture on one hand, and provision of household electrification utilizing agricultural cooperatives as the project owner on the other. The definition of a cooperative according to the International Cooperative alliance is "Cooperatives are people-centred enterprises owned, controlled and run by and for their members to realise their common economic, social, and cultural needs and aspirations" [60]. While cooperatives were initially introduced in Africa as part of colonial economic policies, since the 1980s, structural changes and legal battles have taken place in order to minimize government interference towards independence and further growth [61]. Significant promotion of agricultural cooperative producer structures has been observed and is considered as an important way forward [62]. Agricultural cooperatives can indeed be one of the vehicles for promoting sustainable rural development.

As was showcased in Section 3, if an autonomous electricity system based on photovoltaics is sized to meet the needs of a commercial load, such as milling, large amounts of energy are wasted. This is due to the fact that production of electricity is much cheaper than storing it and it makes economic sense to have increased installed power to meet the load during, for example, cloudy days than to store electricity produced during sunny days. As presented in Figure 8, almost half of the produced electricity is discarded because it does not make economic sense to store it. This discarded electricity can be used to provide Tier 2 household electrification to the villagers. The actual cost for supplying this electricity to households is linked to the microgrid infrastructure. It is also understandable that almost all of the households will be members of the agricultural cooperative of the village. They can decide to use part of the agricultural profits for cross-subsidization of their households' electrification. This is fully in line with the spirit of cooperatives, citing from [60] "Putting fairness, equality and social justice at the heart of the enterprise, cooperatives around the world are allowing people to work together to create sustainable enterprises that generate long-term jobs and prosperity. Cooperatives allow people to take control of their economic future and, because they are not owned by shareholders, the economic and social benefits of their activity stay in the communities where they are established."

It has to be noted at this point that the selected case study applies to Rwanda, but comparable results are expected to be obtained in all East African countries in terms of decentralized maize flour production because the maize markets in these countries have many similarities [41]. Agriculture related productive loads applicable to remote areas of Sub-Saharan Africa, though, are not limited to on-site flour production. Pumping water for irrigation has been proven to be one of the best interventions for increased production and consequent farmers' income increase [63]. Small scale palm oil processing has proven to be profitable [64]. Ice makers can allow communities near the sea or a lake to store fish for longer periods of time; this in turn leads to increased fish sales and, consequently, increased income for these communities [65]. Rice threshers can also be powered by off-grid systems to provide benefits to remote farms [66]. The list of agriculture related productive loads that can be powered by rural electrification system is extensive [33]. This means that the proposed business model can be applied anywhere in Sub-Saharan Africa where an agricultural cooperative is active and profitable post-harvest processing of the agriculture produce is possible.

The proposed business model has further benefits because rural electrification provision companies can offer their services as subcontractors and not have to undertake a much more extended role, as in the JUMEME example presented above, by utilizing available resources more effectively. Moreover, since agricultural electrification investments have really low payback periods (in the above case study, it was less than two years) it can be relatively easy for the cooperatives to get access to development financing and treat the household electrification component as a side-product for the benefit of their community on the path to achieving sustainable development.

6. Conclusions

Electrification of agriculture is one of the main drivers for improving agricultural production and income of local farmers. Such investments have very low payback periods for the developing world and also present multiple site-benefits, such as reduced immigration to the big cities and abroad, creation of new jobs, increase of income, improvement of living conditions, and, ultimately, poverty alleviation.

The electrification of households is a challenging investment, as can be seen in Section 3.1. This is why the electrification of productive uses of energy has been proposed throughout the last decade in order to make both the electrification investments viable and improve the socio-economic conditions of these areas. Even with new digital technologies and business models developed in the last few years, it is very difficult to deploy viable microgrid investments through only commercial lending.

With the aid of digital technologies, it is easy to understand that the profits presented by agricultural electrification are so high, they can actually cross-subsidize household electrification. If the agricultural cooperatives and businesses becomes the main investor in a household electrification activity and this is considered a part of an agricultural electrification investment, it is possible to see higher electrification rates in the short term. This approach is fully in line with the guidelines for institutional and policy frameworks for microgrids that have been adopted by the African Union Commission and are also fully in line with the policy and regulatory frameworks of many Sub-Saharan African countries.

This paper provides justification for why such business models need to be developed further, applied, pilot tested, and evaluated in Sub-Saharan Africa.

Author Contributions: Conceptualization, G.K.; methodology, G.K. and A.T.B.; software, G.K.; validation, G.K. and D.B.; investigation, G.K.; resources, G.K. and A.T.B.; data curation, G.K. and D.B.; writing—original draft preparation, G.K.; writing—review and editing, D.B. and A.T.B.; visualization, G.K. and A.B.; supervision, D.B. All authors have read and agreed to the published version of the manuscript.

Funding: This research received no external funding.

Conflicts of Interest: The authors declare no conflict of interest.

Abbreviations

AFD	Agence Française de Développement
CAPEX	Capital Expenditure
DESCO	Distributed energy services company
ESMAP	Energy Sector Management Assistance Program
EU	European Union
FAO	Food and Agriculture Organization/
FAOSTAT	Food and Agriculture Organization Statistics
GHG	Greenhouse Gas
LCOE	Levelized cost of electricity
OECD	Organization for Economic Co-operation and Development
OPEX	Operational Expenditure
SDG	Sustainable Development Goal
SOC	State of Charge
UN	United Nations
WBG	World Bank Group

References

1. IEA. *World Energy Outlook 2018*; OECD/IEA: Paris, France, 2018.
2. Silberston, A. Economies of Scale in Theory and Practice. *Econ. J.* **1972**, *82*, 369–391. [CrossRef]
3. Marshall, A. *The Principles of Economics*; McMaster University Archive for the History of Economic Thought: London, UK, 1890.

4. Chen, M.; Zhang, H.; Liu, W.; Zhang, W. The Global Pattern of Urbanization and Economic Growth: Evidence from the Last Three Decades. *PLoS ONE* **2014**, *9*, e103799. [CrossRef] [PubMed]

5. Martínez-Zarzoso, I.; Maruotti, A. The impact of urbanization on CO2 emissions: Evidence from developing countries. *Ecol. Econ.* **2011**, *70*, 1344–1353. [CrossRef]

6. Zhou, L.; Dickinson, R.E.; Tian, Y.; Fang, J.; Li, Q.; Kaufmann, R.K.; Tucker, C.J.; Myneni, R.B. Evidence for a significant urbanization effect on climate in China. *Proc. Natl. Acad. Sci. USA* **2004**, *101*, 9540–9544. [CrossRef] [PubMed]

7. Bao, C.; Chen, X. The driving effects of urbanization on economic growth and water use change in China: A provincial-level analysis in 1997–2011. *J. Geogr. Sci.* **2015**, *25*, 530–544. [CrossRef]

8. McDonald, R.I.; Kareiva, P.; Forman, R.T.T. The implications of current and future urbanization for global protected areas and biodiversity conservation. *Biol. Conserv.* **2008**, *141*, 1695–1703. [CrossRef]

9. Hassell, J.M.; Begon, M.; Ward, M.J.; Fèvre, E.M. Urbanization and Disease Emergence: Dynamics at the Wildlife–Livestock–Human Interface. *Trends Ecol. Evol.* **2017**, *32*, 55–67. [CrossRef]

10. Anríquez, G.; Stamoulis, K. Rural Development and Poverty Reduction: Is Agriculture still the Key? *Electron. J. Agric. Dev. Econ.* **2007**, *4*, 5–46. Available online: http://www.fao.org/3/a-ah885e.pdf (accessed on 3 February 2020).

11. Moseley, M. *Rural Development: Principles and Practice*; SAGE: London, UK, 2003. [CrossRef]

12. Rowley, T.D. *Rural Development Research: A Foundation for Policy*; Greenwood Press: Westport, CT, USA, 1996.

13. Ward, N.; Brown, D.L. Placing the Rural in Regional Development. *Reg. Stud.* **2009**, *43*, 1237–1244. [CrossRef]

14. Brundtland, G.H.; Khalid, M.; Agnelli, S.; Al-Athel, S.; Chidzero, B. *Our Common Future*; Oxford University Press: Oxford, UK, 1987.

15. Vélez Echeverry, S.; Burnett, B.; Diniz, J.; Els, R. Rural Electrification and Sustainable Development in South America Afro Descendant's Communities: A Comparison between the Hinterlands of Brazil and Suriname. In Proceedings of the 9th Conference on Sustainable Development of Water, Energy and Environment System, 20–27 September 2014; Available online: https://www.researchgate.net/publication/277815804_Rural_Electrification_and_Sustainable_Development_in_South_America_Afro_Descendant\T1\textquoterights_Communities_a_Comparison_Between_the_Hinterlands_of_Brazil_and_Suriname (accessed on 10 April 2020).

16. Barron, M.; Torero, M. Household electrification and indoor air pollution. *J. Environ. Econ. Manag.* **2017**, *86*, 81–92. [CrossRef]

17. Buyinza, F.; Kapeller, J. *Household Electrification and Education Outcomes: Panel Evidence from Uganda*; ICAE Working Paper Series, No. 85; Institute for Comprehensive Analysis of the Economy (ICAE), Johannes Kepler University Linz: Linz, Austria, 2018; Available online: http://hdl.handle.net/10419/193624 (accessed on 3 February 2020).

18. Khandker, S.; Barnes, D.H.S. *Welfare Impacts of Rural Electrification: A Panel Data Analysis from Vietnam*; Economic Development and Cultural Change, University of Chicago Press: Chicago, IL, USA, 2013; Volume 61, pp. 659–692.

19. Bos, K.; Chaplin, D.; Mamun, A. Benefits and challenges of expanding grid electricity in Africa: A review of rigorous evidence on household impacts in developing countries. *Energy Sustain. Dev.* **2018**, *44*, 64–77. [CrossRef]

20. Lewis, J.; Severnini, E. Short- and long-run impacts of rural electrification: Evidence from the historical rollout of the U.S. power grid. *J. Dev. Econ.* **2019**, *143*, 102412. [CrossRef]

21. Ding, H.; Qin, C.; Shi, K. Development through electrification: Evidence from rural China. *China Econ. Rev.* **2018**, *50*, 313–328. [CrossRef]

22. Zeyringer, M.; Pachauri, S.; Schmid, E.; Schmidt, J.; Worrell, E.; Morawetz, U.B. Analyzing grid extension and stand-alone photovoltaic systems for the cost-effective electrification of Kenya. *Energy Sustain. Dev.* **2015**, *25*, 75–86. [CrossRef]

23. Bhatia, M.; Angelou, N. *Beyond Connections—Energy Access Redefined*; ESMAP Technical Report 008/15; World Bank: Washington, DC, USA, 2015; Available online: https://openknowledge.worldbank.org/handle/10986/24368 (accessed on 3 February 2020).

24. Palit, D.; Chaurey, A. Off-grid rural electrification experiences from South Asia: Status and best practices. *Energy Sustain. Dev.* **2011**, *15*, 266–276. [CrossRef]

25. Kyriakarakos, G.; Papadakis, G. Multispecies Swarm Electrification for Rural Areas of the Developing World. *Appl. Sci.* **2019**, *9*, 3992. [CrossRef]
26. Dagnachew, A.G.; Lucas, P.L.; Hof, A.F.; Gernaat, D.E.H.J.; de Boer, H.-S.; van Vuuren, D.P. The role of decentralized systems in providing universal electricity access in Sub-Saharan Africa–A model-based approach. *Energy* **2017**, *139*, 184–195. [CrossRef]
27. Chaurey, A.; Kandpal, T.C. A techno-economic comparison of rural electrification based on solar home systems and PV microgrids. *Energy Policy* **2010**, *38*, 3118–3129. [CrossRef]
28. OECD. Detailed aid statistics: Total official development financing ODF. OECD International Development Statistics (database): 2020. Available online: https://doi.org/10.1787/data-00071-en (accessed on 10 April 2020).
29. International Finance Corporation, World Bank Group. *Benchmarking Mini-grid DESCOs 2017 Update—Summary of Findings*; IFC: Washington, DC, USA, 2018.
30. Kyriakarakos, G.; Papadakis, G. Microgrids for Productive Uses of Energy in the Developing World and Blockchain: A Promising Future. *Appl. Sci.* **2018**, *8*, 580. [CrossRef]
31. Cabraal, R.A.; Barnes, D.F.; Agarwal, S.G. Productive uses of energy for rural development. *Annu. Rev. Environ. Resour.* **2005**, *30*, 117–144. [CrossRef]
32. Murabula, H.M.; Kanyua, K.E. Murabula Harrison Masiga; Kanyua, K.E. Factors influencing performance of private electricity mini grid projects in Kenya: A case of Kirinyaga county. *Int. J. Adv. Res. Manag. Soc. Sci.* **2017**, *6*, 62–89.
33. GIZ. *Photovoltaics for Productive Use Applications*; GIZ: Eschborn, Germany, 2016.
34. Chakravorty, U.; Emerick, K.; Ravago, M.-L. Lighting up the Last Mile: The Benefits and Costs of Extending Electricity to the Rural Poor. Available online: http://dx.doi.org/10.2139/ssrn.2851907 (accessed on 10 April 2020).
35. Pode, R.; Pode, G.; Diouf, B. Solution to sustainable rural electrification in Myanmar. *Renew. Sustain. Energy Rev.* **2016**, *59*, 107–118. [CrossRef]
36. Yadoo, A.; Cruickshank, H. The value of cooperatives in rural electrification. *Energy Policy* **2010**, *38*, 2941–2947. [CrossRef]
37. Bijaoui, I. Generators of People's Economy. In *Multinational Interest & Development in Africa: Establishing A People's Economy*; Bijaoui, I., Ed.; Springer International Publishing: Cham, Switzerland, 2017; pp. 101–177. [CrossRef]
38. Contejean, A.; Verin, L. *Making Mini-Grids Work—Productive Uses of Electricity in Tanzania*; IIED Working Paper; IIED: London, UK, 2017; Available online: https://pubs.iied.org/pdfs/16632IIED.pdf (accessed on 2 February 2020).
39. Caselli, G.; Vallin, J.; Wunsch, G. *Demography: Analysis and Synthesis, Four Volume Set: A Treatise in Population*; Elsevier Science; Academic Press: Cambridge, MA, USA, 2005.
40. Batirbaev, P.; Kim, T.; Ma'ani, R.; Shim, R.; Singer, J.; Snyder, M.; Yawson, F. Maize in Rwanda: A Value Chain Analysis; UNIDO: 2013. Available online: https://open.unido.org/api/documents/5328232/download/Maize%20in%20Rwanda%20-%20A%20Value%20Chain%20Analysis (accessed on 2 February 2020).
41. Daly, J.; Hamrick, D.; Gereffi, G.; Guinn, A. *Maize Value Chains in East Africa; Center on Globalization, Governance & Competitiveness*; Duke University: Durham, NC, USA, 2016.
42. Kornher, L. *Maize markets in Eastern and Southern Africa (ESA) in the Context of Climate Change—Background Paper for The State of Agricultural Commodity Markets (SOCO) 2018*; FAO: Rome, Italy, 2018; Available online: http://www.fao.org/3/CA2155EN/ca2155en.pdf (accessed on 2 February 2020).
43. Blodgett, C.; Dauenhauer, P.; Louie, H.; Kickham, L. Accuracy of energy-use surveys in predicting rural mini-grid user consumption. *Energy Sustain. Dev.* **2017**, *41*, 88–105. [CrossRef]
44. Hartvigsson, E.; Ahlgren, E.O. Comparison of load profiles in a mini-grid: Assessment of performance metrics using measured and interview-based data. *Energy Sustain. Dev.* **2018**, *43*, 186–195. [CrossRef]
45. CLASP. Global LEAP Awards. Available online: https://globalleapawards.org/tvs (accessed on 3 February 2020).
46. Laender, D. *Powering Health: Electrification Options for Rural Health Centers*; USAID: Washington, DC, USA, 2018. Available online: http://www.poweringhealth.org/Pubs/PNADJ557.pdf (accessed on 2 February 2020).
47. IRENA. *Renewable Energy Benefits: Decentralised Solutions in the Agri-Food Chain*; IRENA: Abu Dhabi, UAE, 2016.
48. USAID. *Rwanda Power Africa Factsheet*; USAID: Washington, DC, USA, 2018. Available online: https://www.usaid.gov/sites/default/files/documents/1860/Rwanda_-_November_2018_Country_Fact_Sheet.pdf (accessed on 2 February 2020).

49. Steerneman, M. Energy efficiency: How to save energy in a mill today? In Proceedings of the European Flour Millers' Conference 2013, Brussels, Belgium, 16 May 2013.

50. UNFCCC. *Standardized Baseline: Rwanda Grid Emission Factor*; UNFCCC: Bonn, Germany, 2015. Available online: https://cdm.unfccc.int/sunsetcms/storage/contents/stored-file-20161025142637883/ASB0017.pdf (accessed on 2 February 2020).

51. United States Environmental Protection Agency. *Greenhouse Gas Emissions from a Typical Passenger Vehicle*; United States Environmental Protection Agency: Washington, DC, USA, 2018. Available online: https://nepis.epa.gov/Exe/ZyPDF.cgi?Dockey=P100U8YT.pdf (accessed on 2 February 2020).

52. Trócaire. *Analysis of National and Regional Agricultural Trade in Maize, Soybeans Andwheat: A Focus on Rwanda*; Trócaire: Kildare, Ireland, 2014; Available online: https://www.trocaire.org/sites/default/files/resources/policy/a-focus-on-rwanda.pdf (accessed on 2 February 2020).

53. Yang, F.; Yang, M. Rural electrification in sub-Saharan Africa with innovative energy policy and new financing models. *Mitig. Adapt. Strateg. Glob. Chang.* **2018**, *23*, 933–952. [CrossRef]

54. IEG—World Bank Group. *Reliable and Affordable Off-Grid Electricity Services for the Poor: Lessons from World Bank Group Experience*; World Bank Group: Washington, DC, USA, 2016; Available online: http://documents.worldbank.org/curated/en/360381478616068138/pdf/109573-WP-PUBLIC.pdf (accessed on 2 February 2020).

55. ACP-EU Energy Facility. *Thematic Fiche No. 7 "Sustainability—Business Models for Rural Electrification"*; ACP-EU Energy Facility: Brussels, Belgium, 2012; Available online: https://europa.eu/capacity4dev/file/10582/download?token=yK25eQn9 (accessed on 2 February 2020).

56. Martinez, N.; Oliver, P.; Trowbridge, A. *Cost-Benefit Analysis of Off-Grid Solar Investments in East Africa*; USAID: Washington, DC, USA, 2017. Available online: https://www.usaid.gov/sites/default/files/documents/1865/Cost-Benefit-Analysis-Off-Grid-Solar-Investments-East-Africa.pdf (accessed on 2 February 2020).

57. Butare, A.; Kyriakarakos, G. *Guidelines for Institutional and Policy Model for Micro-/Mini-Grids*; African Union Commission: Addis Ababa, Ethiopia, 2018.

58. Madriz-Vargas, R.; Bruce, A.; Watt, M. A Review of Factors Influencing the Success of Community Renewable Energy Minigrids in Developing Countries. In Proceedings of the 2015 Asia-Pacific Solar Research Conference, Brisbane, Australia, 8–9 December 2015.

59. AFD. *Digital Energy Facility for the Promotion of Energy Transition and Energy Access*; AFD: Paris, France, 2019; Available online: https://europa.eu/capacity4dev/file/89985/download?token=r0boUBTB (accessed on 2 February 2020).

60. International Co-operative Allianc. What is a Cooperative? Available online: https://www.ica.coop/en/cooperatives/what-is-a-cooperative (accessed on 27 March 2020).

61. Boadu, F.O. Chapter 8—Cooperatives in Sub-Saharan Africa. In *Agricultural Law and Economics in Sub-Saharan Africa*; Boadu, F.O., Ed.; Academic Press: San Diego, CA, USA, 2016; pp. 263–281. [CrossRef]

62. Moyo, S. *Family Farming in Sub-Saharan Africa: Its Contribution to Agriculture, Food Security and Rural Development*; FAO, 2016; Available online: http://www.fao.org/3/a-i6056e.pdf (accessed on 2 February 2020).

63. Burney, J.A.; Naylor, R.L. Smallholder Irrigation as a Poverty Alleviation Tool in Sub-Saharan Africa. *World Dev.* **2012**, *40*, 110–123. [CrossRef]

64. Ohimain, E.; Izah, S.; Dorcas, A.; Cletus, I. Small-Scale Palm Oil Processing Business in Nigeria; A Feasibility Study. *Greener J. Bus. Manag. Stud.* **2014**, *4*, 070–082. [CrossRef]

65. Briganti, M.; Vallve, X.; Alves, L.; Pujol, D.; Cabral, J.; Lopes, C. Implementation of a PV Rural Micro Grid in the Island of Santo Antão (Cape Verde) with an Individual Energy Allowance Scheme for Demand Control. In Proceedings of the 27th European Photovoltaic Solar Energy Conference and Exhibition, Frankfurt, Germany, 24–28 September 2012. [CrossRef]

66. Itodo, I.N.; Aju, A.S.E. Design of a Solar Photovoltaic System to Power a Rice Threshing Machine. *Niger. J. Sol. Energy* **2015**, *26*, 109–116. Available online: http://sesn-ng.com/journal_view.php?id=156 (accessed on 27 March 2020).

Article

Fuzzy Cognitive Map-Based Sustainable Socio-Economic Development Planning for Rural Communities

Konstantinos Papageorgiou [1,*], **Pramod K. Singh** [2], **Elpiniki Papageorgiou** [3,4], **Harpalsinh Chudasama** [2], **Dionysis Bochtis** [4,*] **and George Stamoulis** [1]

[1]	Computer Science Department, University of Thessaly, 35100 Lamia, Greece; georges@e-ce.uth.gr
[2]	Institute of Rural Management Anand (IRMA), Gujarat 388001, India; pramod@irma.ac.in (P.K.S.); harpalsinh@irma.ac.in (H.C.)
[3]	Faculty of Technology, University of Thessaly, Geopolis Campus, 41500 Larisa, Greece; elpinikipapageorgiou@uth.gr
[4]	Institute for Bio-Economy and Agri-Technology (iBO), Center for Research and Technology–Hellas (CERTH), 6th km Charilaou-Thermi Rd, GR 57001 Thermi, Thessaloniki, Greece
*	Correspondence: konpapageorgiou@uth.gr (K.P.); d.bochtis@certh.gr (D.B.)

Received: 9 December 2019; Accepted: 26 December 2019; Published: 30 December 2019

Abstract: Every development and production process needs to operate within a circular economy to keep the human being within a safe limit of the planetary boundary. Policymakers are in the quest of a powerful and easy-to-use tool for representing the perceived causal structure of a complex system that could help them choose and develop the right strategies. In this context, fuzzy cognitive maps (FCMs) can serve as a soft computing method for modelling human knowledge and developing quantitative dynamic models. FCM-based modelling includes the aggregation of knowledge from a variety of sources involving multiple stakeholders, thus offering a more reliable final model. The average aggregation method for weighted interconnections among concepts is widely used in FCM modelling. In this research, we applied the OWA (ordered weighted averaging) learning operators in aggregating FCM weights, assigned by various participants/ stakeholders. Our case study involves a complex phenomenon of poverty eradication and socio-economic development strategies in rural areas under the DAY-NRLM (*Deendayal Antyodaya Yojana*-National Rural Livelihoods Mission) in India. Various scenarios examining the economic sustainability and livelihood diversification of poor women in rural areas were performed using the FCM-based simulation process implemented by the "FCMWizard" tool. The objective of this study was three-fold: (i) to perform a brief comparative analysis between the proposed aggregation method called "OWA learning aggregation" and the conventional average aggregation method, (ii) to identify the significant concepts and their impact on the examined FCM model regarding poverty alleviation, and (iii) to advance the knowledge of circular economy in the context of poverty alleviation. Overall, the proposed method can support policymakers in eliciting accurate outcomes of proposed policies that deal with social resilience and sustainable socio-economic development strategies.

Keywords: circular economy; sustainable socio-economic development; quality of life; poverty alleviation; participatory modelling; ordered weighted averaging; aggregation

---

1. Introduction

Worldwide, the impacts of various anthropogenic activities, such as increased industrialisation, pollution, deforestation, and overconsumption, are causing destruction and overexploitation of natural resources and have been recognised as the most significant risk not only for the environment but also for human health and well-being [1–3]. The overexploitation of natural resources has grown rapidly in the last two decades, and global resource supply chains have become extremely complex. This has resulted in increasing environmental pressures and impacts [2]. If humankind continues to live on the edge of or outside of the ecological limits, it will be much more difficult to achieve equity, justice, prosperity, well-being, and healthy quality of life for everyone at the global and local levels [4–6].

The need for humanity to remain within the safe operating space of planetary boundaries and the need to eradicate poverty and accelerate sustainable socio-economic development are linked by the concept of "safe and just space for humanity" [7–9]. Therefore, a global shift from a linear economy towards a circular economy is needed, in which socio-institutional change along with resource efficiency and innovative product design [10,11] contribute to economic development and human well-being, with reduced pressures and impacts on the environment. The circular economy approach aims at continuous economic development to achieve waste minimisation, energy efficiency, and environmental conservation, without posing significant challenges to the environment and natural resources [12]. The circular economy is a competitive environmental strategy to production processes and economic activity that allows resources to maintain their highest value while benefiting the business and society as a whole with better supply chain, low volatility of resource prices, better customer relations, improved services, and new employment opportunities [12,13]. The key considerations in the implementation of a circular economy are to refuse, reduce, rethink, repair, restore, remanufacture, and reuse resources [10,11,14], and pursue longevity, renewability, replaceability, and upgradability for resources and products that are used.

The sustainable development goal "one" of the United Nations seeks to eradicate poverty and subsequent inequalities in all forms, leaving no one behind. Owing to the multi-dimensional nature of poverty, its eradication involves complex interactions within socio-economic systems. Across the world, several development projects [15,16] and poverty alleviation programmes [17,18] have been implemented, which are primarily aimed at socio-economic development and poverty reduction of poor and vulnerable communities. The *Deendayal Antyodaya Yojana*–National Rural Livelihoods Mission (DAY-NRLM), a centrally sponsored programme of the Government of India, is one of the world's largest poverty eradication programmes that aims to eliminate rural poverty in the country through the promotion of multiple livelihoods for each rural poor household. The programme follows a participatory and community-demand-driven approach that focusses largely on the eradication of poverty and pays special attention to the development of social resilience and sustainable socio-economic development.

This study tried to identify an ideal strategy for sustainable socio-economic development planning for rural communities in developing economies by evaluating the DAY-NRLM programme. For this, we used fuzzy cognitive mapping—a participatory modelling approach that allows for the exploration and management of various interrelationships between human, environmental, and socio-economic systems. The main principle behind using participatory modelling is that the resulting policy can effectively support a range of social, economic, and environmental goals. The systems-based integrated and participatory modelling approaches allow the stakeholders to contribute to the development of a system model, as well as support decision-makers' understanding of the system being examined and all its aspects, by participating in the scenario development process and identifying correct strategies within the domain. Therefore, it is safe to say that the contribution that stakeholders can offer to a given problem, usually related to support for decision-making, policies, regulation, or management, is valuable [19]. The fuzzy cognitive mapping approach allows stakeholders to collaboratively develop a "cognitive map" (i.e., a weighted, directed graph), which represents the perceived causal structure of their system [20]. Fuzzy cognitive map (FCM) [21] is a collection of concepts and causal relationships between them, in the form of a directed graph that can model a real-world system. Because FCMs can easily incorporate human knowledge and adapt to any given domain [22], they have been extensively applied in many research fields and areas of application [23,24]. Their simple model structure gives them considerable awareness and broad research interest [25].

This study also explored a new method for aggregating individual FCMs using the application of ordered weighted averaging (OWA) operators. The aggregation of individual FCM models must be taken into account when a variety of sources need to be included in the modelling procedure of a given system [23]. Individual FCMs, constructed by experts and/or stakeholders, can be aggregated to produce a combined FCM that will incorporate the knowledge from all the different experts and/or stakeholders involved in the FCM construction process. The main objective behind aggregating individual FCMs is to improve the reliability of the overall model and make them less susceptible to potentially inaccurate knowledge of a single expert and/or stakeholders or knowledge inconsistency among the participants. Regarding the combination of multiple FCMs into a single collective model, among the techniques that can be found in the related literature, two methods are widely used in real-life problems [26]—the weighted average and the OWA method introduced by Yager [27]. The weighted average method is considered as the benchmark method for FCM aggregation purposes, where the final FCM model is built by averaging numerical values for a given interconnection [28]. However, there are certain limitations to implementing this methodology. For example, when a large number of stakeholders/participants are asked to assign values on the relationships of a given system, significant deviations can arise between these values, which shows an inconsistency of knowledge among the participants that leads to an inaccurate overall FCM. On the other hand, the application of OWA operators in the aggregation of individual FCMs has a limited presence in the literature. In particular, OWA operators in an FCM framework were introduced by Zhenbang and Lihua [29], who highlighted the ability of OWA operators to simulate the various AND/OR relationships between the concepts and studied the OWA aggregation under different conditions. An OWA operator based on distance was examined by Leyva-Vazquez et al. [30] to rank the scenarios depending on the decision-makers' risk preferences.

The study used the context of the DAY-NRLM programme, which predominantly focuses on livelihood enhancement through building self-managed and sustainable community institutions of "the rural poor women". These community institutions are expected to overcome their social, financial, and economic exclusion. The broader objectives of the programme involve social mobilisation, institution building, enrolment of women in social security schemes and entitlements, socio-economic inclusion, sustainable livelihoods, capacity building, promotion of economic stability, improvement of social resilience, and skill development, aimed at elimination of rural poverty. Ultimately these outcomes will lead to the reduced socio-economic poverty and improved quality of life.

The paper is structured as follows: Section 2 describes the problem along with the methods used in this study. This section includes in particular the basic principles of FCMs, the specifications of the OWA operators, and the proposed algorithm based on the weights of the OWA operator to aggregate FCM weights. This section also deals with the quasi-qualitative method for constructing the collective FCM from groups, whereas in Section 3, the aggregation results produced from the application of the OWA method in the developed FCMs, the sensitivity analysis, and the scenario analysis are provided. In the end, Section 4 summarises the conclusions of this study.

2. Methodology

2.1. Overview of Fuzzy Cognitive Maps

FCMs are soft computing techniques that combine fuzzy logic and neural networks [31]. They were first introduced by Kosko [21,32] as fuzzy-graph structures for representing causal reasoning. They provide a more flexible and natural ability for knowledge representation and reasoning, which are essential to intelligent systems [33,34].

A set of nodes (concepts), representing the key elements of the given system, and directed arcs (links), defining the causal relationships between the nodes, form an FCM [35]. Each concept is indicated by $C_i, i = 1, \ldots, N$ (where N is the number of concepts in the problem domain) and is also specified by an activation value Ai in [0, 1]. The links are labeled with fuzzy weights in the interval of

[0, 1] or [−1, +1], which show the strength of the impact between the concepts [36]. Weights of links are associated with a weight value matrix E NxN, where each element of the matrix w_{ij} takes values in [−1, +1]. There are three types of weights:

- $w_{ij} = 0$ indicates no causality between concepts;
- $w_{ij} > 0$ indicates a causal increase (i.e., C_j increases as C_i increases, and C_j decreases as C_i decreases);
- $w_{ij} < 0$ indicates a causal decrease (i.e., C_j decreases as C_i increases, and C_j increases as C_i decreases).

Word weights like "little" or "somewhat" can be used instead of numeric values [37]. Figure 1 shows an example of a FCM with its corresponding adjacency weight matrix.

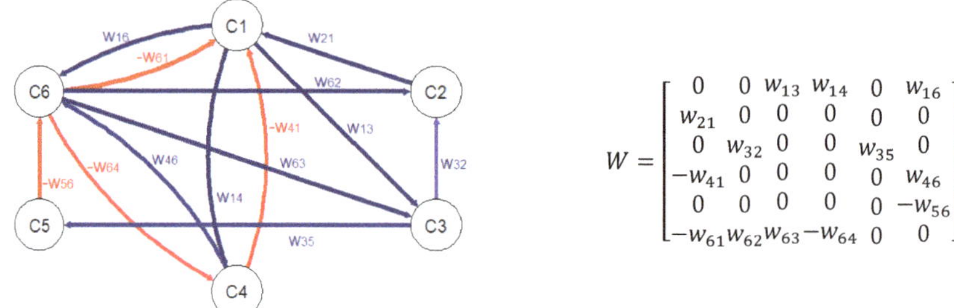

$$W = \begin{bmatrix} 0 & 0 & w_{13} & w_{14} & 0 & w_{16} \\ w_{21} & 0 & 0 & 0 & 0 & 0 \\ 0 & w_{32} & 0 & 0 & w_{35} & 0 \\ -w_{41} & 0 & 0 & 0 & 0 & w_{46} \\ 0 & 0 & 0 & 0 & 0 & -w_{56} \\ -w_{61} & w_{62} & w_{63} & -w_{64} & 0 & 0 \end{bmatrix}$$

Figure 1. Fuzzy cognitive map (**left**) and the correspondent weight adjacency matrix (**right**), showing the positive and negative causal influences.

An FCM is not a static representation of the world; FCM's graph structure facilitates causal reasoning, where calculations can be made to assess the consequences of a specific system state. In [38], *auto-associative neural network* mechanisms were used to study the system dynamics of FCM models and produce projections for different possible scenarios.

We used a multi-step FCM development process, as shown in Figure 2.

Figure 2. A visual illustration of the fuzzy cognitive map (FCM) development process.

2.2. Obtaining Cognitive Maps from the Participants

This study adopted a mixed-concept design approach for obtaining cognitive maps involving open-concept design followed by closed-concept design. Expert-based FCM was obtained using open-concept design. A total of 31 national and state-level implementers identified 20 categories of main concepts and 129 sub-concepts. The visual representation of this expert-based FCM model is depicted in Figure 3, which will be further compared with the models derived from the aggregation processes in order to help this study meet its goals.

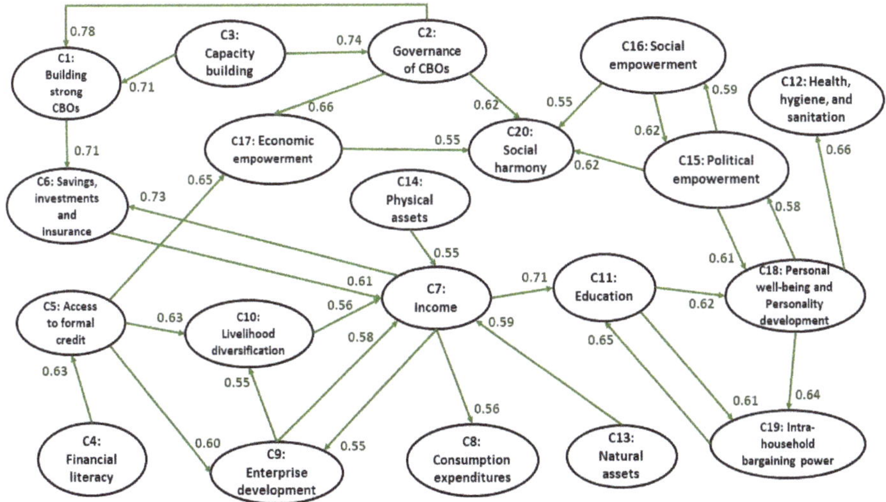

Figure 3. Expert-based FCM model. **Note:** SHGs federate into VO and VOs federate into CLF. These together called CBOs (community-based organisations).

We prepared a protocol/instrument based on the expert-based FCM model drawing all the 20 categories of main concepts and 129 sub-concepts on the paper.

Four different groups of women functionaries of DAY-NRLM of Jammu and Kashmir state of India—The Self-Help Groups (SHGs), the Village Organization (VOs), the Cluster Level Federations (CLFs), and the Community Resource Persons (CRPs) constructed FCMs in groups of 4–5 members. The SHGs group consists of 36 participants, the VOs have 52 participants, and the CLFs comprise 60 participants, whereas the CRPs include 31 participants. We generated a list of SHGs, VOs, and CLFs, to be interviewed for the FCM exercise, using "Microsoft Excel RAND function" from the 10 districts of Jammu and Kashmir, 5 districts from each region. In total, 179 individual fuzzy cognitive maps were collected during the FCM exercise from approximately 600 participants coming from 10 districts of the state of Jammu and Kashmir in India.

Every key concept was formed from a list of certain sub-concepts that define it well. The sub-concepts were introduced by the experts and policymakers to offer a deeper understanding of the examined problem, making it clear to the participants. Table 1 illustrates all the sub-concepts for each key concept.

Table 1. The sub-concepts list for each of the key concepts.

Key Concept	Sub-Concepts	
C1: Building strong CBOs (Community-based Organizations)	Building Self-Help Groups (SHGs)	Building Cluster Level Federations (CLFs)
	Building Village Organization (VOs)	
C2: Governance of CBOs	Practicing panch sutra	Developing repayment culture
C3: Capacity building	Training on SHG concept and management	Training to SHG and VO on bookkeeping
	Training on VO concept and management	

Table 1. *Cont.*

Key Concept	Sub-Concepts	
C4: Financial literacy	Analytical literacy	Institutional literacy
	Functional literacy	Adoption of better financial technology
	Technical literacy	
C5: Access to formal credit	Inter-loaning	Banking awareness
	High-cost loan into low-cost loan	SHG–bank linkage
C6: Savings, investments, and insurance	SHG savings	Health insurance
	Regular savings accounts (individual or joint)	Vehicle insurance
	Life insurance	Livestock insurance
C7: Income	Income from agriculture/horticulture	Income from self-employment
	Income from animal husbandry	Income from wage labour
C8: Consumption expenditures	Usage of freeze, mixer, etc.	Expenditure on more nutritious food
	More expenditure on clothes	More expenditure on health
	Usage of bed, sofa, etc.	More expenditure on education
	Usage of TV, radio, tape recorder, etc.	
C9: Enterprise development	Enterprise development	Skill development training
	Enterprise management	Guidance from experts/professionals
	Decision-making with regard to enterprise	Value addition
	Fostering entrepreneurship	Market identification
	New livelihoods activities/interventions	Market linkages
C10: Livelihood diversification	Agriculture	Self-employment
	Horticulture	Wage labour
	Livestock rearing	
C11: Education	Better implementation of Aanganwadi scheme	Better avenues for higher education for children
	Increased access to education for children	Easy access to mid-day meals
	Improved quality of education	Easy access to information about schooling
	Increased literacy among children	
	Reduced school drop-out rate	
C12: Health, hygiene, and sanitation	Reduced open defecation	Reduced epidemic
	Better sanitation and hygiene	Reduced malnutrition among children
	Increased access to hospitals	Increased nutritional security
	Vaccination	
C13: Natural assets	Management of land	Management of livestock
C14: Physical assets	Productive assets	Asset in the name of women
	Consumptive assets	Community owned assets

Table 1. *Cont.*

Key Concept	Sub-Concepts	
C15: Political empowerment	Political inclusion	Participation in Mahila Mandal
	Political justice	Participation in Mother's Committee
	Awareness of rights	Participation in Watershed Committee
	Exercising rights	Approaching to higher officials in block and district for exercising rights
	Initiatives focusing on women development	Participation in Social Audit
	Participation in political groups/parties	Participation in Vigilance Committee
	Participation in Village Development Committee	Participation in Gram Sabha
	Participation in Village Education Committee	Role in policy making
	Participation in Village Health Committee	
C16: Social empowerment	Universal social mobilization	Increased social development
	Universal social inclusion	Increased social awareness
	Improved social behavior	Increased social welfare
	Increased social values	
C17: Economic empowerment	Financial self-sufficiency	Increased savings
	Economic decision-making	
C18: Personal well-being and personality development	Adoption of agricultural machines/equipment, adoption of better cooking methods/equipment	Sense of authority in public space
	Better health and hygiene awareness	Recognition in family and society
	Access to basic infrastructure	Power in household decision making
	Increased women literacy	Increased gender equality
	Increased women safety	Voicing against ill treatments
	Increased self-confidence	Individual values/integrity
	Engagement with household level issues	Women's identity
	Engagement with community level issues	Women's independence
		Women's mobility
		Own perceptions towards changes
C19: Intra-household bargaining power	Mediation in household/family disputes	Access to credit
	Seeking benefit under government scheme	Admissions to schools/colleges
	Obtaining ration card	Admissions to hospitals
	Lodging police complaint	
C20: Social harmony	Community cohesion	Conflict avoidance
	Community support	Increased tolerance
	Reduced social tension	Solidarity

Meanwhile, participants of all the four groups (SHG, VO, CLF, and CRP) were asked to assign a numerical value (degree of significance) to every sub-concept on a scale of 1–10 in order to assess the impact of each sub-concept to the corresponding key concept. An overall average degree of significance was calculated for every key concept to estimate the significance of each concept in the examined FCM model. This process also helped us in the next step, with the selection of the most important concepts for policy-making for the scenario analysis.

Table 2 illustrates an example regarding the mean values of the degree of significance for the three sub-concepts of the key concept C1, from each group of the Kashmir region, along with the calculated average value (degree of significance) that corresponds to the key concept C1.

Table 2. Degree of significance for the sub-concepts of C1, for Kashmir. CRP: Community Resource Persons.

Key Concept	Sub-Concept	SHG	VO	CLF	CRP	Average
C1: Building Strong CBOs						8.87
	Building SHGs	8.3	8.5	9.1	9.6	8.875
	Building VOs	8.0	8.4	8.6	9.8	8.70
	Building CLFs	7.9	9.1	9.3	9.8	9.025

2.3. Coding Individual Cognitive Maps into an Adjacency Matrix

Individual fuzzy cognitive maps were coded into separate excel sheets to form adjacency matrices [39,40]. The values were coded into the square matrix only when a connection existed between two concepts [39]. Weights given to each link were then normalised between 0 and +1 (the value 7 was normalised to 0.7) for coding into the adjacency matrix [40–42].

2.4. Aggregating Individual Fuzzy Cognitive Maps from Each Group to Produce Aggregated FCMs

All coded maps from each group were aggregated using the two aggregation methods, the weighted average, and the OWA, to make collective fuzzy cognitive maps. Two collective FCMs were created for every group of participants for each aggregation method, named as the average-FCM and OWA-FCM.

2.4.1. Average Aggregation

We aggregated individual FCMs to form collective-FCM through matrix addition [39,41,43–45]. The collective-FCM represents the perception of all members of a particular stakeholder group.

Considering n stakeholders assign a weight value w_{ij}, between the nodes C_i and C_j on individual FCMs, then the aggregated weight $w_{ij}^{(ave)}$ between the average value of the n weights w_{ij}:

$$w_{ij}^{(ave)} = \frac{w_{ij}^{(1)} + w_{ij}^{(2)} + \cdots + w_{ij}^{(n)}}{n}. \tag{1}$$

2.4.2. OWA Aggregation

An OWA operator [19] of dimension n is a mapping:

$$f : R^n \rightarrow R,$$

that has an associated weighting vector W

$$W = [w_1 \ w_2 \ \ldots \ w_n]^T, \tag{2}$$

such that

$$\sum_i w_i = 1; \ w_i \in [0, 1], \tag{3}$$

where

$$f(a_1, \ldots, a_n) = \sum_{j=1}^{n} w_j \, b_j \tag{4}$$

where b_j is the jth largest element of the collection of the aggregated objects a_1, a_2 ... , a_n. The function value $f(a_1, \ldots , a_n)$ determines the aggregated value of arguments a_1, a_2 ... , a_n.

A fundamental aspect of the OWA operator is the re-ordering step, in particular, an argument a_i is not associated with a particular weight w_i but rather a weight w_i is associated with a particular ordered position i of the arguments. A known property of the OWA operators is that they include the *Max*, *Min*, and arithmetic mean operators for the appropriate selection of the vector W:

$$\text{i} \quad \text{For } W = \begin{bmatrix} 1 \\ 0 \\ \vdots \\ 0 \end{bmatrix}, \ f(a_1, \ldots, a_n) = \underset{i}{\text{Max}} \ a_i,$$

$$\text{ii} \quad \text{For } W = \begin{bmatrix} 0 \\ 0 \\ \vdots \\ 1 \end{bmatrix}, \ f(a_1, \ldots, a_n) = \underset{i}{\text{Min}} \ a_i,$$

$$\text{iii} \quad \text{For } W = \begin{bmatrix} 1/n \\ 1/n \\ \vdots \\ 1/n \end{bmatrix}, \ f(a_1, \ldots, a_n) = \tfrac{1}{n} \sum_{i=1}^{n} a_i \, .$$

It can be easily seen [46] that the OWA operators are aggregation operators, satisfying the commutativity, monotonicity, and idempotency properties and are bounded by the *Max* and *Min* operators, for OWA operators:

$$\underset{i}{\text{Min}} \ a_i \ \leq f(a_1, \ldots, a_n) \ \leq \underset{i}{\text{Max}} \ a_i. \tag{5}$$

Because this class of operators runs between the *Max* (*or*) and the *Min* (*and*), the author of [24] introduced a measure to characterise the type of aggregation being performed for a particular value of the weighting vector. This measure, called the *orness measure* of the aggregation, is defined as:

$$\textit{Orness} \ (W) = \frac{1}{n-1} \sum_{i=1}^{n} (n-1) w_i. \tag{6}$$

As suggested by Kosko [24], this measure, which lies in the unit interval, characterises the degree to which the aggregation is like an *or* (*Max*) operation. It can be shown as follows:

$$\textit{orness} \ ([1 \ 0 \ldots 0]^T) \ = \ 1, \tag{7}$$

$$\textit{orness} \ ([0 \ 0 \ldots 1]^T) \ = \ 0, \tag{8}$$

$$\textit{orness}\left(\left[\frac{1}{n} \ \frac{1}{n} \ldots \frac{1}{n} \right]^T \right) = 0.5. \tag{9}$$

Therefore, the, *Min*, and arithmetic mean operators can be regarded as OWA operators with degree of orness, respectively, 1, 0, and 0.5.

2.4.3. Learning OWA Operators for Aggregating FCM Weights

This study took into consideration a previously conducted preliminary research [47] about an alternative FCM aggregation method using OWA operators and made it one step further. Specifically, learning OWA operators' weights were introduced for aggregating FCM connections/links defined by multiple experts and/or stakeholders. This work focused on developing an alternative aggregation methodology for the FCM modelling, in order to fill the absence of learning OWA operators in aggregation of weights in FCMs, considering that FCMs have been proposed as a unique methodology able to aggregate diverse sources of knowledge to represent a "scaled-up" version of individuals' knowledge and beliefs [39,48]. Moreover, learning operators were used in this study for defining weights among the concepts, so that the studying of the OWA aggregation approach would be strengthened in terms of performance. This method has particularly broad applicability and had high effectiveness when a large number of participants/stakeholders were present. A complex real-life strategic decision-making problem was studied, in which the proposed methodology was applied in order to examine/validate this approach. Specifically, the studied problem dealt with the aggregation and modeling of communities' perception, as well as scenario analysis using the FCM-based simulation process implemented by the new FCMWizard tool [31].

In this study, we present an algorithm that can be used for aggregating weights assigned by experts/stakeholders' opinion in designing FCMs. The proposed algorithm learns the weights associated with a particular use of the OWA operator from a group of experts and/or stakeholders of the specific scientific domain. The OWA weights can be obtained through the following procedure:

At first, experts' opinions are considered as argument values $(a_{k1}, a_{k2} \ldots, a_{kn})$.

i   Step 1: Generate slightly different parameters ρ for each argument which represents the optimism of the decision-maker, $0 \leq \rho \leq 1$.

ii   Step 2: Calculate the aggregated values for each sample using the Hurwics method, according to which the aggregated value d obtained from a tuple of n arguments, $a_1, a_2, \ldots, a_n$, is defined as a weighted average of the *Max* and *Min* values of that tuple.

$$\rho \operatorname*{Max}_i a_i + (1 - \rho)\operatorname*{Min}_i a_i = d. \tag{10}$$

iii   Step 3: Reorder the objects $a_{k1}, a_{k2} \ldots, a_{kn}$.

iv   Step 4: Calculate the current estimate of the aggregated values d_k

$$\hat{d}_k = b_{k1}\ w_1 +\ b_{k2}\ w_2 + \ldots +\ b_{kn}\ w_n \tag{11}$$

with initial values of the OWA weights $w_1 = 1/n$.

v   Step 5: Calculate the total $\hat{d}_k$ for each i. The parameters λ_i determine the weights of OWA and are updated with the propagation of the error $\hat{d}_k - d_k$ between the current estimated aggregated value and the actual aggregated value [39].

vi   Step 6: Calculate the current estimates of the λ_i

$$\lambda_i(l+1) = \lambda_i(l) - \beta w_i(l)(b_{ki} - \hat{d}_k)(\hat{d}_k - d_k) \tag{12}$$

with initial values $\lambda_i(0){=}0, i = (1, \ldots, n)$, and a learning rate of $\beta = 0.35$.

vii   Step 7: Use $\lambda_i, i = (1, \ldots, n)$,, to provide a current estimate of the weights

$$w_i = \frac{e^{\lambda_i(l)}}{\sum_{j=1}^n e^{\lambda_i(l)}}, i = (l, n) \tag{13}$$

viii   Step 8: Update wi and $\hat{d}_k$ at each iteration until the estimates for all the λ_i converge to, that is, $\Delta = |\lambda(l+1) - \lambda(l)|$, are small.

In what follows, an explanatory paradigm considering three experts will better illustrate the proposed FCM construction approach in an environmental domain.

As mentioned above, the experts' opinions are considered as argument values $a_{k1}, a_{k2} \ldots, a_{kn}$, and the weight between two concepts as a sample. Zero values for weights were not considered in the aggregation process. An FCM model consisting of 7 concepts and 14 weighted connections among concepts was selected to show how the above steps are implemented (Table 3).

Table 3. The explanatory paradigm for aggregating three experts' opinions on agriculture.

Sample Weight	Expert 1	Expert 2	Expert 3	Aggregated Value
C1–C7	0.43	0.50	0.58	0.45
C2–C1	0.57	0.60	0.68	0.58
C2–C7	0.57	0.60	0.68	0.58
C–C2	−0.30	−0.35	−0.25	−0.32
C3–C7	−0.39	−0.25	−0.30	−0.32
C4–C5	−0.32	−0.40	−0.47	−0.38
C4–C7	−0.43	−0.30	−0.30	−0.35
C5–C4	−0.37	−0.40	−0.45	−0.37
C5–C7	0.68	0.60	0.50	0.53
C6–C2	0.58	0.55	0.70	0.56
C6–C7	0.55	0.50	0.65	0.53
C7–C1	0.22	0.37	0.33	0.23
C7–C2	0.67	0.70	0.75	0.67
C7–C5	0.54	0.58	0.47	0.48

We calculated the aggregated values using various values for parameter ρ within ρ $(0.01 \leq \rho \leq 0.2)$. For example, $\rho = 0.153; 0.131; 0.181; 0.075; 0.055$.

Using *min* and *max* values of ρ, the aggregated value of weight was calculated as follows.

$$0.153(0.58) + (1 - 0.153)(0.43) = 0.45.$$

We initialised $\lambda_i(0) = 0$, $i = (1, n)$, $\beta = 0.35$, and $w_1 = w_2 = w_3 = 0.33$. The estimated values of λ_i after 108 iterations were

$$\lambda_1 = 0.63, \; \lambda_2 = -0.19, \; \lambda_3 = 0.82.$$

The following OWA weights were calculated considering the above λ_i:

$$w_1 = 0.146, \; w_2 = 0.227, \; w_3 = 0.626.$$

We followed the same process for parameter ρ for $0.3 < \rho < 0.5$ and $0.5 < \rho < 0.7$.

Table 4 depicts the calculated values for OWA aggregated weights for all interrelationships among FCM concepts, as well as the deviations between the benchmark weight Wb (average method) and the Wowa, weight produced by learning OWA operators.

Table 4. Weights produced by learning OWA (ordered weighted averaging) operators.

Weight	Average Weight (Wb)	Weight by 0.01 < ρ < 0.2	Weight by 0.3 < ρ < 0.5	Weight by 0.5 < ρ < 0.7	ΔW (Deviation in Aggregated Value)
C1–C7	0.50	0.47	0.53	0.58	0.03
C–C1	0.38	0.33	0.40	0.45	0.05
C2–C7	0.62	0.59	0.64	0.68	0.03
C3–C2	−0.30	−0.32	−0.28	−0.25	0.02
C3–C7	−0.31	−0.35	−0.29	−0.25	0.04
C4–C5	−0.40	−0.43	−0.37	−0.32	0.03
C4–C7	−0.34	−0.38	−0.32	−0.30	0.04
C5–C4	−0.41	−0.43	−0.39	−0.37	0.02
C5–C7	0.59	0.55	0.62	0.68	0.04
C6–C2	0.61	0.58	0.57	0.70	0.03
C6–C7	0.57	0.53	0.59	0.65	0.04
C7–C1	0.31	0.27	0.33	0.37	0.04
C7–C2	0.71	0.69	0.72	0.75	0.02
C7–C5	0.53	0.5	0.55	0.58	0.03

2.5. Visual Interpretation of Collective FCM

The collective FCMs were analysed using the FCMWizard software tool (www.fcmwizard.com) [31]. The tool includes modelling and visualisation capabilities for the consensus FCM models, depicting the connections among the factors and also reflecting the importance of different concepts within various asset classes [43]. The average-FCM models designed by each group (SHG, VO, CLF, CRP) and produced by the FCMWizard tool are illustrated in the figures below (Figures 4–7).

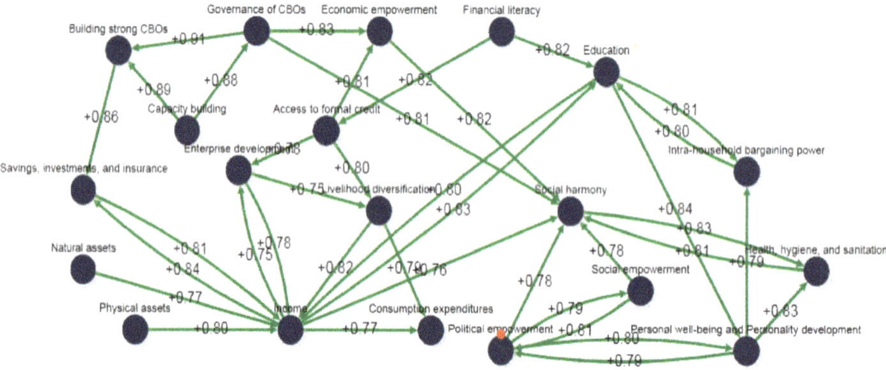

Figure 4. Fuzzy cognitive map of SHG group of DAY-NRLM (Jammu-Kashmir National Rural Livelihoods Mission) programme.

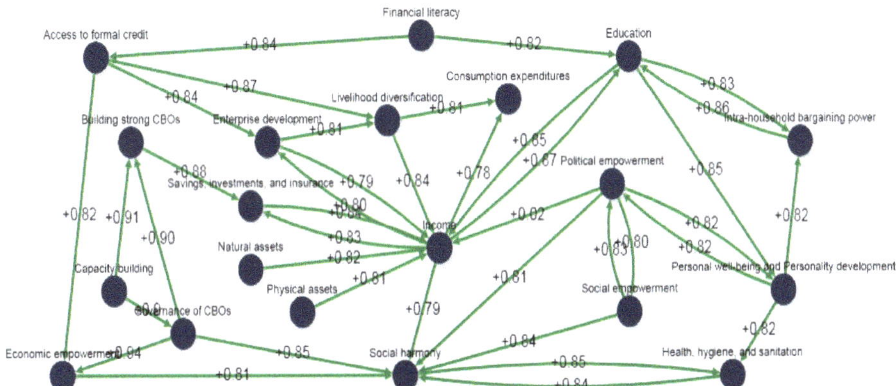

Figure 5. Fuzzy cognitive map of VO group of DAY-NRLM programme.

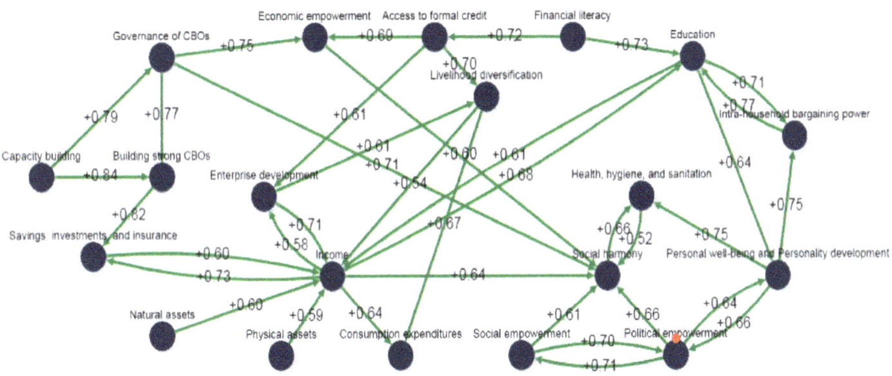

Figure 6. Fuzzy cognitive map of CLF group of DAY-NRLM programme.

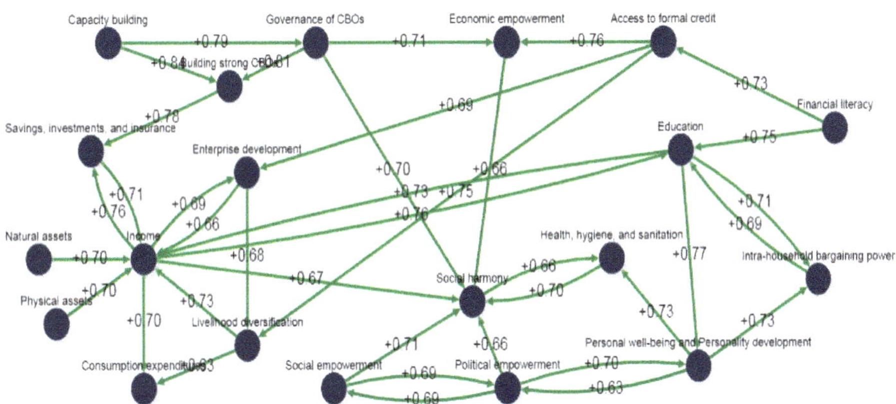

Figure 7. Fuzzy cognitive map of CRP group of DAY-NRLM programme.

2.6. FCM-Based Simulations

Typically, an FCM of n concepts could be represented mathematically by an $n \times n$ weight matrix (W). By feeding the fuzzy cognitive map with an initial stimulus state vector $X^{(k)}$ (state vector at time (k)), it can model the evolution of a scenario over time by evolving forward and letting concepts interacting with one another. Each subsequent value of the concept state $X^{(k+1)}$ can be computed as previous state $X^{(k)}$ and weight matrix multiplication, according to Equation (14):

$$X_i^{(\kappa+1)} = f\left(\sum_{j=1,\ j\neq i}^{n} w_{ji} \times X_j^{\kappa} \right). \tag{14}$$

Based on the literature, two other equations were proposed for FCM inference, the modified Kosko (Equation (15)) and the rescaled Kosko (Equation (16)):

$$X_i^{(\kappa+1)} = f\left(X_i(t) + \sum_{\substack{j=1 \\ j \neq i}}^{n} X_j(t){\cdot}w_{j,i} \right), \tag{15}$$

$$X_i^{(\kappa+1)} = f\left(\left(2 \times X_i^{\kappa} - 1\right) + \sum_{j=1,\ j\neq i}^{n} w_{ji} \times \left(2 \times X_j^{\kappa} - 1\right) \right), \tag{16}$$

where $X_i^{(\kappa+1)}$ is the value of concept C_i at simulation step $\kappa + 1$, $X_j^{(\kappa)}$ is the value of concept C_j at the simulation step κ, w_{ji} is the weight of the interconnection between concept C_j and concept C_i, and $f(\cdot)$ is the threshold transfer function used to retain the values within the range of [0, 1] or [−1, +1]. Generally, the most commonly used transfer function is Sigmoid [49], as shown by Equation (17).

$$f(x) = \frac{1}{1 + e^{-\lambda x}} \tag{17}$$

where λ is a real positive number ($\lambda > 0$), which determines the steepness of the continuous function f, and x is the value $X_i^{(\kappa)}$ for a given iteration.

The simulation stops when the system reaches equilibrium, that is, a limit vector is reached as $X^t = X^{t-1}$ or when $X^t - X^{t-1} \leq e$, where e is a residual, describing the minimum error difference among the subsequent concepts. Its value depends on the application type and it is typically set to 0.001.

3. Results and Discussion

3.1. Characteristics of the Key Concepts of the DAY-NRLM Programme

When there are a large number of concepts that need to be studied individually, then it is necessary to keep only the most influential ones. The filtering technique of key concepts is common in scenario planning and helps linking storylines to the quantitative model, as well as to pay attention to pivotal concepts that can influence, directly or not, the outcome of the examined system, or even significantly change its balance [50]. Key concepts were mainly identified by the experts in the FCM-based scenario analysis or emerged by certain characteristics of the studied model. There were three indicators, based on the connection weights, which help researchers to recognise the important key concepts of the system—indegree (weight of inbound links), outdegree (weight of outbound links), and degree centrality. The first two indicate to what extent a concept is a transmitter (influential) or receiver (dependent). This is similar to the bi-dimensional categorisation of influence-dependence axes in cross-impact analysis [51]. Degree centrality is the relative importance of a concept within the FCM structure, which is calculated by the sum of the corresponding absolute indegree and outdegree causal weights [52]. These calculated indices for the collective average-FCM, along with the concepts identified previously, are summarised in Table 5. Additionally, the overall specifications of the above FCM model are presented in Table 6.

Table 5. Finalised concepts, their description, and type with three major indices values (indegree, outdegree, and degree centrality) for the collective average-FCM.

Concepts	Description	Indegree	Outdegree	Degree Centrality	Betweenness Centrality	Closeness Centrality	Type
C1: Building strong CBOs	Build competence and confidence	0.90	1.80	2.60	11.83	8.58	Ordinary
C2: Governance of CBOs	Organize the goals, roles, and responsibilities of CBOs	2.60	0.90	3.40	40.50	9.91	Ordinary
C3: Capacity building	Obtain, improve, and retain skills, knowledge, and other resources for their jobs	1.80	0.00	1.80	0.00	7.36	Transmitter
C4: Financial literacy	Possession of skills and knowledge to make effective decisions with all the financial resources	1.60	0.00	1.60	7.50	8.41	Transmitter
C5: Access to formal credit	Demand for loans provided by formal banking institutions	2.40	0.80	3.20	23.33	9.83	Ordinary
C6: Savings, investments, and insurance	Increasing income, reducing expenses, protection from financial loss	0.80	1.70	2.50	25.83	9.33	Ordinary
C7: Income	Wages, salaries, profits, and other forms of earnings	4.00	4.80	8.80	201.3	13.83	Ordinary
C8: Consumption expenditures	Spending by households on goods and services	0.00	1.60	1.60	0.00	8.91	Receiver
C9: Enterprise development	Provide contributions to develop business sustainability	1.50	1.50	3.10	4.66	9.75	Ordinary
C10: Livelihood diversification	Construct a diverse portfolio of social activities to improve the standards of living	1.60	1.60	3.20	9.66	10.25	Ordinary
C11: Education	Acquisition of knowledge, skills, values, beliefs, and habits	2.50	2.50	4.90	53.58	10.58	Ordinary
C12: Health, hygiene, and sanitation	Practices that contribute to good health and keep our environment healthy	0.80	1.60	2.40	6.83	8.83	Ordinary
C13: Natural assets	Augmentation of natural resources, etc.	0.80	0.00	0.80	0.00	8.24	Transmitter

Table 5. *Cont.*

Concepts	Description	Indegree	Outdegree	Degree Centrality	Betweenness Centrality	Closeness Centrality	Type
C14: Physical assets	Water supply and irrigation, infrastructure development	0.80	0.00	0.80	0.00	8.24	Transmitter
C15: Political empowerment	Political inclusion, political justice	2.40	1.60	3.90	29.33	10.83	Ordinary
C16: Social empowerment	Participation in various village-level committees, universal social mobilization, social inclusion	1.60	0.80	2.40	0.00	8.41	Ordinary
C17: Economic empowerment	Savings, financial self-sufficiency, etc.	0.80	1.70	2.50	22.49	9.75	Ordinary
C18: Personal well-being and personality development	Experience of health, happiness, and prosperity Improving behavior and attitude that makes a person distinctive	2.40	1.60	4.00	21.16	9.50	Ordinary
C19: Intra-household bargaining power	Decisions regarding the household unit, such as whether to spend or save, to study or work	0.80	1.60	2.40	0.00	7.94	Ordinary
C20: Social harmony	The promotion of ethnic cohesion and peace	0.80	4.80	5.50	101.6	12.16	Ordinary

Table 6. Specifications of the FCM model.

Index	Value
Positive (negative) connections	59 (76)
Density	0.1026
Hierarchy index	0.0256
Average degree centrality	3.07
Average betweeness centrality	27.9875
Average closeness centrality	9.5366

3.2. Characteristics of the Sub-Concepts of the DAY-NRLM Programme

Every participant was given a list of sub-concepts for each key concept (as presented in Table 1) to assign a degree of significance on the scale of 1–10, where 10 is the most significant sub-concept and 1 the least significant. This procedure was followed for two regions, Jammu and Kashmir. Table 7 includes the key concepts with the respective mean values for both regions, as well as the overall average value of the degree of significance, whereas the next figure (Figure 8) illustrates these values in a graph for better visual interpretation of the results.

Table 7. Mean values of significance for two regions and the average degree of significance.

Key Concepts	Degree of Significance (Mean Value) with Respect to the Region		Average Degree of Significance
	Kashmir	Jammu	
C1: Building strong CBOs	8.79	9.43	9.11
C2: Governance of CBOs	8.92	9.41	9.16
C3: Capacity building	8.69	8.95	8.82
C4: Financial literacy	7.73	6.98	7.35
C5: Access to formal credit	8.18	8.74	8.46
C6: Savings, investments, and insurance	5.97	8.03	7.00
C7: Income	6.62	7.07	6.84
C8: Consumption expenditures	5.47	6.79	6.13
C9: Enterprise development	5.76	6.61	6.18
C10: Livelihood diversification	7.03	7.72	7.38
C11: Education	7.17	8.05	7.61
C12: Health, hygiene, and sanitation	7.32	8.10	7.71
C13: Natural assets	6.05	6.75	6.40
C14: Physical assets	5.11	6.20	5.66
C15: Political empowerment	6.27	7.21	6.74
C16: Social empowerment	8.25	7.76	8.00
C17: Economic empowerment	8.37	8.06	8.21
C18: Personal well-being and personality development	7.60	8.00	7.80
C19: Intra-household bargaining power	7.20	7.73	7.46
C20: Social harmony	8.26	8.64	8.45

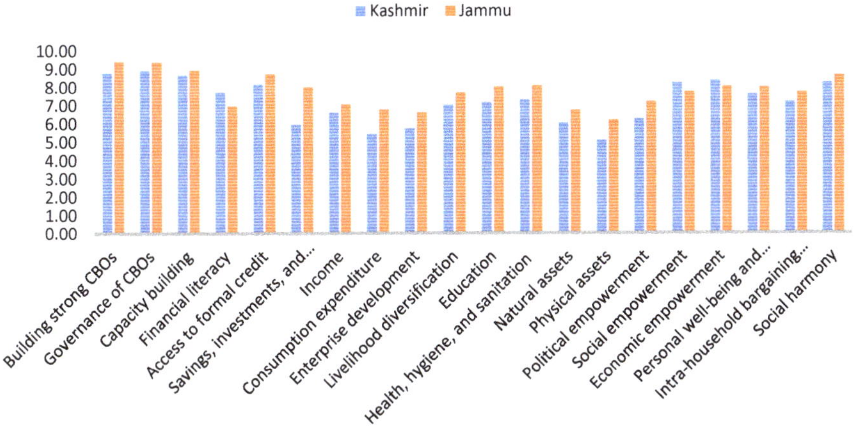

Figure 8. Mean values of significance for the regions of Jammu and Kashmir.

Having a thorough look at the values presented in Table 8, it was observed that the concepts C1, C2, C3, C15, C17, C18, and C20 were among concepts with the highest values for both degree centrality and degree of significance. Moreover, comparing these values with the corresponding degree centrality of the aggregated and the expert-based FCM model, we can verify that the key concepts mentioned earlier can be indeed selected as the most significant ones, and will be further used in the scenario analysis.

Table 8. The average degree of significance and degree centrality values for all key concepts.

Key Concept	Degree of Significance (Average Value)	Degree Centrality (Expert-Based)	Degree Centrality (Aggregated)
C1: Building strong CBOs	9.11	2.2	2.60
C2: Governance of CBOs	9.16	2.7	3.40
C3: Capacity building	8.82	1.5	1.80
C4: Financial literacy	7.35	0.6	1.60
C5: Access to formal credit	8.46	2.5	3.20
C6: Savings, investments, and insurance	7.00	2.0	2.50
C7: Income	6.84	5.4	8.80
C8: Consumption expenditures	6.13	0.6	1.60
C9: Enterprise development	6.18	2.3	3.10
C10: Livelihood diversification	7.38	1.7	3.20
C11: Education	7.61	2.5	4.90
C12: Health, hygiene, and sanitation	7.71	0.7	2.40
C13: Natural assets	6.40	0.6	0.80
C14: Physical assets	5.66	0.6	0.80
C15: Political empowerment	6.74	3.0	3.90
C16: Social empowerment	8.00	1.8	2.40
C17: Economic empowerment	8.21	1.9	2.50
C18: Personal well-being and personality development	7.80	3.1	4.00
C19: Intra-household bargaining power	7.46	1.9	2.40
C20: Social harmony	8.45	2.3	5.50

3.3. Aggregation Results

In this section, the results produced from the application of the two aggregation methods on the FCM models are presented. The FCM models constructed by every participant group (SHG, VO, CLF, and CRP) were aggregated using the two aggregation methods, the average, and the OWA. A collective FCM was produced from each of these methods. The aggregation process was delivered with the help of the OWA tool that was developed for this purpose.

From the comparative analysis that was conducted among the aggregated average-FCM, OWA-FCM, and experts-based FCM, it was observed that the minimum mean deviation value (= 0.12) was located between the OWA-FCM and the experts-based FCM. This means that the OWA-FCM model resembles the structure of the Expert-based FCM and consequently can present a similar performance to the model constructed by the experts.

3.4. Scenario Results

3.4.1. Scenario Development

For the scenario analysis, the researchers identified the most important concepts (called decision concepts) that affected the status of the system being examined. During the FCM exercise, we also recorded the degree of significance of every key concept on the basis of the perception of over 600 participants of SHGs, VOs, CLFs, and CRPs. During the FCM exercise, the participants also identified the most significant concepts in the system. On the basis of the FCM models prepared by the participants, we can infer that social harmony (C20), women's socio-economic empowerment (C16 and C17), and personal well-being and personality development (C18) were the most significant concepts, which are likely to have considerable impacts on the system. The results for the same analysis are presented in Figure 9. The first established approach in scenario planning was the selection of the most important concepts. The seven concepts that were selected are C1—"Building strong CBOs", C2—"Governance of CBOs", C3—"Capacity building", C5—"Access to formal credit", C15—"Political empowerment", C16—"Social empowerment", and C17—"Economic empowerment". These concepts, assigned by the programme participants and implementers, were selected properly as they were among concepts with the highest degree centrality, having both in/out-degree values, whereas their degree of significance was the highest among all key concepts (see Table 8). Thus, they could significantly influence the dynamics of the system. The selected scenarios with their concepts are briefly presented in the following Table 9.

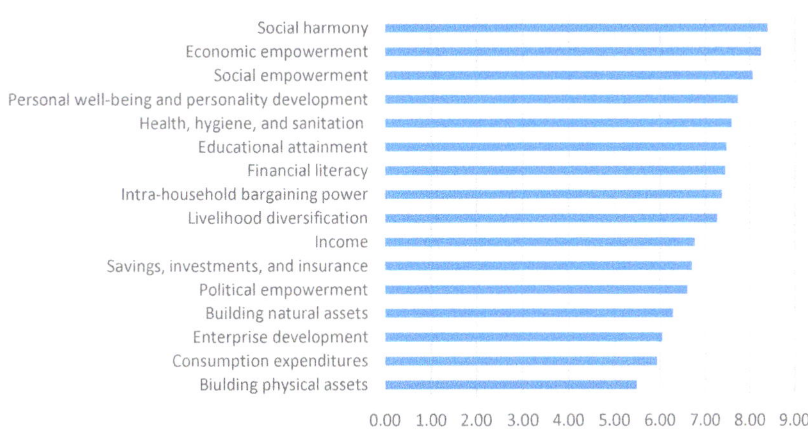

Figure 9. Most significant concepts of the DAY-NRLM (*Deendayal Antyodaya Yojana*-National Rural Livelihoods Mission) programme.

Table 9. The key concepts of each scenario.

Scenarios	Concepts		
Scenario 1 (S1)	C1: Building strong CBOs		
Scenario 2 (S2)	C2: Governance of CBOs		
Scenario 3 (S3)	C3: Capacity building		
Scenario 4 (S4)	C3: Capacity building	C5: Access to formal credit	
Scenario 5 (S5)	C2: Governance of CBOs	C3: Capacity building	
Scenario 6 (S6)	C1: Building strong CBOs	C3: Capacity building	
Scenario 7 (S7)	C15: Political empowerment		
Scenario 8 (S8)	C16: Social empowerment		
Scenario 9 (S9)	C17: Economic empowerment		

- Scenario 1 examined the effects of building CBOs (C1), whereas scenario 2 presented the effects of governance of CBOs (C2) in terms of SHGs, VO, and CLFs.
- Scenario 3 presented the effects of capacity building of the CBOs (C3) in terms of governance and management.
- Scenario 4 highlighted the effects of capacity building of the CBOs (C3) in terms of governance and management, along with access to formal credit (C5) in terms of micro-finance and SHG-bank linkage.
- Scenario 5 showed the effects of governance of CBOs (C2) in terms of SHGs, VO, and CLFs, along with the capacity building of the CBOs (C3) in terms of governance and management.
- Scenario 6 illustrated the effects of building strong CBOs (C1) in terms of SHGs, VO, and CLFs, and capacity building of the CBOs (C3) in terms of governance and management.
- Scenarios 7 to 9 highlighted the effects of political (C15), social (C16), and economic (C17) empowerment of women, respectively. These empowerments represent political inclusion, political justice, participation in various village-level committees, savings, financial self-sufficiency, universal social mobilization, and social inclusion, among others.

3.4.2. FCM-Based Simulations/Scenario Analysis

FCM-based simulations can offer a deeper understanding of the concepts' behavior and their relations in terms of how one concept affects others. The researchers conducted FCM-based simulations/scenario analyses for the respective case study. The simulation process was performed by "clamping" the initial values of the key concepts one-by-one (Equation (14)). This outcome was compared against a baseline scenario where the system (output vector) reached the steady-state through clamping all the initial values to zero. Exploring the dynamic change of concepts' values between the baseline steady-state and outcome of the clamping procedure enabled quantitative interpretation of the impact of the key concepts on the system. The simulation process entailed the application of a sigmoid function with lambda = 1 as a threshold function on the adjacency matrix after it was multiplied with the input vector. The process was iterated until the system reached a steady-state. The FCMWizard, a web-based software tool, was used for the simulation purposes, as it has the unique ability to construct an FCM using data that come from experts or stakeholders' knowledge and can perform simulations for different possible scenarios, in different scientific domains, using a very intuitive graphical user interface [31]. The impact of the conducted scenarios on the selected decision concepts was examined, further identifying which key concepts affect the final deliverables of the program.

The scenario analysis performed simulations for the selected nine scenarios (Table 9). For example, scenario 1 (S1) was devoted to increasing the concept C1—"Building strong CBOs by "clamping" it to one, whereas scenarios 2 (S2) and 3 (S3) studied the effects of the concepts C2—"Governance of CBOs" and C3—"Capacity building" by clamping the values of these concepts to one. The nine scenarios were conducted with the two aggregated collective FCMs, average-FCM and OWA-FCM. The expert-based FCM, which was constructed by the experts, was considered as the benchmark model that would help the researchers to further investigate the usefulness, importance, and superiority of the proposed OWA aggregation method against the average aggregation method. The results for scenario analysis for the two aggregated groups, for both average-FCM and OWA-FCM, as well as the expert-based FCM, were illustrated in the following figures. The scenario results for the expert-based FCM model, along with the corresponding scenario results for the OWA aggregated FCM, are illustrated in Table A1 in Appendix A.

Figure 10 illustrates all three FCMs (average, OWA, and expert-based), the deviation from the steady-state for all concepts after the nine scenarios had been conducted. The following figures (Figures 11 and 12) illustrate the outcomes regarding the percentage of change for certain key concepts, for all performed scenarios, with respect to FCMs.

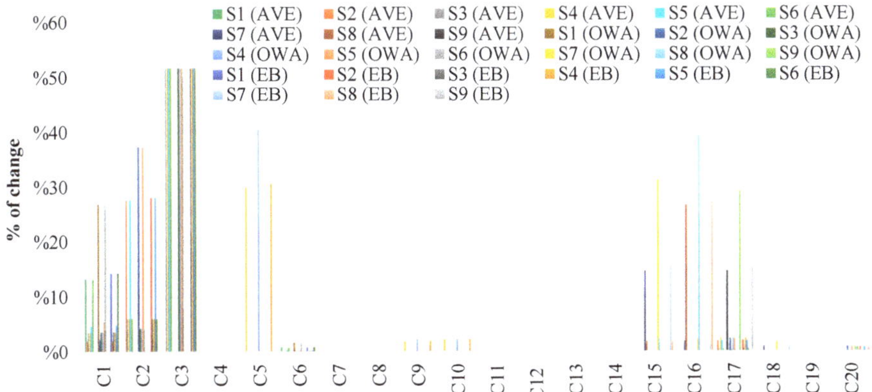

Figure 10. Scenario analysis considering deviations from the steady state for all FCMs considering confidences and links. (AVE is the abbreviation of Average, EB is the abbreviation of Expert-based).

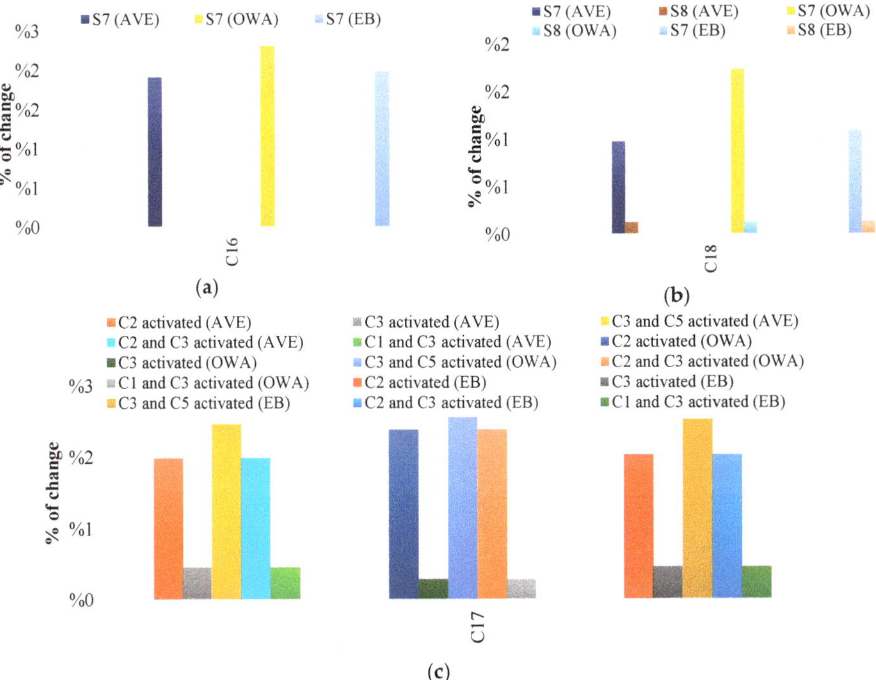

Figure 11. Percentage of change for decision concepts (**a**) C16, (**b**) C18, and (**c**) C17 when all scenario concepts were clamped to one for the expert-based FCM (AVE is the abbreviation of Average, EB is the abbreviation of Expert-based).

Figure 13 depicts the corresponding results of the deviation from the steady-state when FCMs were used in the scenario analysis. Figures 14 and 15 illustrate the outcomes regarding the percentage of change for decision concepts when the concepts for each scenario were clamped to one.

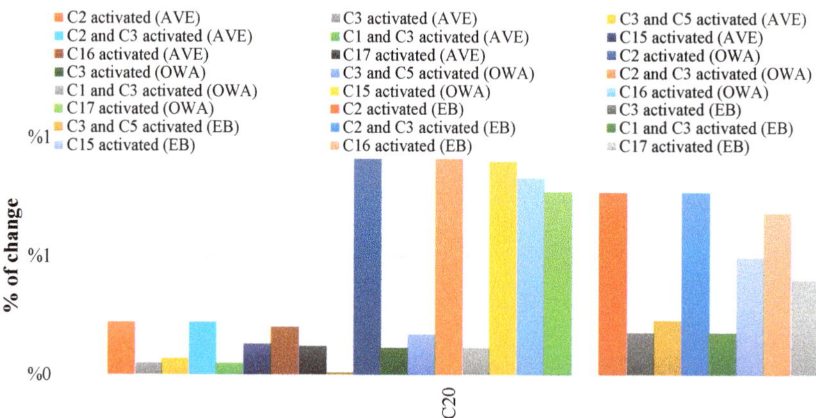

Figure 12. Decision concept C20 (social harmony) percentage of change when the key concepts of each devoted scenario were clamped to one. (AVE is the abbreviation of Average, EB is the abbreviation of Expert-based).

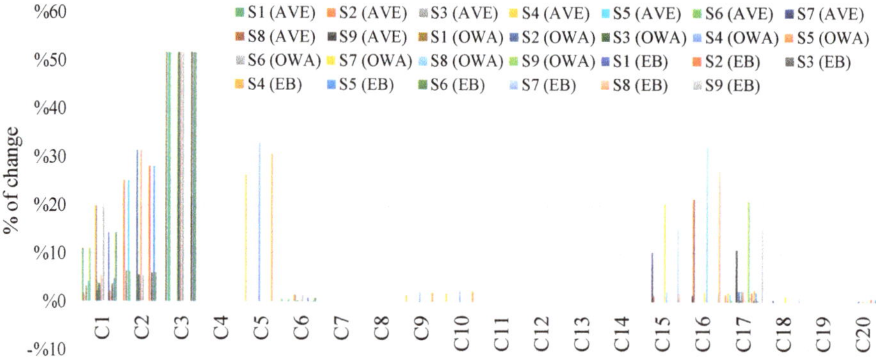

Figure 13. Scenario Analysis for all FCMs (average, OWA, expert) considering links. (AVE is the abbreviation of Average, EB is the abbreviation of Expert-based).

The following table (Table 10) briefly presents the impact that the examined scenarios had on the four decision concepts of the system.

Table 10. Scenarios mainly affecting decision concepts.

Decision Concepts	Scenarios Mainly Affecting Decision Concepts
C16	C15 activated (S7).
C17	C2 activated (S2), C2 and C3 activated (S5), C3 and C5 activated (S5).
C18	C15 activated (S7), C16 activated (S8).
C20	C2 activated (2), C2 and C3 activated (S5), C15 activated (S7), C16 activated (S8), C17 activated (S9).

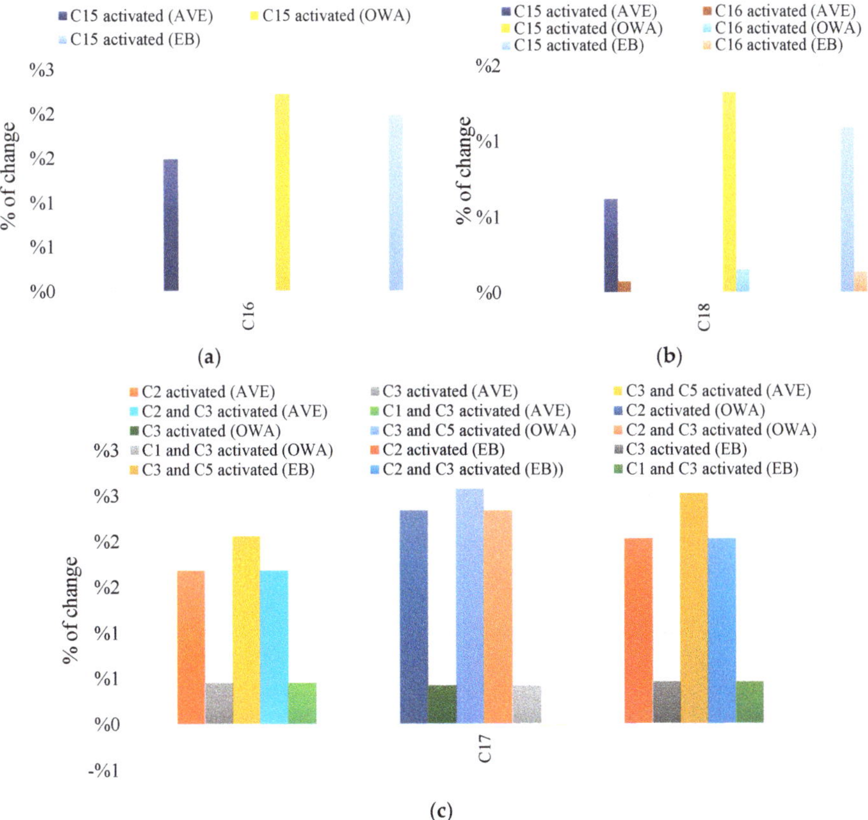

Figure 14. Percentage of change for decision concepts (**a**) C16, (**b**) C18, and (**c**) C17 when all scenario concepts were clamped to one, compared to the initial steady state (baseline scenario), considering links. (AVE is the abbreviation of Average, EB is the abbreviation of Expert-based).

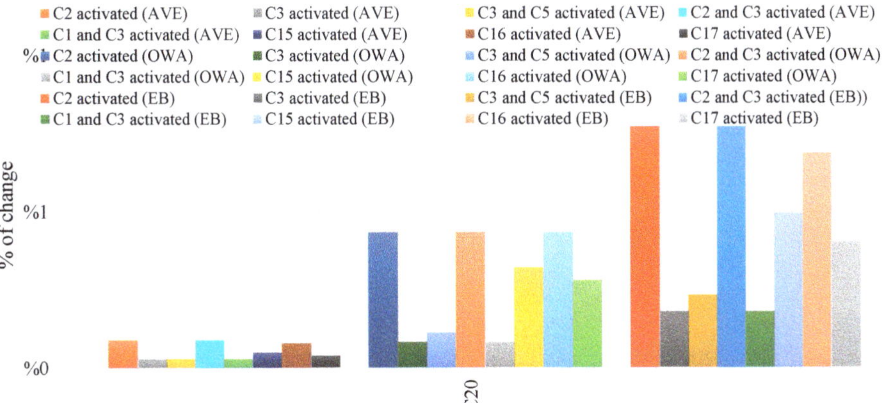

Figure 15. Decision concept C20 (social harmony) percentage of change when the key concepts of each devoted scenario were clamped to one. (AVE is the abbreviation of Average, EB is the abbreviation of Expert-based).

The observations that were drawn from the above figures and tables focus on the following two main points:

- The impact that certain key concepts have on other concepts and how they could affect the examined system, and;
- The performance of both the average and the OWA aggregation methods, comparing their outcomes with the expert-based FCM model. With respect to concepts' potential in affecting the state of the system, the following considerable observations emerged:

  i The decision concept C16—"Social empowerment" was solely affected by C15—"Political empowerment" for all FCMs (average, OWA, and expert). Moreover, women's personal well-being and personality development (decision concept C18) increased when more political and social empowerment (decision concepts C15 and C16) were offered to them.

  ii Furthermore, it was observed that the key concept C2 (Governance of CBOs), as well as the combinations C2 (Governance of CBOs) and C3 (Capacity building), and C3 (Capacity building) and C5 (Access to formal credit), had the highest impact in the decision concept C17—"Economic empowerment" for all collective FCMs, showing a significant increase in C17, particularly when the OWA aggregation method was applied.

  iii It also emerged from the above figures and Table 10, that the increase of social harmony (C20) was directly connected to the increase of the following key concepts: C2—"Governance of CBOs", the combination of C2—"Governance of CBOs" and C3—"Capacity building", as well as the concepts C15—" Political empowerment", C16—"Social empowerment" and C17—"Economic empowerment". Results of FCM-based simulations revealed that impacts of the DAY-NRLM programme could be realised better if strong institutions are built.

  iv Overall, the concepts C2, C3, C5, C15, C16, and C17 had a significant impact in the policy-making and strategic planning of socio-economic sustainability of poor rural communities because they presented considerably higher deviations from the steady-state than the rest of the concepts (see Figures 10 and 13). To further check the validity of the outcomes, a sensitivity analysis regarding the impact that these key concepts had on decision concept C20, for all three FCMs (average, OWA, and expert-based), was conducted, and the corresponding results are presented in Appendix B (Figures A1–A3). There seemed to be an influence from the absence of political, social, and economic empowerment, as well as from the lack of governance of CBOs, which corresponded to the concepts C2, C15, C16, and C17, affecting the community's social harmony (C20).

Concerning the performance of the two examined aggregation methods, the following important observations were extracted in particular after a careful analysis of the tables and figures above.

i The OWA aggregation method seemed to be superior to the average aggregation method, in terms of participatory modelling, when a large number of participants were involved. After a thorough view of Figure 12, Figure 13, Figure 14, Figure 15, where the scenario analysis of the three different FCMs was conducted, it can be concluded that in most cases, the deviations were higher when the OWA aggregation method was applied, compared with those resulting from the average aggregation method.

ii As mentioned before, we considered the expert-based FCM model as the benchmark model, as experts' opinions produced it. When the OWA-FCM model was compared to the expert-based FCM model, we noticed that the results on decision making were better or similar to those regarding the experts' FCM model. Thus, it can be concluded that the OWA-FCM model resembles the structure of the expert-based FCM model.

iii Overall, for analysis and decision-making, the OWA aggregation method was proven to be proper for policy-making and strategic decision planning, considering a large number of maps, outperforming the average aggregation method.

Some limitations of the proposed methodology that need to be considered are (i) the lack of imprecision of human perception in fuzzy form in FCM representation, as this approach cannot deal with aggregating fuzzy values of multiple FCMs into a collective FCM, and (ii) the weakness in coping with complex FCMs where a large number of concepts and weights are assigned by many participants. In this case, it is difficult to accurately define the learning parameters in OWA and handle the aggregated weighted connections.

3.5. Discussion on Scenario Results

The scenarios 1 to 6 examine how certain key concepts of the DAY-NRLM programme such as strong CBOs, good governance within CBOs, better capacity building of communities and CBOs, as well as access to formal credit, would help to achieve the final objectives of the programme, that is, alleviation of socio-economic poverty and better quality of life. In particular, increased access to formal credit and good governance would empower the SHG women economically, politically, and socially, as well as increase social harmony in the community. Consequently, income and savings would increase, and that would lead to an increase in consumption expenditure, livelihood diversification, and enterprise development. Higher-income will lead to better access to education for women and their children, consequently developing their personality, personal well-being, and overall socio-economic status.

Scenarios 7 to 9 highlight the effects of political (C15), social (C16), and economic (C17) empowerment of women, respectively. These empowerments represent political inclusion, political justice, participation in various village-level committees, savings, financial self-sufficiency, universal social mobilization, and social inclusion, among others. These scenarios also show an increase in income and savings, which would further lead to a rise in consumption expenditure, livelihood diversification, and enterprise development. Increased income will lead to better access to education for SHG women and their children; it would consequently develop the personality and personal well-being of SHG women. Better education will improve their intra-household bargaining power and health, hygiene, and sanitation. However, the result showed that empowerment alone is inadequate, and hence building strong CBOs and better access to formal credit are essential.

The outcomes of the scenario analysis highlight the importance of the simultaneous implementation of multiple concepts for the development of SHG members. Enhancing the capacities of SHG members, good governance within the CBOs and micro-finance through high-quality CBOs comprise the main aspects that should be taken into consideration in the examined case. Access to micro-finance and higher income will help community members to diversify their livelihood options and develop small enterprises. As a result, women empowerment and social safety nets will emerge, and women will improve their education, health, and develop their personality. All of the above factors will help poor rural communities and their members to alleviate socio-economic poverty while improving social resilience and promoting economic stability.

However, there is a need to incorporate resource efficiency in local and collective businesses, which can reduce pressures and impacts on the environment while contributing to socio-economic development and human well-being [10,11]. A shift towards the circular economy could translate into significant changes in people's lives [14]. Worldwide, small and medium enterprises are trying to move towards circular business models and solutions; however, the lack of consumer interest and awareness along with the lack of support from demand networks prevent the implementation of green innovations and act as the main obstacle for a transition towards the circular economy [14].

Several concepts identified in this study have the potential to incorporate the circular economy approach. The characteristics of those concepts that can influence communities' perceptions and attitudes towards circular solutions could include (C8) consumption—encouraging a non-materialistic environment among households and communities, supporting decisions to buy refurbished products over new ones, and increasing longevity of purchased products; (C9) enterprise development—building green enterprises, reusing/repairing/recycling resources at various levels, and more focus on product quality and service offering; (C10) livelihood diversification—a shift towards green livelihoods, and building farm and non-farm livelihood portfolios; (C12) natural assets—sustainable use and management of water and land resources, soil nutrient management through organic fertilisers, composting, and mulching, among others, and sustainable livestock management; and (C13) health, hygiene, and sanitation—changing consumer behavior, waste management at various levels, that is, households and industries, through reduction, reusing, and recycling, and wastewater treatment.

4. Conclusions

This study identified the key- and respective sub-concepts aimed at eliminating rural poverty in the developing economy. The CBOs are likely to allow the poor to overcome the social, financial, and economic exclusion responsible for perpetuating poverty. Social mobilisation and building institutions of the poor, including them in conventional financial institutions, the promotion of livelihood diversification, convergence with other development programmes, and social development are critical components designed to address the exclusion of the rural poor, eliminating their poverty and bringing them into the economic and social mainstream. The DAY-NRLM programme created the necessary implementation architecture, financial and management systems, and a conducive environment for the CBO-led model for the promotion of livelihoods and poverty alleviation. The CBOs show potential for long-term sustainability, as the programme has generated large social capital in the form of CRPs and community cadres, and thus it is likely that they will be able to sustain and nurture the community institutions.

The pathways of approaching a circular economy are strongly connected with the existing social, economic, and political systems, which could vary for every country. The following are some recommendations for moving towards a circular economy: (i) introduction of effective policy measures in order to enhance productivity and efficiency of resources; (ii) focus on research and development within enterprises to increase resource productivity and product longevity; (iii) strengthen education to raise awareness about use and limits of resources; (iv) change lifestyles to develop sustainable consumption behavior; and (v) improve communication between policymakers, communities, and businesses/industries.

Furthermore, this study also deals with the contribution of the OWA-FCM aggregation method by learning OWA operator weights in the participatory modelling domain. The innovation of this work lies in the fact that the strengths of the relationship were calculated with the proposed aggregation approach and further compared with the weights of the average method and those assigned by the experts. The results showed that the OWA-FCM resembles the structure of the expert-based FCM and, in most cases, showcased an improved performance compared to the model constructed by the experts. The FCM-based scenarios try to model the situation of livelihood enhancement through building self-managed and sustainable institutional platforms along with promoting social resilience and economic stability of the rural poor women. What makes this study important with regards to the scenario analysis results is that the OWA-FCM method presents similar trends to the other two aggregation methods examined with regards to the impact that certain key concepts such as political, social, and economic empowerment of women can have on other important key concepts such as increasing social harmony in the community. It was well demonstrated in the study that there was consistency among the impacts of the model of certain key concepts on other components, which makes the proposed method significant for policymakers.

Moreover, the proposed aggregation method can be applied by policymakers for strategic decisions in various scientific domains, validating its generic applicability and convenience when a significantly large number of experts and/or stakeholders are involved in designing FCMs. This new aggregation method, when combined with FCM-based simulations, can facilitate the preparation of more appropriate, equitable, and effective policy scenarios and responses, including shifting investment, production, distribution, and consumption towards more sustainable approaches, and the development of better governance capacities at multiple scales.

Author Contributions: Conceptualisation, K.P. and P.K.S.; data collection, P.K.S. and H.C., methodology, K.P., P.K.S., and E.P.; software, K.P.; validation, K.P., P.K.S., and H.C.; investigation, K.P.; resources, P.K.S. and H.C.; data curation, P.K.S.; writing—original draft preparation, K.P. and P.K.S.; writing—review and editing, K.P., P.K.S., E.P., H.C., and D.B.; visualisation, K.P., P.K.S., and E.P.; supervision, E.P., D.B., and G.S. All authors have read and agreed to the published version of the manuscript.

Funding: This research received no external funding.

Acknowledgments: Part of this work has received funding from the Ministry of Rural Development, Government of India, for providing financial support. We sincerely thank the participants who took part in the research.

Conflicts of Interest: The authors declare no conflict of interest.

Appendix A

Table A1. Scenario results (initial and final value) for each concept, for the expert-based FCM.

Key Concept	Scenario 1		Scenario 2		Scenario 3		Scenario 4		Scenario 5		Scenario 6		Scenario 7		Scenario 8		Scenario 9	
	Initial Value	Final Value	Initial Value	Final Value	Initial Value	Final Value	Initial Value	Final Value	Initial Value	Final Value	Initial Value	Final Value	Initial Value	Final Value	Initial Value	Final Value	Initial Value	Final Value
C1	1	1	0	0.7764	0	0.9072	0	0.9072	0	0.9072	1	1	0	0.7763	0	0.776	0	0.7763
C2	0	0.659	1	1	0	0.8281	0	0.8281	1	1	0	0.828	0	0.659	0	0.659	0	0.659
C3	0	0	0	0	1	1	1	1	1	1	1	1	0	0	0	0	0	0
C4	0	0	0	0	0	0	0	0	0	0	0	0	0	0	0	0	0	0
C5	0	0.659	0	0.659	0	0.659	1	1	0	0.659	0	0.659	0	0.659	0	0.659	0	0.659
C6	0	0.8959	0	0.8959	0	0.9051	0	0.9052	0	0.9051	0	0.905	0	0.8959	0	0.896	0	0.8959
C7	0	0.9537	0	0.9537	0	0.954	0	0.954	0	0.954	0	0.954	0	0.9537	0	0.954	0	0.9537
C8	0	0.8718	0	0.8718	0	0.8718	0	0.8719	0	0.8718	0	0.872	0	0.8718	0	0.872	0	0.8718
C9	0	0.8562	0	0.8562	0	0.8563	0	0.8563	0	0.8563	0	0.856	0	0.8562	0	0.856	0	0.8562
C10	0	0.8516	0	0.8516	0	0.8516	0	0.8517	0	0.8516	0	0.852	0	0.8516	0	0.852	0	0.8516
C11	0	0.8908	0	0.8908	0	0.8908	0	0.8908	0	0.8908	0	0.891	0	0.8908	0	0.891	0	0.8908
C12	0	0.8912	0	0.8912	0	0.8914	0	0.8914	0	0.8914	0	0.891	0	0.8912	0	0.891	0	0.8912
C13	0	0	0	0	0	0	0	0	0	0	0	0	0	0	0	0	0	0
C14	0	0	0	0	0	0	0	0	0	0	0	0	0	0	0	0	0	0
C15	0	0.8663	0	0.8663	0	0.8663	0	0.8663	0	0.8663	0	0.866	1	1	0	0.866	0	0.8663
C16	0	0.7851	0	0.7851	0	0.7851	0	0.7851	0	0.7851	0	0.785	0	0.7852	1	1	0	0.7851
C17	0	0.8483	0	0.8483	0	0.8641	0	0.8641	0	0.8641	0	0.864	0	0.8483	0	0.848	1	1
C18	0	0.877	0	0.877	0	0.877	0	0.877	0	0.877	0	0.877	0	0.877	0	0.877	0	0.877
C19	0	0.8806	0	0.8805	0	0.8806	0	0.8806	0	0.8806	0	0.881	0	0.8806	0	0.881	0	0.8805
C20	0	0.9769	0	0.9769	0	0.9794	0	0.9794	0	0.9794	0	0.979	0	0.9769	0	0.977	0	0.9769

Appendix B

Sensitivity Analysis:

The following sensitivity analysis reveals the relative changes in social harmony (C20) under different values of key concepts. It was observed that there was a strong influence from the absence of political and social empowerment, as well as from the absence of governance of CBOs. This conveyed the notion that the key factors affecting social harmony were C2 and C15–C17. Sensitivity analysis was also performed for the key concepts C11 and C12, but no influences were observed from their deviations. This means that they were reluctant in playing a significant role in the decision concept of social harmony.

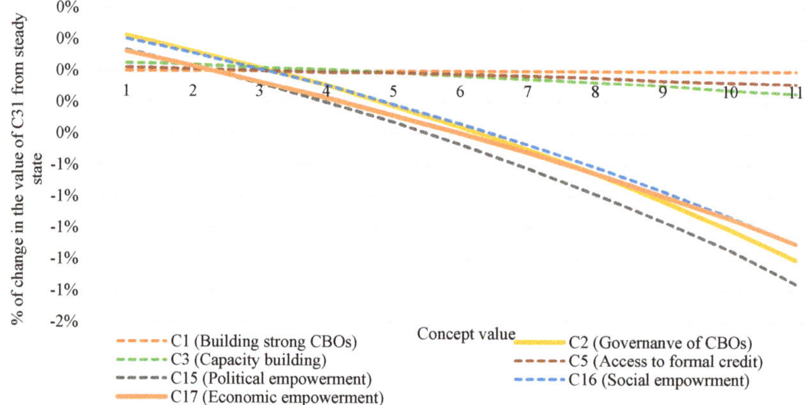

Figure A1. Sensitivity analysis for the concept C20 regarding the average FCM.

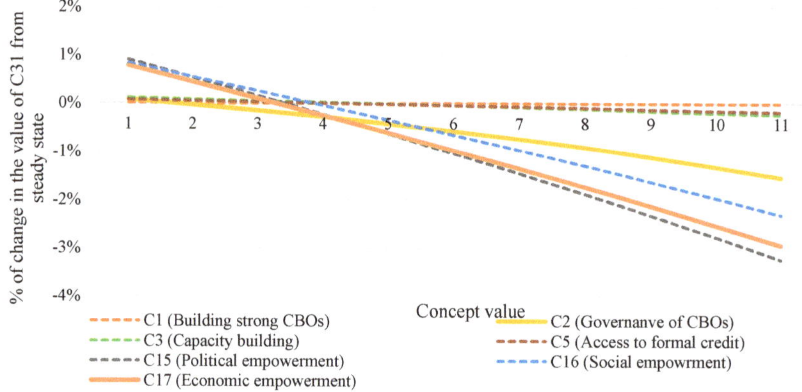

Figure A2. Sensitivity analysis for the concept C20 regarding the OWA-FCM.

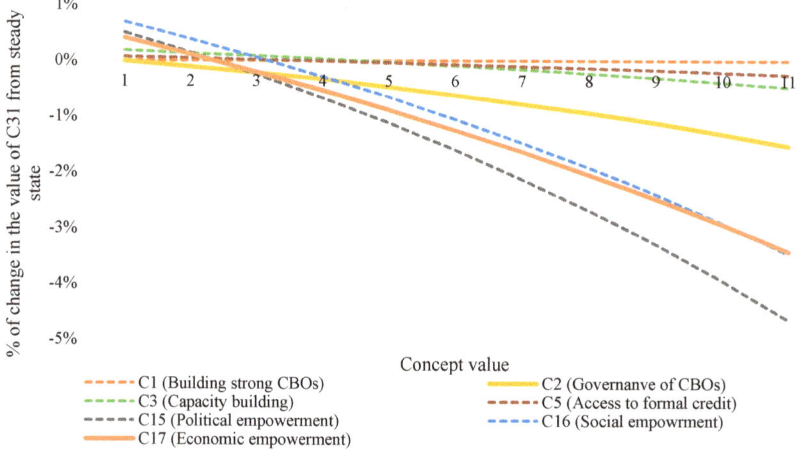

Figure A3. Sensitivity analysis for the concept C20 regarding the initial expert-based FCM.

References

1. IPCC. Global warming of 1.5 °C. In *An IPCC Special Report on the Impacts of Global Warming of 1.5 °C Above Pre-Industrial Levels and Related Global Greenhouse Gas Emission Pathways, in the Context of Strengthening the Global Response to the Threat of Climate Change, Sustainable Development, and Efforts to Eradicate Poverty*; Masson-Delmotte, V., Zhai, P., Pörtner, H.-O., Roberts, D., Skea, J., Shukla, P.R., Pirani, A., Moufouma-Okia, W., Péan, C., Pidcock, R., Eds.; IPCC: Geneva, Switzerland, 2018; in press.
2. Landrigan, P.J.; Fuller, R.; Acosta, N.J.R.; Adeyi, O.; Arnold, R.; Basu, N.; Baldé, A.B.; Bertollini, R.; Bose-O'Reilly, S.; Boufford, J.I.; et al. The Lancet Commission on pollution and health. *Lancet* **2018**, *391*, 462–512. [CrossRef]
3. IPCC. Climate Change 2014: Impacts, Adaptation, and Vulnerability. Part B: Regional Aspects. In *Contribution of Working Group II to the Fifth Assessment Report of the Intergovernmental Panel on Climate Change*; Barros, V.R., Field, C.B., Dokken, D.J., Mastrandrea, M.D., Mach, K.J., Bilir, T.E., Chatterjee, M., Ebi, K.L., Estrada, Y.O., Genova, R.C., Eds.; Cambridge University Press: Cambridge, UK; New York, NY, USA, 2014; p. 688.
4. Steffen, W.; Rockström, J.; Richardson, K.; Lenton, T.M.; Folke, C.; Liverman, D.; Summerhayes, C.P.; Barnosky, A.D.; Cornell, S.E.; Crucifix, M.; et al. Trajectories of the Earth System in the Anthropocene. *Natl. Acad. Sci. Proc. Natl. Acad. Sci.* **2018**, *115*, 8252–8259. [CrossRef] [PubMed]
5. Steffen, W.; Richardson, K.; Rockstrom, J.; Cornell, S.E.; Fetzer, I.; Bennett, E.M.; Biggs, R.; Carpenter, S.R.; de Vries, W.; de Wit, C.A.; et al. Planetary boundaries: Guiding human development on a changing planet. *Science* **2015**, *347*, 1259855. [CrossRef] [PubMed]
6. Steffen, W.; Persson, A.; Deutsch, L.; Zalasiewicz, J.; Williams, M.; Richardson, K.; Crumley, C.; Crutzen, P.; Folke, C.; Gordon, L.; et al. The Anthropocene: From global change to planetary stewardship. *Ambio* **2011**, *40*, 739–761. [CrossRef]
7. Algunaibet, I.M.; Pozo, C.; Galán-Martín, Á.; Huijbregts, M.A.J.; Dowell, N.M.; Guillén-Gosálbez, G. Powering sustainable development within planetary boundaries. *Energy Environ. Sci.* **2019**, *12*. [CrossRef]
8. O'Neill, D.W.; Fanning, A.L.; Lamb, W.F.; Steinberger, J.K. A good life for all within planetary boundaries. *Nat. Sustain.* **2018**, *1*, 88–95. [CrossRef]
9. Raworth, K.A. *Safe and Just Space for Humanity: Can We Live Within the Doughnut*; Oxfam: Oxford, UK, 2012; Available online: https://www.oxfam.org/sites/www.oxfam.org/files/dp-a-safe-and-just-space-for-humanity-130212-en.pdf (accessed on 30 November 2019).
10. Potting, J.; Hekkert, M.P.; Worrell, E.; Hanemaaijer, A. *Circular Economy: Measuring Innovation in the Product Chain*; PBL Netherlands Environmental Assessment Agency: Hague, The Netherlands, 2017; Available online: http://www.pbl.nl/sites/default/files/cms/publicaties/pbl-2016-circular-economy-measuringinnovation-in-product-chains-2544.pdf (accessed on 30 November 2019).
11. Stahel, W.R. The circular economy. *Nature* **2016**, *531*, 435. [CrossRef]
12. Singh, M.P.; Chakraborty, A.; Roy, M. Developing an extended theory of planned behavior model to explore circular economy readiness in manufacturing MSMEs, India. *Resour. Conserv. Recycl.* **2018**, *135*, 313–322. [CrossRef]
13. Kok, L.; Wurpel, G.; Ten Wolde, A. *Unleashing the Power of the Circular Economy*; Report by IMSA Amsterdam for Circle Economy; The Circle Economy: Amsterdam, The Netherlands, 2013; p. 48. Available online: http://www.imsa.nl/uploads/Unleashing_the_Power_of_the_Circular_Economy-Circle_Economy.pdf (accessed on 20 December 2019).
14. Camacho-Otero, J.; Boks, C.; Pettersen, I.N. Consumption in the Circular Economy: A Literature Review. *Sustainability* **2018**, *10*, 2758. [CrossRef]
15. United Nations. Transforming Our World: The 2030 Agenda for Sustainable Development. The UN General Assembly. A/RES/70/1. 2015. Available online: https://www.refworld.org/docid/57b6e3e44.html (accessed on 30 November 2019).
16. World Bank. *The World Bank: Annual Report 2015*; The World Bank: Washington, DC, USA, 2015.
17. Yalegama, S.; Chileshe, N.; Ma, T. Critical success factors for community-driven development projects: A Sri Lankan community perspective. *Int. J. Proj. Manag.* **2016**, *34*, 643–659. [CrossRef]
18. Chakrabarti, A.; Dhar, A. Social funds, poverty management and subjectification: Beyond the World Bank approach. *Camb. J. Econ.* **2013**, *37*, 1035–1055. [CrossRef]
19. Voinov, A.; Bosquet, F. Modeling with stakeholders. *Environ. Model. Softw.* **2010**, *25*, 1268–1281. [CrossRef]

20. Penn, A.S.; Knight, C.J.; Lloyd, D.J.; Avitabile, D.; Kok, K.; Schiller, F.; Woodward, A.; Druckman, A.; Basson, L. Participatory development and analysis of a fuzzy cognitive map of the establishment of a bio-based economy in the Humber region. *PLoS ONE* **2013**, *8*, e78319. [CrossRef] [PubMed]
21. Kosko, B. Fuzzy cognitive maps. *Int. J. Man Mach. Stud.* **1986**, *24*, 65–75. [CrossRef]
22. Stach, W. Learning and Aggregation of Fuzzy Cognitive Maps—An Evolutionary Approach. Ph.D. Thesis, University of Alberta, Edmonton, AB, Canada, 2010. Available online: http://gradworks.umi.com/NR/62/NR62921.html (accessed on 3 December 2019).
23. Aguilar, J. A survey about fuzzy cognitive Maps papers. *Int. J. Comput. Cogn.* **2005**, *3*, 27–33.
24. Kosko, B. *Fuzzy Engineering*; Prentice-Hall: New York, NY, USA, 1997.
25. Papageorgiou, E.; Stylios, C. *Fuzzy Cognitive Maps. Handbook of Granular Computing*; John Wiley and Son Ltd.: Hoboken, NJ, USA; Publication Atrium: Chichester, UK, 2008; pp. 755–774.
26. Mezei, J.; Sarlin, P. Aggregating expert knowledge for the measurement of systemic risk. *Decis. Support Syst.* **2016**, *88*, 38–50. [CrossRef]
27. Yager, R.R. On ordered weighted averaging aggregation operators in multi-criteria decision making. *IEEE Trans. Syst. Man Cybern.* **1988**, *18*, 183–190. [CrossRef]
28. Gray, S.; Gagnon, A.; Gray, S.; O'Dwyer, B.; O'Mahony, C.; Muir, D.; Devoy, R.J.N.; Falaleeva, M.; Gault, J. Are coastal managers detecting the problem? Assessing stakeholder perception of climate vulnerability using Fuzzy Cognitive Mapping. *Ocean Coast Manag.* **2014**, *94*, 74–89. [CrossRef]
29. Zhenbang, L.; Lihua, Z. Advanced Fuzzy Cognitive Maps Based on OWA Aggregation. *Int. J. Comput. Cogn.* **2007**, *5*, 31–34.
30. Leyva-Vázquez, M.; Pérez-Teruel, K.; John, R. A Model for Enterprise Architecture Scenario Analysis Based on Fuzzy Cognitive Maps and OWA Operators. In Proceedings of the IEEE Conference, Cholula, Mexico, 26–28 February 2014.
31. Papageorgiou, E.; Papageorgiou, K.; Dikopoulou, Z.; Mouhrir, A. A web-based tool for Fuzzy Cognitive Map Modeling. In Proceedings of the 9th International Congress on Environmental Modelling and Software. (iEMSs), Fort Collins, CO, USA, 25–29 June 2018.
32. Papageorgiou, E. A new methodology for Decisions in Medical Informatics using fuzzy cognitive maps based on fuzzy rule-extraction techniques. *Appl. Soft Comput.* **2011**, *11*, 500–513. [CrossRef]
33. Kosko, B. *Neural Networks and Fuzzy Systems*; Prentice Hall: Englewood Cliffs, NJ, USA, 1992.
34. Miao, Y.; Liu, Z.; Siew, C.; Miao, C. Dynamical cognitive network–An extension of fuzzy cognitive map. *IEEE Trans. Fuzzy Syst.* **2001**, *9*, 760–770. [CrossRef]
35. Papakostas, G.A.; Polydoros, A.S.; Koulouriotis, D.E.; Tourassis, V.D. Training Fuzzy Cognitive Maps by using Hebbian learning algorithms: A comparative study. In Proceedings of the IEEE International Conference on Fuzzy Systems, Taipei, Taiwan, 27–30 June 2011; pp. 851–858.
36. Papageorgiou, E.I.; Kontogianni, A. Using fuzzy cognitive mapping in environmental decision making and management: A methodological primer and an application. In *International Perspectives on Global Environmental Change*; Young, S., Ed.; InTech: London, UK, 2012; pp. 427–450.
37. Papageorgiou, E.I.; Salmeron, J.L. Methods and Algorithms for Fuzzy Cognitive Map-based Modeling. In *Fuzzy Cognitive Maps for Applied Sciences and Engineering*; Springer: Berlin/Heidelberg, Germany, 2014; Chapter 1; pp. 5–24.
38. Groumpos, P.; Stylios, C.D. Modeling supervisory control systems using fuzzy cognitive maps. *ChaosSolitons Fractals* **2000**, *11*, 329–336. [CrossRef]
39. Özesmi, U.; Özesmi, L.O. Ecological models based on people's knowledge: A multi-step fuzzy cognitive mapping approach. *Ecol. Model.* **2004**, *176*, 43–64. [CrossRef]
40. Diniz, F.H.; Kok, K.; Hoogstra-Klein, M.; Arts, B. Mapping future changes in livelihood security and environmental sustainability based on perceptions of small farmers in the Brazilian Amazon. *Ecol. Soc.* **2015**, *20*, 26. [CrossRef]
41. Singh, P.K.; Papageorgiou, K.; Chudasama, H.; Papageorgiou, E.I. Evaluating the Effectiveness of Climate Change Adaptations in the World's Largest Mangrove Ecosystem. *Sustainability* **2019**, *11*, 6655. [CrossRef]
42. Brown, R.G. *Smoothing, Forecasting and Prediction of Discrete Time Series*; Prentice-Hall: Englewood Cliffs, NJ, USA, 1963.
43. Singh, P.K.; Chudasama, H. Pathways for drought resilient livelihoods based on people's perception. *Clim. Chang.* **2017**, *140*, 179–193. [CrossRef]

44. Singh, P.K.; Chudasama, H. Assessing impacts and community preparedness to cyclones: a fuzzy cognitive mapping approach. *Clim Chang.* **2017**, *143*, 337–354. [CrossRef]
45. Singh, P.K.; Nair, A. Livelihood vulnerability assessment to climatic variability and change using fuzzy cognitive mapping approach. *Clim Chang.* **2014**, *127*, 475–491. [CrossRef]
46. Tsadiras, A. Comparing the inference capabilities of binary, trivalent and sigmoid fuzzy cognitive maps. *Inf. Sci.* **2008**, *178*, 3880–3894. [CrossRef]
47. Papageorgiou, K.; Papageorgiou, E.I.; Singh, P.K.; Stamoulis, G. A software tool for FCM aggregation employing credibility weights and learning OWA operators. In Proceedings of the 10th International Conference on Information Intelligence, Systems and Applications (IISA), Patras, Greece, 15–17 July 2019; pp. 15–17. [CrossRef]
48. Papageorgiou, K.; Papageorgiou, E.; Mouhrir, A.; Stamoulis, G. Exploring OWA Operators For Aggregating Fuzzy Cognitive Maps. In Proceedings of the 12th International Conference on Information, Technology in Agriculture, Food and the Environment (EFITA), Rhodes island, Greece, 27–29 June 2019.
49. Kosko, B. Adaptive inference in fuzzy knowledge networks. In Proceedings of the First International Conference on Neural Networks, San Diego, CA, USA, 21–24 June 1987; Volume 2, pp. 261–268.
50. Amer, M.; Daim, T.U.; Jetter, A. Technology roadmap through fuzzy cognitive map-based scenarios: The case of wind energy sector of a developing country. *Technol. Anal. Strateg. Manag.* **2016**, *28*, 131–155. [CrossRef]
51. Alipour, M.; Hafezi, R.; Amer, M.; Akhavan, A. A new hybrid fuzzy cognitive map-based scenario planning approach for Iran's oil production pathways in the postesanction period. *Energy* **2017**, *135*, 851e64. [CrossRef]
52. Gray, S.; Gray, S.; Zanre, E. Fuzzy Cognitive Maps as representations of mental models and group beliefs: Theoretical and technical issues. In *Fuzzy Cognitive Maps for Applied Sciences and Engineering —From Fundamentals to Extensions and Learning Algorithms*; Elpiniki, I.P., Ed.; Springer Publishing: Heidelberg, Germany, 2014; pp. 29–48.

 *sustainability*

Article

Agricultural Workforce Crisis in Light of the COVID-19 Pandemic

Dionysis Bochtis [1,*]**, Lefteris Benos** [1]**, Maria Lampridi** [1]**, Vasso Marinoudi** [2]**, Simon Pearson** [2] **and Claus G. Sørensen** [3]

1   Institute for Bio-Economy and Agri-Technology (IBO), Centre of Research and Technology-Hellas (CERTH), 6th km Charilaou-Thermi Rd, GR 57001 Thessaloniki, Greece; e.benos@certh.gr (L.B.); m.lampridi@certh.gr (M.L.)
2   Lincoln Institute for Agri-Food Technology (LIAT), University of Lincoln, Lincoln LN6 7TS, UK; vm352@efb.gr (V.M.); SPearson@lincoln.ac.uk (S.P.)
3   Department of Engineering, Aarhus University, DK-8000 Aarhus C, Denmark; claus.soerensen@eng.au.dk
*   Correspondence: d.bochtis@certh.gr

Received: 10 August 2020; Accepted: 1 October 2020; Published: 5 October 2020

Abstract: COVID-19 and the restrictive measures towards containing the spread of its infections have seriously affected the agricultural workforce and jeopardized food security. The present study aims at assessing the COVID-19 pandemic impacts on agricultural labor and suggesting strategies to mitigate them. To this end, after an introduction to the pandemic background, the negative consequences on agriculture and the existing mitigation policies, risks to the agricultural workers were benchmarked across the United States' Standard Occupational Classification system. The individual tasks associated with each occupation in agricultural production were evaluated on the basis of potential COVID-19 infection risk. As criteria, the most prevalent virus transmission mechanisms were considered, namely the possibility of touching contaminated surfaces and the close proximity of workers. The higher risk occupations within the sector were identified, which facilitates the allocation of worker protection resources to the occupations where they are most needed. In particular, the results demonstrated that 50% of the agricultural workforce and 54% of the workers' annual income are at moderate to high risk. As a consequence, a series of control measures need to be adopted so as to enhance the resilience and sustainability of the sector as well as protect farmers including physical distancing, hygiene practices, and personal protection equipment.

Keywords: coronavirus; occupational health and safety; food security; resilience; control measures

---

1. Introduction

1.1. The Pandemic Background

On 30 January 2020, the World Health Organization (WHO) triggered their highest alert by announcing the coronavirus disease (COVID-19) as a public-health emergency of international concern. On 11 March 2020, COVID-19 was declared a pandemic. As the director-general of WHO explained: *"CO stands for corona, VI for virus, D for disease and 19 for the year the outbreak was first identified"*. COVID-19 is the infectious disease resulting from the severe acute respiratory syndrome coronavirus 2 (SARS-CoV-2) [1]. The virus can be transmitted during close contact between people via small respiratory droplets produced when an infected individual speaks, sneezes, or coughs. Furthermore, these droplets can contaminate surfaces. Common symptoms include dry cough and fever or mild symptoms such as nasal congestion, sore throat, loss of smell or taste as well as toes and fingers discoloration [1]. The virus can be asymptomatic, making COVID-19 control extremely challenging, as it can be passed on by individuals who might not notice that they have been infected. At the global level, governments have taken precaution measures to "flatten the curve", such as quarantine, lockdown, the isolation of infected individuals, travel restrictions, border shutdowns and social distancing [2,3]. However, these actions proved to have a detrimental effect on the economy leading to the economic recession and crisis [4].

COVID-19 has severely tested the resilience of supply chains. The effects of COVID-19 on agriculture, as in any sector, have not been manifested in full, while currently a second wave of the virus is impacting many countries. Key impacts on the food system up to now include the general population panic shopping and warehousing of durable food, including pasta, flour, beans and rice [5]. This led to empty shelves at supermarkets. Afraid of running out of domestic supplies, some countries were cautious and decided to close their borders. For example, Russia, Kazakhstan and Serbia temporarily banned exports of key staple foods [6,7]. In the same vein, European Union (EU) countries, such as France, suggested closing the borders of Europe until October 2020 [8]. According to the Food and Agricultural Organization of the United Nations (FAO), from the beginning of May (2020), the international prices of the major staple commodities, such as wheat and maize have dropped. Conversely, rice is the only staple product whose price has risen. This is attributed to export restrictions of Vietnam, which is a key supplier, until 1 May 2020 [9]. On the other hand, great disturbance of supply chains as a consequence of population "lockdowns" has provoked a global decline in demand across the food service sector, such as restaurants, open markets, catering and hotels [10]. Effective closure of food service segments has impacted all businesses across the supply chain including farms which provide the primary produce. To make matters worse, transport restrictions have hindered farmers' and fishers' ability to access markets, hence, limiting their productive capacities [9]. Disturbances downstream from farms can also cause accumulative surpluses, putting extra pressure on storage facilities, especially for highly perishable commodities.

1.2. Pandemic Effects on Agriculutral Sector

COVID-19 has impacted the agricultural workforce, especially the pool of seasonal agricultural workers. These are often migrant workers, typically employed in the crop harvesting, who use highly dexterous and physical skills [11,12]. Lockdowns and restrictions in the mobility of workers across borders contributed to labor shortages, mainly in countries that rely on seasonal workers. However, the ability of an agricultural system to exploit workers that can travel between workplaces constitutes a fundamental condition for its sustainability [13]. Unfortunately, emergency travel bans considerably decreased the available workforce. Moreover, no certainty exists that seasonal workers would like to work in countries that have been infected by COVID-19. Additionally, it was noted that many native workers fell ill or took care of sick members of the family or children, due to the closing of schools, further impacting the availability of seasonal personnel [14]. These consequences have particularly affected vegetable and fruit producers as well as garden nurseries and horticulture [15]. However, for many crops, the harvesting season is fixed and a deficiency of labor can result in production shortages in the food market and higher prices, making markets even more unforeseeable [16].

Owing to disruptions in logistics and transport services, COVID-19 lockdowns also impacted the provision of key intermediate products for farmers, such as pesticides, fertilizers and seeds. Additional supply chain checks and procedures resulted in delays to the transit of these products. Shortages or high prices in personal protection equipment from COVID-19 infection, such as hand sanitizers and face masks, caused additional delays and problems [5]. A representative example was China, where pesticide production declined suddenly after production plants shut down. Delays to the transport of these intermediate products can disturb supply chains for extended periods from 2020 and beyond [17].

In a nutshell, lockdown measures to contain the spread of COVID-19 caused a cascading effect on agricultural supply chains, especially of perishable products. In particular, a considerable decline in labor productivity, higher labor and transport costs, substantial income losses for farmers, food shortages and an increase in perishable products' prices, like vegetables and fruits for consumers, was observed during the first weeks [8]. As a means to document the existing situation in the agricultural sector, predicting the potential effects of the COVID-19 pandemic and suggesting measures to mitigate them, several studies have been conducted. Some of them analyzed solely the impact of the virus on agricultural production regarding the first infected by COVID-19 countries, namely China [18,19] and later Italy [7,20–22]. In contrast, some studies dealt with the agricultural sector of countries, which were later infected by the viral pandemic, including countries from the rest of Asia (India [23,24] and Iran [25]), Oceania (Australia [26] and New Zealand [27]), Europe (rest of EU [5,28] and the United Kingdom (UK) [29]) as well as America (Argentine [30], Peru [31], Canada [32–36] and the US [14,37,38]).

1.3. Policy Approaches to Mitigate the Negative Consequences of the COVID-19 Pandemic on Agriculture

Undoubtedly, the agricultural sector is one of the most precarious and unforeseeable sectors. This fact has become even more intense because of the COVID-19 pandemic outbreak. The major short-term concern was to keep farmworkers healthy. For this purpose, farmers, like the entire society, coped with unprecedented measures in order to contain the virus' spread. In a few words, precaution interventions, such as social distancing, travel restrictions, lockdown and self-isolation, proved to have a cascade effect on agriculture, introducing major limitations for farmworkers that have led to potentially devastating consequences. In the aftermath of these measures, the mobility of seasonal workers, especially that of migrant ones, was highly restricted, resulting in delays in harvesting and increased food losses, mostly affecting perishable goods. The governments of developed countries, including the US, UK, France, Germany, Spain and Italy, which highly rely on this labor force, urgently adopted strategies to avoid disturbances owing to the imminent labor shortage. Overall, these strategies seem to be altered from one day to another, mirroring the problem's depth.

In the EU, for example, the policies for mitigating the seasonal workers shortage amid the harvesting period can be briefly analyzed into four axes according to the study of Mitaritonna and Ragot [8]:

- Substituting seasonal migrant labor with domestic workers: websites were created to put unemployed individuals and part-time workers in touch with farmers. As a means to encourage this policy, the workers could combine unemployment benefits with the agricultural wage. Although this measure seemed to be successful at the first stage, attracting a plethora of applicants, the recruitment rates were very low, as the applicants wanted to return to their jobs as soon as possible. In addition to this, extra training costs arose, given that there was an important mismatch of the required skills.
- Applying deviations from labor laws so as to allow agricultural workers to work more: for instance, in France, workers were allowed to work also on Sundays and for more hours. In return, the hours worked further than a defined threshold were paid as overtime.
- Implementing very strict health measures during the reception of the seasonal migrant workers: to that end, seasonal migrants could enter Germany, for example, exclusively by plane and only when they were tested for COVID-19. Afterwards, for the first two weeks they should live and work separately from the other workers. Nevertheless, taking into account the high rates of COVID-19 cases in Europe, they faced the reasonable workers' fear of being infected by the virus if they came to work.
- Regularizing irregular migrants: even though the exact number of them is hard to assess, irregular migrants are working in the agricultural sector. Their assistance in such a labor shortage crisis would definitely be beneficial. However, this constitutes a controversial approach, particularly for countries having very restrictive migration policies.

The aforementioned strategies can help in the emergency situation characterized by a shortage of workers. It is very hard to single out which policy is the most effective, that is to say which combines the overcoming labor shortage with the lower costs.

Another important aspect is that health should be guaranteed throughout the food chain. To achieve this goal, measures like establishing biosecurity arrangements, enacting stricter employee health policies, using cashless transactions and gathering and communicating scientific evidence as soon as possible could help [39]. Limiting the COVID-19 spread in workplaces is another major concern. The example of the virus clusters in the meatpacking industry in California [14] reveals that COVID-19 may spread rapidly, thus reducing the availability of workers and leading to possible local lockdowns. More recently, workers from a meat processing industry in Greece [40] contracted COVID-19, with health officials deciding self-isolation for them in their home and closing the company. One solution to eradicate such problems would be working in shifts in order to avoid crowding with fewer people involved in the process by keeping safe distances and using personal protection like face masks, gloves and antiseptics.

1.4. Aim of the Study

In total, significant progress has been made in identifying the main problems caused by the first emergency measures to prevent the COVID-19 spread and the subsequent chain reaction in the food stocks, demand and prices. However, it is of major importance to assess the potential consequences of the COVID-19 pandemic on agricultural-related occupations. To our knowledge, no study exists in the relative literature on this topic. Towards this direction, the aim of the present investigation is to examine the above consequences and suggest ways to ensure the smooth operation of the agricultural sector in case of a second wave of COVID-19. To this end, since two clear sources of infection occur pertaining to the virus' transmission, namely the close proximity of workers and contaminated surfaces, the individual tasks (based on US Standard Occupational Classification system—SOC) are assessed with respect to these sources. There is an imperative necessity to manage potential COVID-19 resurgence, to protect workers and their jobs, as well as assure food supply and security.

2. Materials and Methods

In the absence of a methodology for such an analysis, a new methodology was developed. To meet this objective, each agricultural occupation was analyzed and characterized based on the individual tasks comprising this occupation and the corresponding potential risks.

For the analysis of the occupations, the employment and salaries data from the US Bureau of Labor Statistics (BLS) were implemented. For the standardization of occupations, the US 2018 Standard Occupational Classification (SOC) system was employed and more specifically, the eight-digit scheme of the Occupational Information Network (O*NET) classification system [41]. Overall, 17 occupations related to agriculture were considered for the current investigation. It should be stressed that the jobs involving aquaculture and logging activities were excluded from the present analysis. The investigated occupations are summarized in Table 1 along with their eight-digit code and the number of tasks they encompass.

In order to assess the effect of the pandemic on the total agricultural workforce and on the total budget allocated to the agricultural occupations' salaries, the distribution of each occupation in the above metrics has to be considered. By processing the data provided by the US Department of Labor statistics (data refer to May 2017), Figure 1a,b presents the annual budget and the workforce distribution among the selected occupations.

Table 1. The O*NET categorization of agricultural occupations and the corresponding codes along with the number of tasks they involve according to [41].

8-Digit O*NET Code	Occupation	No. of Tasks
11-9013.01	Nursery and Greenhouse Managers	20
11-9013.02	Farm and Ranch Managers	26
13-1074.00	Farm Labor Contractors	8
17-2021.00	Agricultural Engineers	13
19-1011.00	Animal Scientists	9
19-1013.00	Soil and Plant Scientists	20
19-4011.01	Agricultural Technicians	25
19-4011.02	Food Science Technicians	15
45-1011.07	First-Line Supervisors of Agricultural Crop and Horticultural Workers	24
45-1011.08	First-Line Supervisors of Animal Husbandry and Animal Care Workers	18
45-2011.00	Agricultural Inspectors	22
45-2041.00	Graders and Sorters, Agricultural Products	5
45-2091.00	Agricultural Equipment Operators	17
45-2092.01	Nursery Workers	21
45-2092.02	Farmworkers and Laborers, Crop	14
45-2093.00	Farmworkers, Farm and Ranch Animals	22
49-3041.00	Farm Equipment Mechanics and Service Technicians	13

With the intention of assessing the risk level of the tasks comprising each occupation, each task was classified with respect to the contamination risk. Four levels of risk were identified, namely: minimal, low, moderate and high, by considering the most prevalent virus transmission mechanisms, namely the proximity of workers and the possibility of touching contaminated surfaces. In brief, a task has:

- Minimal risk—when it does not require the physical presence of the worker in order to be performed. Usually, this level includes office tasks that can be carried out remotely (teleworking potential). Obviously, considering the remote execution of the task, there is no chance of touching contaminated surfaces.

- Low risk—when it requires the physical presence of the worker in the field/office/laboratory. Little contact with other people is required, mostly with the same individuals every day, while there is a little chance of touching contaminated surfaces.

- Moderate risk—when physical human presence is required in the field/office/laboratory. Moderate contact with other people is needed, mostly with the same people every day. Moreover, contact with costumers or workers from different places may be observed, while there is a moderate chance of touching contaminated surfaces.

- High risk—when physical human presence is required in the field/office/laboratory while considerable contact with other people is observed including customers, workers from different places and so on. In addition, there is a high chance of touching contaminated surfaces.

Six assessors, namely the authors of this study, independently assessed the level of risk of each task taking into account their own knowledge and the elaborated knowledge of an interviewed group of agricultural professionals including farm managers, first-line supervisors of crop and animal production, and various agricultural workers on horticulture, nursery, and livestock production. A consensus telemeeting of the assessors was held for the purpose of resolving any disagreement and arriving at the final result. The assessors have proved expertise in various fields of agricultural production including occupational health and safety, automation, operations management, agricultural technologies assessment, and agricultural ergonomics.

As can be seen in Figure 2, for the characterization of each task, the following methodology was implemented. Each individual task of an occupation is uniquely characterized by one of the four defined risk levels. To this end, given the four risk levels, (minimal ($j = 1$), low ($j = 2$), moderate ($j = 3$), or high risk level ($j = 4$)), for each one of the n defined tasks composing an occupation, the grade "1" was assign for the risk level to which the particular task was classified, while the grade "0" was assigned for the other three levels. For example, if the first task ($i = 1$) of an hypothetical occupation was characterized as "moderate", then $X_{1,3} = 1$ and $X_{1,1} = X_{1,2} = X_{1,4} = 0$. After assigning grades to all the tasks of an occupation according to the same procedure, the average grade was calculated providing the percentage of tasks classified at each risk level ($w_j = \frac{1}{n} \sum_{i=1}^{n} X_{i,j} \cdot 100\%$, $j \in [1, 2, 3, 4]$, noting that $\sum_{j=1}^{4} w_j = 100\%$). Finally, by considering the above approach, each occupation receives a weighted risk level characterization of the tasks it involves, which can be illustrated in the form of a single bar chart, as can be seen in the right side of Figure 2.

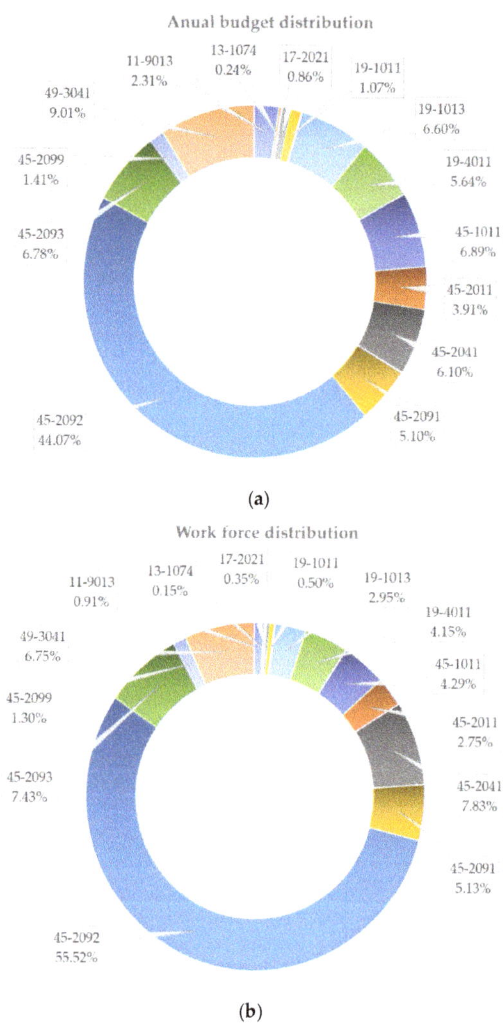

Figure 1. (a) Annual budget and (b) workforce distribution of agricultural occupations; the correspondence between the codes and occupations is shown in Table 1.

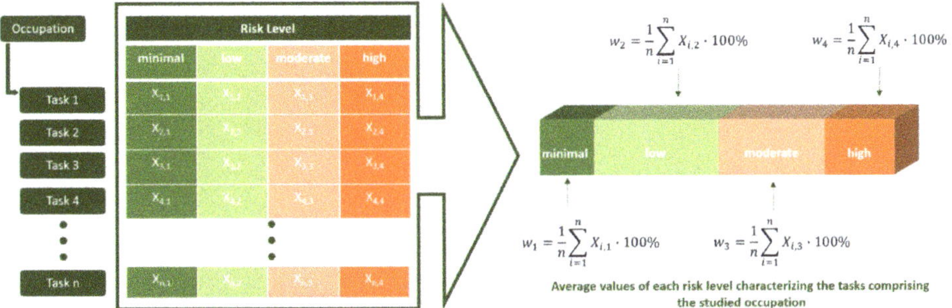

Figure 2. Methodology for the characterization of the risk level of the tasks comprising an occupation.

3. Results

3.1. Contamination Risk Level Distribution in Agricultural Occupations

The aforementioned methodology was implemented for the 17 occupations including in total 292 individual tasks (an elaborate description of them is provided in [41]), which were assigned to different risk levels in conformity with the proposed methodology described in Section 2. The contamination risk level distribution among all the agricultural occupations is illustrated in Figure 3. The high risk occupation was shown to be "Graders and Sorters, Agricultural Products" (45-2041.00), whose responsibilities include the activities of grading, sorting and/or classifying agricultural products by condition, weight, size or color, with 20% of the tasks demonstrating moderate and 80% of the tasks demonstrating high risk. The next occupation under high risk was "Farm Labor Contractors" (13-1074.00), who have the responsibility of recruiting and hiring seasonal agricultural workers. They may also transport workers to the work sites and provide tools and meals for the workers. This physical interaction engages in a high risk of contamination, either due to close contact with agricultural workers or the high chance of touching contaminated surfaces. In particular, out of the tasks of the above occupation, 63% were observed to have high risk, 25% moderate risk, and 13% low risk, while none of them had minimal risk of being affected by COVID-19.

Vulnerable occupations to the COVID-19 infection, although with lower risk levels, were also observed to be "Farmworkers, Farm and Ranch Animals" (45-2093.00), "Farmworkers and Laborers, Crop" (45-2092.02) and "First-Line Supervisors of Agricultural Crop and Horticultural workers" (45-1011.07). In contrast, the less affected occupations were "Food Science Technicians" (19-4011.02) and "Farm Equipment Mechanics and Service Technicians" (49-3041.00).

It can be deduced that occupations that are mostly related to the scientific aspects of agriculture and management, which can be performed remotely or with little contact with other people, have minimal to low risk of contamination (tasks of minimal and low risk level greater than 50% of the total number of tasks). On the other hand, occupations that require many people working at the same time together or require meetings with different people have a moderate to high risk of contamination (tasks of moderate and high risk level greater than 50% of the total number of tasks).

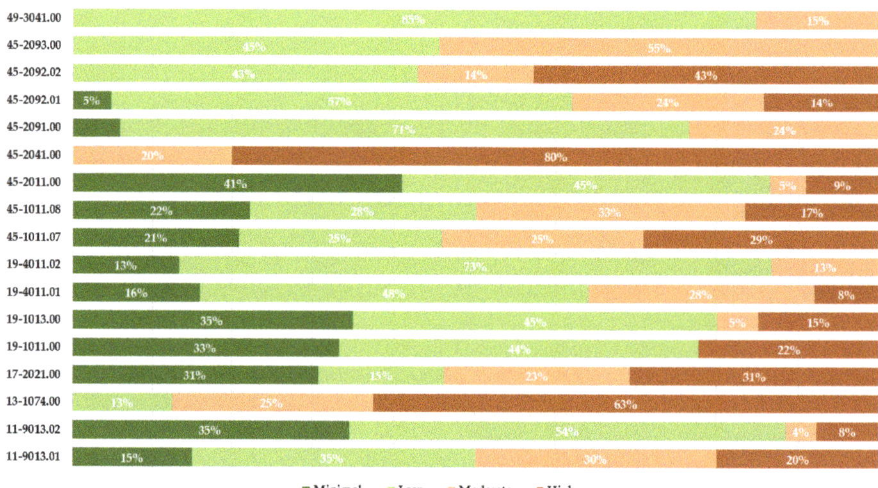

Figure 3. Contamination risk level distribution of each agricultural occupation; the correspondence between the codes and occupations is shown in Table 1.

3.2. Annual Budget and Total Workforce Effect

The worker who is classified in a specific occupation devotes a certain number of working hours to the execution of the individual tasks of the work. Therefore, the risk level of losing working hours can also be expressed as the risk level of the entire job position. Consequently, reducing the results to the workforce, it can be inferred (based on the data presented in Figure 1) that 5% of the total working time of agricultural employees (in the US-based scenario) are at high risk while 45% at moderate risk (Figure 4a). Cumulatively, 50% of the agricultural workforce is at moderate to high risk of contracting the disease in their workplace and the corresponding 50% of workhours are at moderate to high risk of being lost with the eventual consequences to food security and the economy in general. These employees mainly belong in occupations that require many people working at the same time together in close proximity or meeting with different people or/and exchanging tools. These numbers correspond to the 8% and 46% of the annual salaries, respectively (Figure 4b), meaning that 54% of the agricultural annual budget for workers' salaries are at moderate to high risk demonstrating the level of economic insecurity that is related to the pandemic. In contrast, 31% of the workforce time and 27% of the annual income are not expected to be influenced by the pandemic, while 19% of the workforce and annual salaries are at low risk.

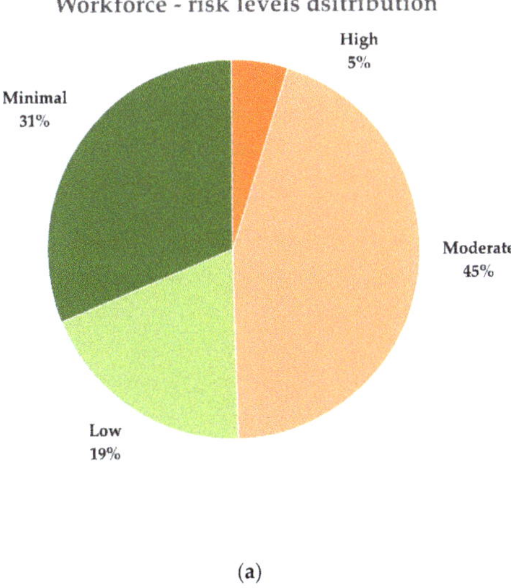

(**a**)

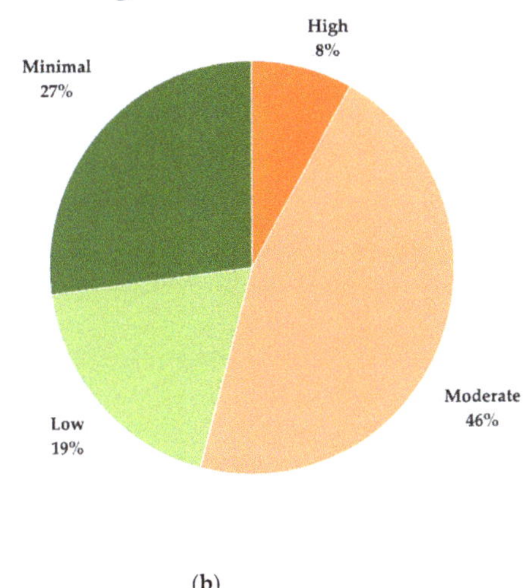

(**b**)

Figure 4. Risk level distribution in (**a**) the workforce and (**b**) the annual budget.

4. Discussion and Conclusions

With the object of assessing the potential impact of the consequences of the COVID-19 pandemic on the agricultural workforce, the well-defined US SOC system was used. In the absence of an existing methodology, the tasks of all SOC occupations were characterized as having minimal, low, moderate or a high risk level of getting the virus. The preliminary results of this study revealed the gravity of the matter, especially for workers of the "Farming, Fishing and Forestry" (45-0000) occupations major group, who work in close proximity conditions. On the other hand, "Management" (11-0000) and "Life, Physical and Social Science" (19-0000)-related occupations present lower risk, as a considerable part of their tasks can be performed remotely. In total, 31% of the workforce in agricultural occupations and 27% of the corresponding annual budget are not anticipated to be affected by COVID-19, while

19% of the workforce and corresponding budget are at low risk. However, it was found that 50% of the agricultural workforce and the 54% of the corresponding salaries are at moderate to high risk, hence, indicating the economic uncertainty associated with the current pandemic.

The adoption of a series of measures that can increase the resilience and sustainability of the sector in such urgent situations are imperative. These measures include different strategies to find seasonal workers, such as substituting seasonal migrant labor with domestic workers, applying deviations from labor laws, implementing very strict health measures during the reception of seasonal migrant workers and regularizing irregular migrants [8]. Another way for the mitigation of the existing problems is to facilitate the transfer of perishable products and logistics by avoiding trade restriction and minimizing trade costs. This would contribute to decrease the food loss. Moreover, facilitating border procedures in essential inputs such as fertilizers, veterinary medicines and pesticides by allowing for digital copies of certificates, for instance, could be beneficial. It would be also very consequential to maintain international markets open and transparent. Transparency can be accomplished through timely market information, which stands for information sharing. As a consequence, panic buying can be alleviated and trust among markets and countries be assured.

As far as the precaution measures pertaining to the spread of the coronavirus are concerned, the policies also implemented in other sectors have already been adopted [42,43]. In summary, control measures to protect workers against COVID-19 on farms include:

- Physical distancing: (a) the limitation of close contact by ensuring a 2 m minimum distance. This distance must also be kept during breaks; (b) the limitation of the number of people working together in one workspace, especially the closed ones like greenhouses, by working in shifts; (c) installing signage for maintaining physical distancing; (d) the use of alternative ways of communication, such as teleconferences and emails instead of face-to-face meetings. If it is necessary, meeting in outdoor spaces is highly recommended; and (e) if social distancing cannot be maintained, face masks must be worn, while their usage, taking off and disposal must follow all the instructions of WHO [44].

- Hygiene practices: (a) all workers should know how to properly wash their hands as well as avoid touching their mouth and nose. Moreover, a personal hand sanitizer should be provided to all workers to prevent the multiple usage of a single one. Frequent hand washing should be encouraged before entering the farm, before and after breaks, or after contact with surfaces and other people; (b) all non-essential visitors must be kept off the farm. Essential visitors, such as those needed for the care of the cleaning facilities and animals, must follow all the above practices concerning both physical distancing and good hygiene. Moreover, visitors should avoid visiting the same washroom facilities with the farm employees; (c) handling packages received at the farm must be left untouched for quite some time or disinfected in order to reduce the possibility of the virus being present on the surfaces; (d) the cleaning frequency of commonly touched surfaces and areas, like machinery, workstations, farm equipment and washrooms, must be increased; and (e) wherever possible, each worker should use their own tool, tractor, etc.

- Other precaution measures: (a) pre-authorizing farm visitors; (b) regularly checking workers for signs of COVID-19, such as shortness of breath, coughing and/or fever. In case someone has any symptom, they must self-isolate and notify their supervisor as soon as possible to call a doctor and provide the COVID-19 testing. If the workers test positive, all the other employees that work in the same environment or came into contact with them must be quarantined. Immediately after, COVID-19 test must also be provided for them in order to protect their families; (b) with the object of restricting the COVID-19 pandemic spread, the farms, as far as possible, need to be isolated so as to minimize the number of infected cases in case of someone contracting COVID-19; (c) farm employers and supervisors must be kept informed, and train workers about how to protect themselves from the coronavirus, well communicating the required measures, and following all national health warning recommendations associated with the COVID-19 pandemic.

The precaution measures are likely to be resorted, considering the precariousness about imminent waves of COVID-19. This pandemic, however, is an opportunity to create a fertile ground for the coordinated efforts of researchers, agricultural practitioners, infectious disease specialists and policymakers. It is anticipated that this preliminary study can serve as a basis for future research concerning integrated strategies for ensuring the smooth operation of the food supply chain, occupational health and jobs' protection.

Author Contributions: Conceptualization, L.B. and D.B.; methodology, L.B., M.L., D.B. and V.M.; investigation, L.B., V.M. and M.L.; writing—original draft preparation, L.B., M.L. and D.B.; writing—review and editing, S.P., C.G.S., and D.B.; visualization, L.B., M.L. and V.M.; supervision, S.M., C.G.S. and D.B. All authors have read and agreed to the published version of the manuscript.

Funding: This research received no external funding.

Conflicts of Interest: The authors declare no conflict of interest.

References

1. Q&A on Coronaviruses (COVID-19). Available online: https://www.who.int/emergencies/diseases/novel-coronavirus-2019/question-and-answers-hub/q-a-detail/q-a-coronaviruses (accessed on 3 August 2020).
2. WHO Situation Report-44. Available online: https://www.who.int/docs/default-source/coronaviruse/%0Asituation-reports/20200304-sitrep-44-covid-19.pdf?sfvrsn=783b4c9d_2 (accessed on 3 August 2020).
3. Coronavirus: Travel Restrictions, Border Shutdowns by Country | Coronavirus Pandemic News | Al Jazeera. Available online: https://www.aljazeera.com/news/2020/03/coronavirus-travel-restrictions-border-shutdowns-country-200318091505922.html (accessed on 3 August 2020).
4. Nicola, M.; Alsafi, Z.; Sohrabi, C.; Kerwan, A.; Al-Jabir, A.; Iosifidis, C.; Agha, M.; Agha, R. The socio-economic implications of the coronavirus pandemic (COVID-19): A review. *Int. J. Surg.* **2020**, *78*, 185–193. [CrossRef] [PubMed]
5. Jámbor, A.; Czine, P.; Balogh, P. The Impact of the Coronavirus on Agriculture: First Evidence Based on Global Newspapers. *Sustainability* **2020**, *12*, 4535. [CrossRef]
6. Russia Wants to Limit Grain Exports to Protect Food Supplies—Bloomberg. Available online: https://www.bloomberg.com/news/articles/2020-03-27/wheat-futures-rise-as-russia-considers-grain-export-quota (accessed on 6 August 2020).
7. Pulighe, G.; Lupia, F. Food First: COVID-19 Outbreak and Cities Lockdown a Booster for a Wider Vision on Urban Agriculture. *Sustainability* **2020**, *12*, 5012. [CrossRef]
8. Mitaritonna, C.; Ragot, L. After Covid-19, Will Seasonal Migrant Agricultural Workers in Europe Be Replaced by Robots? CEPII Policy Brief No. 33. 2020. Available online: http://www.cepii.fr/CEPII/en/publications/pb/abstract.asp?NoDoc=12680 (accessed on 3 October 2020).
9. Q&A: COVID-19 Pandemic—Impact on Food and Agriculture | FAO | Food and Agriculture Organization of the United Nations. Available online: http://www.fao.org/2019-ncov/q-and-a/impact-on-food-and-agriculture/en/ (accessed on 4 August 2020).
10. Prices of Agricultural Commodities Drop 20% Post COVID-19 Outbreak—Rediff Realtime News. Available online: https://realtime.rediff.com/news/india/Prices-of-agricultural-commodities-drop-20-post-COVID19-outbreak/955078599584b749?src=interim_alsoreadimage (accessed on 3 August 2020).
11. Benos, L.; Tsaopoulos, D.; Bochtis, D. A Review on Ergonomics in Agriculture. Part I: Manual Operations. *Appl. Sci.* **2020**, *10*, 1905. [CrossRef]
12. Benos, L.; Tsaopoulos, D.; Bochtis, D. A Review on Ergonomics in Agriculture. Part II: Mechanized Operations. *Appl. Sci.* **2020**, *10*, 3484. [CrossRef]
13. Lampridi, M.; Sørensen, C.; Bochtis, D. Agricultural Sustainability: A Review of Concepts and Methods. *Sustainability* **2019**, *11*, 5120. [CrossRef]
14. Martin, P. COVID-19 and California farm labor. *Calif. Agric.* **2020**, *74*, 67–68. [CrossRef]
15. OECD Fruit and Vegetables Scheme—OECD Fruit and Vegetables Scheme. Available online: https://www.oecd.org/agriculture/fruit-vegetables/ (accessed on 3 August 2020).

16. Seasonal Workers, CAP and COVID-19, Farm to Fork—EURACTIV.com. Available online: https://www.euractiv.com/section/agriculture-food/news/seasonal-workers-cap-and-covid-19-farm-to-fork/ (accessed on 4 August 2020).
17. *COVID-19: Channels of Transmission to Food and Agriculture*; FAO: Rome, Italy, 2020.
18. Pu, M.; Zhong, Y. Rising concerns over agricultural production as COVID-19 spreads: Lessons from China. *Glob. Food Sec.* **2020**, *26*, 100409. [CrossRef]
19. Zhang, S.; Wang, S.; Yuan, L.; Liu, X.; Gong, B. The impact of epidemics on agricultural production and forecast of COVID-19. *China Agric. Econ. Rev.* **2020**. [CrossRef]
20. Cortignani, R.; Carulli, G.; Dono, G. COVID-19 and labour in agriculture: Economic and productive impacts in an agricultural area of the Mediterranean. *Ital. J. Agron.* **2020**, *15*, 172–181.
21. Barcaccia, G.; D'Agostino, V.; Zotti, A.; Cozzi, B. Impact of the SARS-CoV-2 on the Italian Agri-Food Sector: An Analysis of the Quarter of Pandemic Lockdown and Clues for a Socio-Economic and Territorial Restart. *Sustainability* **2020**, *12*, 5651. [CrossRef]
22. Cattivelli, V.; Rusciano, V. Social Innovation and Food Provisioning during Covid-19: The Case of Urban–Rural Initiatives in the Province of Naples. *Sustainability* **2020**, *12*, 4444. [CrossRef]
23. McDonald, A.J.; Balwinder-Singh; Jat, M.L.; Craufurd, P.; Hellin, J.; Hung, N.V.; Keil, A.; Kishore, A.; Kumar, V.; McCarty, J.L.; et al. Indian agriculture, air pollution, and public health in the age of COVID. *World Dev.* **2020**, *135*, 105064. [CrossRef] [PubMed]
24. Kumar, A.; Padhee, A.K.; Kumar, S. How Indian agriculture should change after COVID-19. *Food Secur.* **2020**, *135*, 1–4.
25. Zarei, M.; Rad, A. Covid-19, Challenges and Recommendations in Agriculture. *J. Bot. Res.* **2020**, *2*. [CrossRef]
26. Henry, R. Innovations in Agriculture and Food Supply in Response to the COVID-19 Pandemic. *Mol. Plant* **2020**, *13*, 1095–1097. [CrossRef]
27. Neef, A. Legal and social protection for migrant farm workers: Lessons from COVID-19. *Agric. Human Values* **2020**, *37*, 641–642. [CrossRef]
28. Darnhofer, I. Farm resilience in the face of the unexpected: Lessons from the COVID-19 pandemic. *Agric. Human Values* **2020**, *1*, 3.
29. Phillipson, J.; Gorton, M.; Turner, R.; Shucksmith, M.; Aitken-McDermott, K.; Areal, F.; Cowie, P.; Hubbard, C.; Maioli, S.; McAreavey, R.; et al. The COVID-19 pandemic and its implications for rural economies. *Sustainability* **2020**, *12*, 3973. [CrossRef]
30. Villulla, J.M. COVID-19 in Argentine agriculture: Global threats, local contradictions and possible responses. *Agric. Human Values* **2020**, *1*, 1.
31. Siche, R. What is the impact of COVID-19 disease on agriculture? *Sci. Agropecu.* **2020**, *11*, 3–9. [CrossRef]
32. Barichello, R. The COVID-19 pandemic: Anticipating its effects on Canada's agricultural trade. *Can. J. Agric. Econ. Can. d'agroeconomie* **2020**, *68*, 219–224. [CrossRef]
33. Brewin, D.G. The impact of COVID-19 on the grains and oilseeds sector. *Can. J. Agric. Econ. Can. d'agroeconomie* **2020**, *68*, 185–188. [CrossRef]
34. Cranfield, J.A.L. Framing consumer food demand responses in a viral pandemic. *Can. J. Agric. Econ. Can. d'agroeconomie* **2020**, *68*, 151–156. [CrossRef]
35. Deaton, B.J.; Deaton, B.J. Food security and Canada's agricultural system challenged by COVID-19. *Can. J. Agric. Econ. Can. d'agroeconomie* **2020**, *68*, 143–149. [CrossRef]
36. Richards, T.J.; Rickard, B. COVID-19 impact on fruit and vegetable markets. *Can. J. Agric. Econ. Can. d'agroeconomie* **2020**, *68*, 189–194. [CrossRef]
37. Gunther, A. COVID-19: Fight or flight. *Agric. Human Values* **2020**, *1*, 1.
38. Lal, R. Home gardening and urban agriculture for advancing food and nutritional security in response to the COVID-19 pandemic. *Food Secur.* **2020**, *12*, 1–6. [CrossRef]
39. COVID-19 and the Food and Agriculture Sector: Issues and Policy Responses. Available online: http://www.oecd.org/coronavirus/policy-responses/covid-19-and-the-food-and-agriculture-sector-issues-and-policy-responses-a23f764b/ (accessed on 7 August 2020).
40. Over 100 New Cases of Covid-19 Signals Critical Point in Greece | GreekReporter.com. Available online: https://greece.greekreporter.com/2020/08/01/over-100-new-cases-of-covid-19-signals-critical-point-in-greece/ (accessed on 7 August 2020).

41. Standard Occupational Classification (SOC) System. Available online: https://www.bls.gov/soc/ (accessed on 5 August 2020).
42. 10 Measures to Protect Against COVID-19 on Farms | The Pig Site. Available online: https://thepigsite.com/articles/10-measures-to-protect-against-covid-19-on-farms (accessed on 28 August 2020).
43. COVID-19 Guidance for Farms & Markets | Pasa Sustainable Agriculture. Available online: https://pasafarming.org/covid19/ (accessed on 28 August 2020).
44. When and How to Use Masks. Available online: https://www.who.int/emergencies/diseases/novel-coronavirus-2019/advice-for-public/when-and-how-to-use-masks?gclid=Cj0KCQjw1qL6BRCmARIsADV9Jtao6SCtOBozLUQrHzRRdPAH82L-SMFmQ0rHxRYMxrHKtHvIUCD845IaAusJEALw_wcB (accessed on 28 August 2020).

MDPI

St. Alban-Anlage 66

4052 Basel

Switzerland

Tel. +41 61 683 77 34

Fax +41 61 302 89 18

www.mdpi.com

Sustainability Editorial Office

E-mail: sustainability@mdpi.com

www.mdpi.com/journal/sustainability